MOLECULAR ASPECTS OF NEURODEGENERATION AND NEUROPROTECTION

Editors

Akhlaq A. Farooqui

&

Tahira Farooqui

The Ohio State University

USA

Molecular Aspects of Neurodegeneration and Neuroprotection

Editors: Akhlaq Farooqui and Tahira Farooqui

eISBN: 978-1-60805-092-5

ISBN: 978-1-60805-376-6

© 2011, Bentham eBooks imprint.

Published by Bentham Science Publishers – Sharjah, UAE. All Rights Reserved.

Bentham Science Publishers
Executive Suite Y - 2
PO Box 7917, Saif Zone
Sharjah, U.A.E.
subscriptions@benthamscience.org

Bentham Science Publishers
P.O. Box 446
Oak Park, IL 60301-0446
USA
subscriptions@benthamscience.org

Bentham Science Publishers
P.O. Box 294
1400 AG Bussum
THE NETHERLANDS
subscriptions@benthamscience.org

CONTENTS

FOREWORD

The brain is an extraordinarily complex organ with remarkable capabilities. It undergoes a programmed plan of development over more than two decades and usually continues ability to undergo plasticity associated with learning throughout a lifetime. However, some congenital diseases, injuries to the head, and age-associated diseases produce profoundly negative impact on cognitive function. Understanding these neurodegenerative diseases and finding preventions/cures for them is an important goal of current research. It is becoming increasingly evident that neuroinflammation, oxidative stress, perturbed Ca^{2+} homeostasis, and apoptosis are closely associated with the pathogenesis of neurological diseases. This E-book presents up-to-date, cutting edge, and comprehensive information on the molecular aspects of neurodegeneration and neuroprotection in neurological disorders. Neurotraumatic diseases are characterized by acute neuroinflammation and oxidative stress that develop rapidly due to rapid depletion of ATP; sudden loss of ion homeostasis; accumulation of eicosanoids and platelet activating factor; and the release of proinflammatory cytokines. These processes result in rapid cell death that may occur in hours to days. In contrast, in neurodegenerative diseases oxygen and nutrients continue to be available to the nerve cells. ATP levels and ionic homeostasis are maintained at a level compatible with cell viability but not necessarily at optimum level. The interplay between oxidative stress and neuroinflammation occurs at a slow rate, leading to a neurodegenerative process that takes several years to develop. Thus, in neurodegenerative diseases chronic inflammation and oxidative stress linger for years, causing continued small insult to the brain tissue that accumulates over many years and ultimately reaches the threshold of detection many years after the onset of the neurodegenerative diseases.

Editors are known for their work on neurodegeneration and neuroprotection. They have taken great care in selecting topics on which progress has been made recently. They have done a commendable job in putting together this book and in writing the perspective. Chapters within this E-book are characterized by uniformity of style and simple and clear presentations. Topics addressed include the involvement of arachidonic acid derived lipid mediators (eicosanoids) and docosahexaenoic acid-derived lipid mediators (docosanoids) in neurodegeneration and neuroprotection, contribution of platelet activating factor, complement and hypoxic injury in neurodegeneration, biomarkers for oxidative stress in neurodegenerative diseases, and involvement of oxidative stress in Parkinson disease and schizophrenia. The subject matter develops logically and progresses smoothly from one topic to another. The book contains extensive bibliography. Useful references will help readers in pursuing those areas of interest that are beyond the scope of this book. To aid comprehension, a large number of figures and line diagrams of signal transduction pathways is presented.

This E-book can be used as supplemental text for a range of neuroscience courses. Clinicians will find this E-book useful for understanding molecular aspects of lipid mediators involved in neuroinflammation and oxidative stress in neurological disorders. It is anticipated that senior neuroscientists may find inspiration from this E-book to overcome problems encountered in their research on lipid mediators associated with neuroinflammation and oxidative stress. Students may gain insight into the difficulties experienced in their research on lipid mediators in brain.

Lane J. Wallace, Ph.D
Chair and Professor
Division of Pharmacology, College of Pharmacy
The Ohio State University, Columbus, Ohio 43210 USA

PREFACE

American population is aging. The Census Bureau predicts that the nation will have more than 1 million centenarians in 2050, up from 71,000 today. The number of Americans afflicted with stroke, neurodegenerative diseases, and neuropsychiatric disorders is also increasing. These diseases are caused by acute or progressive loss of neurons resulting in brain dysfunction. Known risk factors for stroke, neurodegenerative diseases, and neuropsychiatric disorders include increasing age, genetic polymorphisms, endocrine conditions, oxidative stress, inflammation, excitotoxicity, hypertension, diabetes, infection, tumors, immune and metabolic conditions, and chemical exposure. According to the NINDS, approximately 50 million Americans are affected by stroke, neurodegenerative diseases and neuropsychiatric disorders each year. The number of people affected with above neurological disorders will double every 20 years, and will cost the U.S. economy billions of dollars each year in direct health care costs and lost opportunities. It is estimated that $100 billion per year is spent on Alzheimer disease alone. In addition to the financial costs, there is an immense emotional burden on patients, their relatives, and caregivers. It is also predicted as the number of senior citizens grows, costs to treat neurological disorders to the society will increase significantly.

Significant progress has been made on molecular aspects of neurodegeneration and neuroprotection in acute neural trauma and neurodegenerative diseases. This information is scattered throughout the literature in the form of original papers, reviews, and some edited books. The purpose of this E-book is to present readers with cutting edge and comprehensive information on molecular aspects of neurodegeneration and neuroprotection in a manner that is useful not only to students and teachers but also to researchers and physicians. This E-book has 11 chapters by leading researchers in the field of neurodegeneration and neuroprotection. Chapter 1, 2, and 3 describe neurochemical aspects of neuroinflammation and biomarkers for oxidative stress. Chapter 4 deals that molecular mechanism of kainic acid neurotoxicity in the brain. Chapter 5, 6, and 7 describe the involvement of mitochondrial signaling in Alzheimer disease and hypoxic injury, molecular aspects of dopamine-mediated neurodegeneration, and role of complement in neurodegeneration and neuroinflammation. Chapter 8 describes the role of platelet activating factor in neurological disorders. Chapter 9 provides information on pathogenesis of schizophrenia. Chapter 10 describes involvement of fatty acids and their metabolites in neuroprotection in brain, and finally chapter 11 deals with the perspective and directions for future development on various aspects of neurodegeneration and neuroprotection.

The writing style and demonstrated ability of various authors to present complicated information on neurodegeneration and neuroprotection makes this book particularly accessible to neuroscience graduate students, teachers, and fellow researchers. This E-book can be used as supplement text for a range of neuroscience courses. Clinicians and pharmacologists will find this book useful for understanding molecular basis of neurodegeneration in neurodegenerative diseases. Editors of this E-book have tried to ensure uniformity in the mode of presentation as well as a logical progression from one topic to another and have made sure that authors provide extensive referencing. Editors hope that their attempts to integrate and consolidate knowledge of molecular aspects of neurodegeneration and neuroprotection in neurological disorders will provide the basis for more dramatic advances and developments in signal transduction processes associated with molecular mechanisms of neurodegeneration and neuroprotection.

Akhlaq A. Farooqui
Tahira Farooqui

CONTRIBUTORS

Bonifati D.M.

Department of Neurological disorders
Santa Chiara Hospital
Largo Medaglie d'Oro 1
Trento
Italy.

Cardoso S.

Department of Zoology
Faculty of Sciences and Technology
University of Coimbra
Coimbra
Portugal

Carvalho C.

Department of Zoology
Faculty of Sciences and Technology
University of Coimbra
Coimbra
Portugal

Coppedè F.

Department of Neuroscience
University of Pisa, Via Roma
Pisa
Italy

Correia S.C.

Department of Zoology
Faculty of Sciences and Technology
University of Coimbra
Coimbra
Portugal

Das U.N.

Jawaharlal Nehru Technological University, / Kakinada, India and UND Life Sci / Shaker Heights, Ohio
USA

Farooqui A.A.

Department of Molecular and Cellular Biochemistry
The Ohio State University
Columbus, Ohio
USA

Farooqui T.

Department Entomology/Center for Molecular Neurobiology
The Ohio State University, Columbus, Ohio
USA

Hirashima Y.

Physiological Chemistry

Faculty of Pharmaceutical Sciences
Teikyo University, Sagamiko
Sagamihara
Kanagawa
Japan

Joshi S.R.

Interactive Research School for Health Affairs
Bharati Vidyapeeth
Pune
India

Kale A.

Interactive Research School for Health Affairs
Bharati Vidyapeeth
Pune
India

Kishore U.

Laboratory of Human Immunology and Infection Biology
Biosciences Division Brunel University
Uxbridge West London UB8 3PH
United Kingdom

Mahadik S.P.

Department of Psychiatry and Health Behavior
Medical College of Georgia
Augusta GA
USA

Migliore L.

Department of Human and Environmental Sciences
University of Pisa, Via S. Giuseppe 22,
Pisa
Italy

Moreira P.I.

Institute of Physiology
Faculty of Medicine,
University of Coimbra
Coimbra
Portugal

Nayak A.

Laboratory of Human Immunology and Infection Biology
Biosciences Division, Brunel University
Uxbridge West London UB8 3PH
UK

Ong W.Y.

Department of Anatomy
National University of Singapore
Singapore

Perry G.

UTSA Institute for Neuroscience and Department of Biology
College of Sciences, University of Texas
San Antonio, Texas
USA

Santos R.X.

Department of Zoology
Faculty of Sciences and Technology
University of Coimbra
Coimbra
Portugal

Smith M.A.

Department of Pathology
Case Western Reserve University
Cleveland, Ohio
USA

Wallace L.J.

Division of Pharmacology
College of Pharmacy
The Ohio State University
Columbus, Ohio
USA

Zhu X.

Department of Pathology
Case Western Reserve University
Cleveland, Ohio
USA

ACKNOWLEDGEMENTS

This book would not have been possible without the contribution of all authors and co-authors, whom we owe a great debt of gratitude. We would like to thank Bentham Science Publishers, Director Mahmood Alam and Assistant Manager Sara Moqeet, for their support and efforts.

Akhlaq A. Farooqui
Tahira Farooqui

CHAPTER 1

Arachidonic Acid and Docosahexaenoic Acid-derived Lipid Mediators in Brain

Akhlaq A. Farooqui[*]

Department of Molecular and Cellular Biochemistry, The Ohio State University, Columbus, Ohio 43210, USA

Abstract: Docosahexaenoic acid and arachidonic acid are major polyunsaturated fatty acids in neural membrane glycerophospholipids. Docosahexaenoic acid and arachidonic acid are metabolically and functionally distinct molecules that have opposing physiological functions. Docosahexaenoic acid is metabolized to docosanoids, whereas arachidonic acid is metabolized to eicosanoids. Like their precursors, docosanoids and eicosanoids are different types of lipid mediators, which play important and opposing roles in modulating inflammatory reactions, oxidative stress, neuroprotection, and neurodegeneration. Increase in levels of eicosanoids occurs in acute neural trauma (stroke and traumatic injury to brain and spinal cord) and neurodegenerative diseases (Alzheimer disease, Parkinson disease, and Huntington disease), whereas consumption of DHA increases levels of docosanoids, which have antioxidant, anti-inflammatory, and anti-apoptotic properties. Synthesis of docosanoids is an endogenous neuroprotective mechanism against acute neural trauma and neurodegenerative diseases.

Keywords: Arachidonic acid; docosahexaenoic acid; eicosanoids; docosanoids; eicosapentaenoic acid; oxidative stress; inflammation; apoptosis; neurodegenerative diseases

INTRODUCTION

Docosanoids are a group of endogenous oxygenated metabolites derived from docosahexaenoic acid (DHA). docosanoids include at least two familes of lipid mediators namely, D series resolvins and docosatrienes (neuroprotectins) [1-3]. As a characteristic feature, docosatrienes contain conjugated triene structures (Fig. 1). In contrast, eicosanoids (prostanoids) is a group of endogenous oxygenated metabolites derived from arachidonic acid (ARA) and eicosapentaenoic acid (EPA). Eicosanoids (prostanoids) not only include prostaglandins (PGs), leukotrienes (LTs) and thromboxanes (TXs), but also hepoxilins (HXs), lipoxins (LXs), and epoxyeicosatrienoic acids (EET) [4-6].

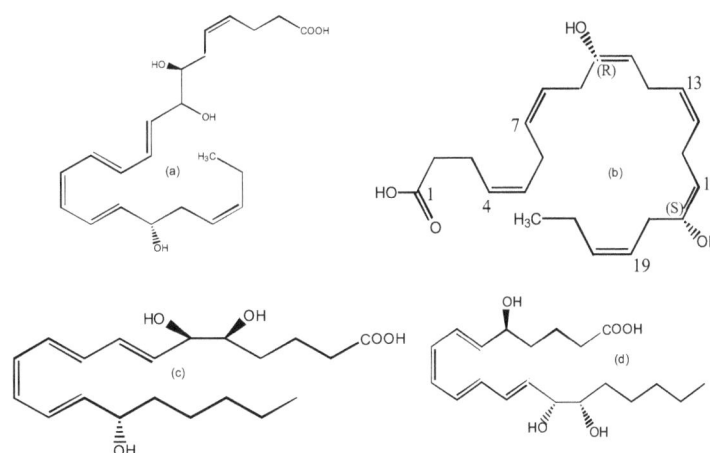

Figure 1: Chemical structures of some docosanoids and eicosanoids. Resolvin D_1 (a); neuroprotectin D_1 (b); lipoxin A_4 (c); lipoxin B_4 (d).

Oxidation of EPA generates less active eicosanoids. DHA and ARA are essential fatty acids that are obtained from diet. Present day Western diet contains very high amounts of saturated fat and with ARA to DHA ratio of about 15-

*Address **correspondence to:** Department of Molecular and Cellular Biochemistry, The Ohio State University, Columbus, OH 43210, USA; Tel.: (614) 488-0361; E-mail: farooqui.1@osu.edu

20:1. The Paleolithic diet on which human beings have evolved, and lived for most of their existence, contained a ratio of 1:1, and was high in fiber, rich in fruits, vegetables, lean meat, and fish [7].

Changes in eating habits (natural versus processed food enriched in vegetable oil) has changed ARA:DHA ratio in favor of ARA. High intake of ARA enriched foods elevates levels of PGs, LTs, and TXs, and upregulates the expression of pro-inflammatory cytokines, such as TNF-α, and IL-1β. ARA-enriched diet mediates elevated levels of proinflammatory cytokines promoting chronic visceral (cardiovascular, inflammatory and autoimmune) diseases as well as neurological disorders (stroke, Alzheimer disease, AD; Parkinson disease, PD; and Huntington disease, HD). In contrast, consumption of DHA-enriched diet has anti-inflammatory and antioxidant effects and exerts cardioprotective, anti-carcinogenic, immunosuppressive, and neuroprotective effects [7, 8, 9]. Perhaps, these effects of DHA-enriched diet are due to elevated levels of docosanoids that have antioxidant and antiapoptotic effects in neural and non-neural tissues [3, 9].

The purpose of this chapter is to describe and compare the properties of eicosanoids and docosanoids in brain. It is hoped that this commentary would not only initiate more studies on signal transduction processes associated with docosanoids, but also will jump start studies on interactions between eicosanoid and docosanoid metabolism in brain.

SYNTHESIS OF EICOSANOIDS AND DOCOSANOIDS IN BRAIN

In neural membrane glycerophospholipids, DHA and ARA are esterified at the sn-2 position of glycerol moiety. ARA is evenly distributed in gray and white matter and among the different cell types in brain tissue. The turnover rates of ARA and DHA in neural membrane phospholipids are rapid and energy consuming [10]. The deacylation/reacylation cycle regulates turnover of ARA and DHA containing glycerophospholipids in brain. The release of DHA from PlsEtn is catalyzed by plasmalogen-selective-phospholipase A_2 (PlsEtn-PLA$_2$). This enzyme has been purified and characterized from bovine brain, rabbit heart, and kidney [11-14]. Similarly, PtdSer is also hydrolyzed by a phospholipase, which has not yet been characterized [15]. The liberation of ARA from PtdCho is catalyzed by cytosolic phospholipase A_2 (cPLA$_2$). Three oxidases: cyclooxygenases (COXs) that produce prostaglandins, lipoxygenases (LOXs) that generate hydroxy derivatives and leukotrienes, and epoxygenases (EPOXs) that form epoxyeicosatrienoic products, oxidize ARA after its release from neural membrane PtdCho by the action of cPLA$_2$. In contrast, COXs do not oxidize DHA. The action of 15-LOX like enzyme on DHA generates 10, 17S-docosatrienes, 17S-resolvins, and neuroprotectins [4, 16-18]. Although, considerable information is available on COX, LOX, and EPOX enzymes that convert ARA into eicosanoids, but very little is known about 15-LOX that oxidizes DHA. Thus, detailed investigations are needed on isolation and characterization of 15-LOX in brain tissue.

NEUROCHEMICAL AND PHYSIOLOGICAL ACTIVITIES OF EICOSANOIDS IN BRAIN

As stated above, COX and LOX oxidize ARA into PGs, LTs, and TXs [4]. Because of their amphiphilic nature, these lipid mediators can cross neural cell membranes and leave the cell in which they are synthesized to act on neighboring cells [19]. Eicosanoids act through their receptors that are located on plasma and nuclear membranes. These receptors modulate signal transduction pathways and gene transcription. Among eicosanoids, PGs are potent autocrine and paracrine oxygenated lipid mediators of ARA metabolism that contribute appreciably to physiological and pathophysiological responses in brain (Table **1**). Collective evidence indicates that the PGs play an important role in brain diseases including ischemic injury and several neurodegenerative diseases. PG signaling is mediated by interactions with four distinct G protein-coupled receptors, EP_1, EP_2, EP_3, and EP_4, which are differentially expressed on neuronal and glial cells throughout the central nervous system. EP_1 activation mediates Ca^{2+}-dependent neurotoxicity in ischemia. EP_2 activation is associated with microglial-mediated paracrine neurotoxicity as well as suppression of microglia internalization of aggregated neurotoxic peptides [20].

EP_2 receptor activation induces BDNF secretion through stimulation of cyclic AMP dependent signaling involving cAMP-dependent protein kinase (PKA). This signaling may be associated with neurotoxicity or neuroprotection in microglial cells and astrocytes. In contrast, EP_4 utilizes phosphatidylinositol 3-kinase (PtdIns3K) as well as PKA. In addition, it is also shown that the EP_4 receptor, but not the EP_2, can activate the extracellular signal-regulated kinases (ERKs) 1 and 2 by way of PtdIns3K leading to the induction of early growth response factor-1 (EGR-1), a transcription factor traditionally associated with wound healing. The induction of EGR-1 expression may have significant implications for the role of PGE_2 in inflammatory processes [6, 20]. EP_3 receptor signals are associated with the inhibition of adenylyl cyclase via Gi activation, and in Ca^{2+}-mobilization through Gβγ from Gi. Along with Gi activation, the EP_3 receptor can stimulate cAMP formation via G(s) activation. EP_3 receptor can augment Gs-coupled receptor-mediated adenylyl cyclase activity. At least three isoforms of EP_3 are generated through alternative RNA splicing from a single gene and differ only in the efficiency of G protein activation and in the specificity of coupling to G proteins. Some PGs modulate neurotransmitters, whereas others regulate circulatory function. TXA_2 is a potent vasoconstrictor and produces vasospasm, whereas PGI_2 has opposing effects (Fig. **2**) [4]. PGs, LTs, and TXs are involved in many processes including fever, sensitivity to pain, sleep, inflammation and oxidative stress. The generation of eicosanoids under pathological conditions is associated with the modulation of cerebrovascular blood flow. Their active production by circulating cells such as platelets and leukocytes induces alterations in the microcirculation and ultimately to CNS dysfunction [4]. High levels of PGs have degenerative effects on differentiated murine neuroblastoma cells in cultures. In vivo, PGs are involved in the regulation of cytokines and maintenance of inflammatory and oxidative stress cascades [4].

Table 1: Activities of eicosanoids and docosanoids in neural and non-neural tissues.

Parameter	Eicosanoids	Docosanoids
Aggregation	Proaggregatory	Antiaggregatory
Thrombotic activity	Prothrombotic	Antithrombotic
Inflammation	Proinflammatory	Antithrombotic
Lipid status	Hyperlipidemic	Hypolipidemic
Heart rhythm	Proarrhythmic	Antiarrhythmic
Effect on excitotoxicity	Proexcitotoxic	Antiexcitotoxic

Prostacyclin, a major prostaglandin derived from enzymic oxidation of ARA at the nuclear and endoplasmic reticular membrane level. They act through their surface as well as nuclear receptors in physiological and pathological processes. Prostacyclins interact with peroxisomal proliferator activated receptors and modulate many biological functions including lipid and energy metabolism, oxidative stress, and inflammation responses. Similarly, leukotrienes are lipid mediators of ARA metabolism, involved in autocrine and paracrine signaling [4]. Lipoxins (LXA_4, LXB_4, 15 epi-LXA_4, and and 15 epiLXB_4), a group of trihydroxytetraene eicosanoids, are derived from ARA (Fig. **1**). They are generated by the action of LOX on hydroperoxyeicosatetraenoic acid (HPETE) and hydroxyeicosatetraenoic acid (HETE). LXs are involved in the resolution of acute inflammation, a neuroprotective process that separates healthy neural cells from injured cells. LXs promote resolution by modulating key steps in leukocyte trafficking and preventing neutrophil-mediated acute tissue injury [18, 2 1]. LXs mediate a number of processes, including regression of pro-inflammatory cytokine production, inhibition of cell proliferation, and stimulation of phagocytosis of apoptotic leukocytes by macrophages [22]. LXs interact with ALX and LXA receptors [23, 24, 25] and mediate their signaling. LXA_4 is a major LXs which binds to G protein coupled with LXA_4 receptors that have been identified in astrocytes and microglia, supporting the view that these glial cells may be a target for LX action in the brain. Quantitative RT-PCR studies indicate that LXA_4 retards the IL-1β-induced stimulation of IL-8 and ICAM-1 expression in 1321N1 human astrocytoma cell [22]. Furthermore, LXA_4 reduces the expression of IL-1β-induced IL-8 protein levels. LXA4 inhibits IL-1β-induced degradation of I-κBα, and the activation of an NF-κB modulated reporter gene construct. Collective evidence suggests that LXA_4 exerts anti-inflammatory effects in 1321N1 astrocytoma cells at least in part via an NF-κB-dependent mechanism [22]. Aspirin mediates the generation of lipoxins and promotes resolution of inflammatory reaction [24, 25].

Various leukotrienes (leukotriene C_4, LTC_4; leukotriene D4, LTD4; and leukotriene (E_4 LTE_4) activate contractile and inflammatory processes through specific interaction with putative seven transmembrane spanning receptors that couple to G proteins and subsequent intracellular signaling pathways. Cysteinyl leukotrienes (CysLTs) are potent mediators of inflammation derived from ARA by the concerted actions of 5-LOX, 5-LOX-activating protein (FLAP), leukotriene C_4 synthase, and additional downstream enzymes. CysLTs are generated by macrophages, eosinophils, mast cells, and other inflammatory cells activate 3 different high-affinity CysLT receptors: $CysLT_1R$,

CysLT$_2$R, and GPR 17. Cys-LTs play a role in a number of inflammatory conditions including cerebrovascular and cardiovascular diseases [26]. Leukotrienes and lipoxins modulate neural stem cell (NSC) functions. Thus, LTB$_4$ and LXA$_4$ modulate proliferation and differentiation in murine embryo brain NSC. LTB$_4$ stimulates NSCs proliferation by interacting with LTB$_4$ receptors, and this process can be blocked by LTB$_4$ receptor antagonist. In contrast, LXA$_4$, and its aspirin-triggered stable analog of LXA$_4$, 15-epi-LXA$_4$, inhibit the growth of NSC at very low concentrations. Collective evidence suggests that LTB$_4$ and LXA$_4$ are closely associated with the regulation of NSC proliferation and differentiation [27].

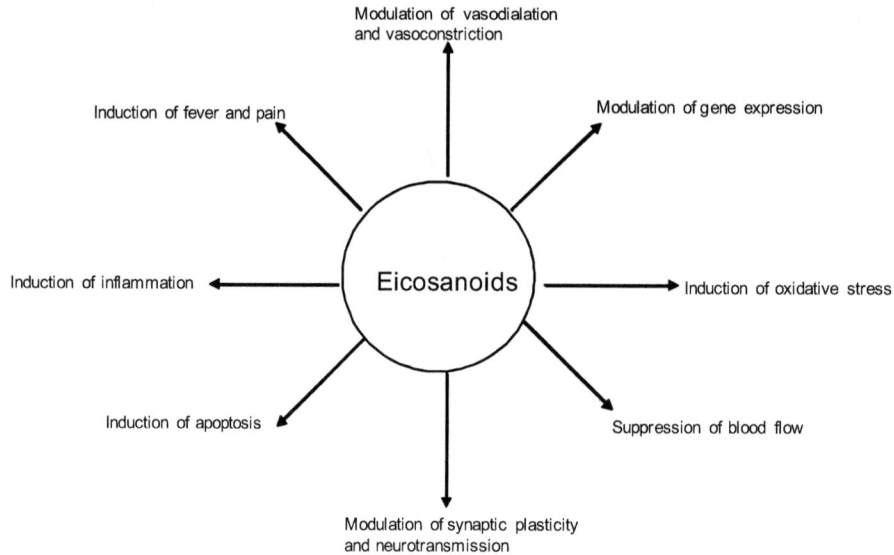

Figure 2: Roles of eicosanoids in brain.

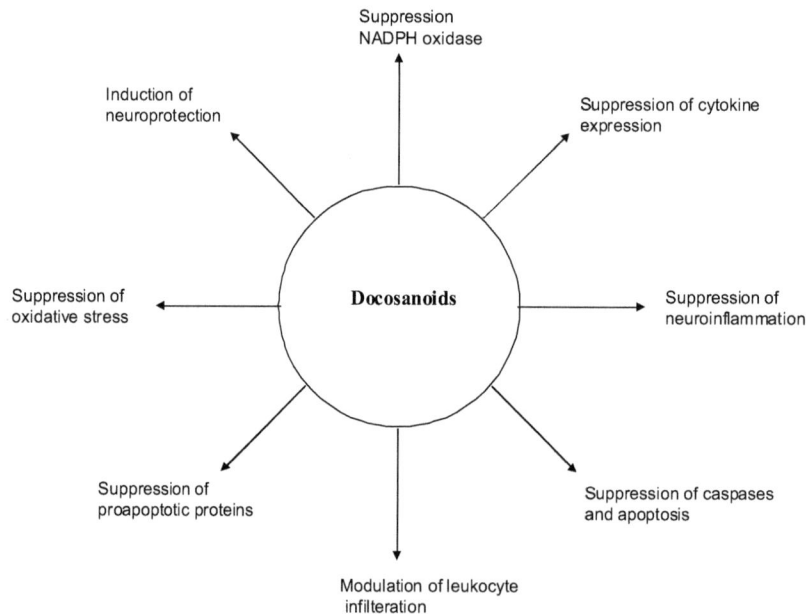

Figure 3: Roles of docosanoids in brain

Lipoxins also participate in modulation of nociception. LXA$_4$ receptors are expressed on spinal astrocytes. The delivery of LXA$_4$ and its stable analogs to spinal cord attenuates inflammation-mediated pain process [28]. Furthermore, activation of extracellular signal-regulated kinase and c-Jun N-terminal kinase in astrocytes play an important role in spinal nociceptive process, which is attenuated in the presence of lipoxins. This observation suggests the possibility that lipoxins regulate spinal nociceptive processing though their actions upon astrocytic activation. Targeting mechanisms that downregulate the spinal consequences of persistent peripheral inflammation provide a novel endogenous mechanism by which chronic pain may be controlled [28].

NEUROCHEMICAL AND PHYSIOLOGICAL ACTIVITIES OF DOCOSANOIDS IN BRAIN

Oxidation of DHA by a 15-LOX-like enzyme generates 10,17S-docosatrienes and 17S-resolvins [1-3, 9, 24]. These second messengers not only antagonize the effects of eicosanoids, but also modulate leukocyte trafficking and down-regulate the expression of cytokines in glial cells. They are collectively called as docosanoids. They possess potent anti-inflammatory, antioxidant, neuroprotective and pro-resolving properties (Fig. **3**) [1-3, 9, 18, 24].

17S D SERIES RESOLVINS IN BRAIN

15-LOX catalyzed oxidation of DHA generates 17-hydroxydocosahexaenoic acid, which is further metabolized to resolvins of the D series (RvDs) (Fig. **1**) and protectins neuroprotectin D_1/ protectin D_1 (NPD_1/PD_1) [3, 29]. Aspirin impinges on these systems, triggering formation of the epimeric 17R-series RvDs--denoted as 'aspirin-triggered-RvDs'--which possess bioactivity *in vivo* equivalent to that evoked by their 17S-series counterparts (i.e. RvDs). They include RvD_1, RvD_2, RvD_3, RvD_4, RvD_5, and RvD_6. These metabolites exert potent agonist actions on macrophages and vascular endothelial cells that can control the magnitude of the local inflammatory response. They act through specific receptors found in neural and non-neural cells. These receptors are called as resolvin D receptors ($resoDR_1$) resolvin E receptors ($resoER_1$) [29]. These receptors have neither been fully characterized in non-neural tissues nor in brain. Thus, detailed investigations on enzymic oxidation of DHA by 15-LOX like enzymic activity and generation of $resoDR_1$ and $resoER_1$ in neural and non-neural cells are urgently required.

DOCOSATRIENES IN BRAIN

Oxidation of DHA by 15-lipoxygenase-like enzyme also generates of NPD_1 (10,17S-docosatriene) (Fig. **4**). In vitro, the synthesis of this lipid mediator is stimulated by calcium ionophore A23187 and IL-1β, or the supply of DHA. Stimulation of PLA_2 in ischemic/ reperfusion injury and epileptic seizures is accompanied by an increase in unesterified pool of DHA. Some DHA is incorporated into neural membranes through the reacylation process and remaining DHA is converted to NPD_1. The infusion of NPD_1 during reperfusion produces neuro-protective effects by counteracting polymorphonuclear neurophil infiltration, proinflammatory gene expression and infarct size [3]. Increase in NF-κB binding and induction of COX-2 by ischemia reperfusion/injury is prevented by NPD_1. NPD_1 also stimulates the expression of anti-apoptotic Bcl-2 proteins (Bcl-2 and BclxL) and blocks the expression of pro-apoptotic proteins, Bax and Bad. NPD_1 reduces caspase-3-mediated oxidative stress [1-3]. Although, the molecular mechanisms and sequence of events associated with above processes are not known, but it is well known that Bcl-2 family proteins regulate apoptotic signaling by modulating the release of cytochrome c and activating caspase-3 in mitochondria. In retina, NPD_1 generation is induced by pigment epithelium-derived factor [30].

Collectively, these studies suggest that NPD_1 promotes homeostatic regulation of the integrity of neural cells through the modulation of multiple signaling pathways and down-regulation of proinflammatory gene expression [3] particularly during oxidative stress and neuroinflammatory events in retinal and neurodegenerative diseases [3]. NPD_1 promotes neural cell survival not only through the induction of antiapoptotic and neuroprotective gene expression programs suppressing amyloid ss peptide (Ass), composed of 42 amino acids (Ass42) generation and its neurotoxicity, but also by retarding inflammatory reactions [31]. NPD_1 contents are reduced in the CA1 subfield of the hippocampal region from Alzheimer disease patients and soluble amyloid precursor protein-α and growth factor enhance the synthesis of NPD_1 [3, 32]. Receptors for NPD_1 have not been characterized in brain tissue, but based on retinal epithelium studies, it is suggested that neurotrophins (growth factors) may act as agonists for NPD_1 [31] (Fig. **4**). Signal transduction mechanisms linked with NPD_1 receptor have not been described. It is also proposed that the generation of resolvins and neuroprotectins may be an internal protective mechanism for preventing apoptotic cell death-mediated brain damage [1, 3, 24] and NPD_1 through its receptors may modulate gene expression programs that are neuroprotective through down-regulation of pro-apoptotic and pro-inflammatory factors and upregulation of Bcl-2-family anti-apoptotic proteins, which are crucial modulators of cell survival [3, 9, 31].

Docosanoids retard inflammation and oxidative stress through several mechanisms including upregulation of γ-glutamyl-cysteinyl ligase and glutathione reductase activities, inhibition of p65 subunit transcription factor NF-κB, preventing cytokine secretion, blocking the synthesis of eicosanoids, blocking toll like receptor-mediated activation of macrophages, and reducing leukocyte trafficking [6]. It is also proposed that DHA enrichment in Neuro 2A cells prevents staurosporine-mediated apoptotic cell death in a PtdSer-and PtdIns3-Kinase-dependent manner [33].

EICOSAPENTAENOIC ACID DERIVED LIPID MEDIATORS IN THE BRAIN

The action of COX enzymes on EPA generates the 3-series of prostaglandins and thromboxanes and the 5-series of leukotrienes (Fig. **5**). These eicosanoids have different biological properties than the corresponding eicosanoids produced by the metabolism of ARA.

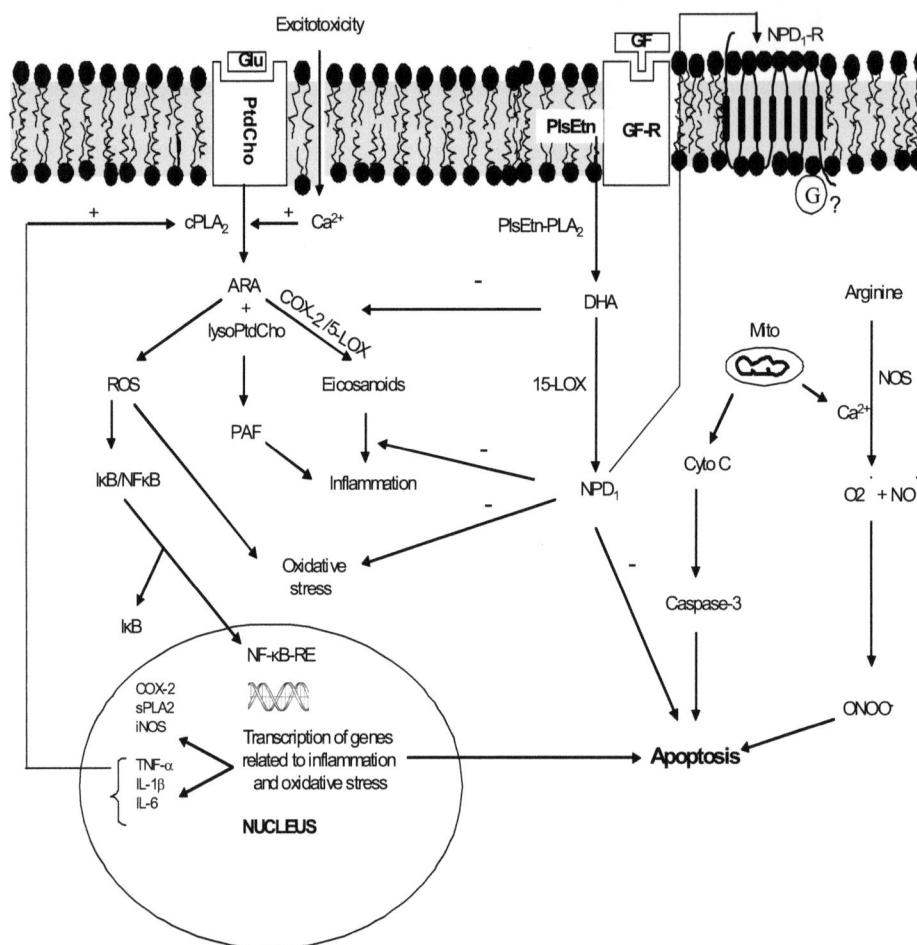

Figure 4: Hypothetical diagram showing the generation of neuroprotectin D_1 from DHA and modulation of inflammation, oxidative stress, and apoptosis by neuroprotectin D_1. Glutamate (Glu); cytosolic phospholipase A_2 (cPLA$_2$); plasmalogen-selective phospholipase A_2 (PlsEtn-PLA$_2$); secretory phospholipase A_2 (sPLA$_2$); phosphatidylcholine (PtdCho); lyso-phosphatidylcholine (lyso-PtdCho); ethanolamine plasmalogen (PlsEtn); platelet activating factor (PAF); arachidonic acid (ARA); docosahexaenoic acid (DHA); cyclooxygenase-2 (COX-2); 5-lipoxygenase (5-LOX); 15-lipoxygenase (15-LOX); nitric oxide synthase (NOS); lysophosphatidylcholine (lyso-PtdCho); rective oxygen species (ROS); neuroprotectin D_1 (NPD$_1$); neuroprotectin D_1 receptor (NPD$_1$-R); growth factor (GF); growth factor receptor (GF-R); nitric oxide (NO); superoxide (O_2^-); tumor necrosis factor-α (TNF-α); interleukin-1β (IL-1β); interleukin-6; and nuclear factor κB-response element (NF-κB).

For example, TXA$_3$ is less active than TXA$_2$ in aggregating platelets and constricting blood vessels [34, 35]. Substrate specificities on oxidation of fatty acids by COX enzymes and PG endoperoxide H synthases indicate that EPA and ARA compete for the same enzymes. However, some differences have been reported. Under optimal conditions purified COX-1 oxygenates EPA with only 10% of the efficiency of ARA, and EPA significantly inhibits

ARA oxygenation by COX-1 [27]. Two- to 3-fold higher activities or potencies with 2-series versus 3-series eicosanoids have been reported for COX-2, PGD synthases, microsomal COX-1 and EP_1, EP_2, EP_3, and FP receptors [27]. Surprisingly, oxygenation of ARA by COX-2 is only modestly inhibited by EPA and TxA_3 have equal affinity for TP-α receptor. These studies indicate that increasing glycerophospholipid EPA/AA ratios in cells may slow-down eicosanoid signaling with the largest effects being on COX-1 pathway [27].

In non-neural cells, oxidation of EPA by 15-LOX like enzyme initiates the synthesis of resolvins of the E series [36, 37]. Thus, resolvin E_1 (RvE_1, 5S, 12R, 18R)-trihydroxy-6Z,8E,10E,14Z,16E-eicosapentaenoic acid) is an oxidized product derived from eicosapentaenoic acid that displays potent anti-inflammation/pro-resolution actions *in vivo* [36]. RvE_1 is an initial metabolite. The conversion of RvE_1 to the oxo product inactivates RvE_1. RvE_1 blocks the activation of NF-κB by TNF-α. Antiinflammatory action of RvE_1 has been studied in human polymorphonuclear leukocyte (PMN) [37]. Human PMN show specific binding with RvE_1. [^{3}H]RvE_1 specific binding to human PMN can be displaced by leukotriene B_4 (LTB_4) and U-75302, a LTB_4 receptor 1, (BLT_1) antagonist, but not by chemerin peptide, a ligand specific for another RvE1 receptor ChemR23. Recombinant human BLT_1 show specific binding with [^{3}H]RvE_1. RvE_1 selectively blocks adenylate cyclase with BLT_1, but not with BLT_2 [37]. RvE_1 partially mediates calcium mobilization, and retards subsequent stimulation by LTB_4. RvE_1 also attenuates LTB_4-mediated NF-κB stimulation in BLT_1-transfected cells. *In vivo* anti-inflammatory actions of RvE_1 are markedly down-regulated in BLT_1 knockout mice when given at low doses in peritonitis. In contrast, RvE_1 at higher concentrations significantly prevents PMN infiltration in a BLT_1-independent manner [37]. RvE_1 also bind with ChemR23 receptors. Treatment of dendritic cells with small interference RNA specific for ChemR23 blocks RvE_1-mediated regulation of IL-12 demonstrating novel counter regulatory responses in inflammation initiated by EPA-derived RvE_1 and its receptor. Collectively, these studies suggest that RvE_1 binds to BLT_1 as a partial agonist, potentially serving as a local damper of BLT_1 signals on leukocytes along with other receptors (e.g., ChemR23-mediated counterregulatory actions) to mediate the resolution of inflammation [36, 37].

Studies on RvE_1 metabolome (metabolic products derived from RvE_1) indicate that RvE_1 is metabolized to several novel products by human polymorphonuclear leukocytes and whole blood as well as in murine inflammatory exudates, spleen, kidney, and liver [38]. These products include 19-hydroxy-RvE_1, 20-carboxy-RvE_1, and 10,11-dihydro-RvE_1. They are less active than RvE_1 and produce reduce bioactivity in vivo. RvE_1 increases macrophage phagocytosis, a proresolving activity which is reduced by its metabolic inactivation. These results indicate that during inflammation and controlled resolution process, specific tissues inactivate RvE_1 and allow the co- ordinated return to homeostasis. Moreover, the RvE_1 metabolome may serve as a biomarker of these processes [38].

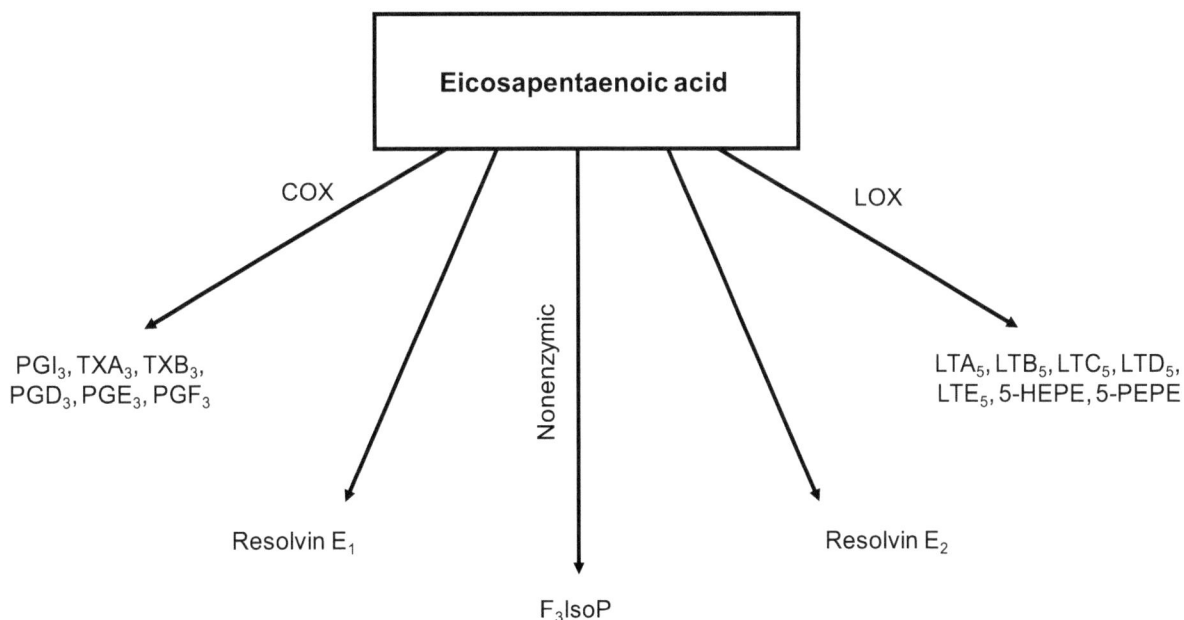

Figure 5: Enzymic and non-enzymic oxidation products of eicosapentaenoic acid.

Isolation, identification, and bioactions of another resolvin, 5S,18-dihydroxy-eicosapentaenoic acid, (resolvin E_2, RvE$_2$) have also been reported [39]. RvE$_2$ blocks zymosan-induced polymorphonuclear (PMN) leukocyte infiltration, and elicits potent anti-inflammatory properties in murine peritonitis. Similar to RvE$_1$, human recombinant 5-lipoxygenase generates RvE$_2$ from a common precursor of E series resolvins, namely, 18-hydroxyeicosa- pentaenoate (18-HEPE). Collective evidence suggests that RvE$_2$, together with RvE$_1$ may contribute to the beneficial actions of n-3 fatty acids in human diseases. It is stated that the 5-lipoxygenase in human leukocytes is a pivotal enzyme that generates both pro- and anti-inflammatory chemical mediators.

EPA modulates lipoprotein metabolism and as stated above, EPA downregulates the synthesis of pro inflammatory cytokines (IL-1β and TNF-α). The molecular mechanism associated with in down-regulation of cytokine secretion by EPA is not fully understood. However, inhibition of the TNF-α expression may be closely associated with downregulation. EPA appears to prevent NF-κB activation by preventing the phosphorylation of I-κB-α. In addition, suppression of ARA-derived eicosanoids by EPA may also play an important role because the EPA competes with ARA for COX and LOX-mediated oxidation [4, 40].

DOCOSANOIDS AND NEUROPROTECTION

The discovery of resolvins and neuroprotectins is a very important from therapeutic point of view. These metabolites antagonize the effects of eicosanoid, modulate leukocyte trafficking, and down-regulate the expression of cytokines in glial cells and modulate interactions among neurons, astrocytes, oligodendrocytes, microglia, and cells of the microvasculature [1, 2, 3, 9, 23]. Based on above studies, it is proposed that generation of resolvins and NPD$_1$ and resolvins is an internal neuroprotective mechanism for preventing brain tissue damage [3, 24, 32]. In addition, n-3 fatty acids also downregulate NF-κB activity [5], and suppress genes induced by pro-inflammatory cytokines and other mediators released by glial cells [5, 41 42]. The effectiveness of n-3 fatty acids in protecting against acute neural trauma depends on their ability to cross the blood brain barrier, their potential in terms of subcellular distribution in mitochondria, plasma membrane, and cytoplasm. A clearer appreciation of the potential therapeutic ability of n-3 fatty acids would emerge only when *in vivo* importance of interactions among excitotoxicity, neuro-inflammation, and oxidative stress is realized and are fully understood at the molecular level [5, 42, 43].

CONCLUSIONS

ARA and DHA are essential fatty acids that belong to n-6 and n-3 fatty acid families respectively. These fatty acids differ from each other in their metabolism as well as in their neurochemical effects. ARA is a precursor of proinflammatory eicosanoids except lipoxins that are anti-inflammatory. ARA metabolism generates ROS and many of its lipid mediators induce apoptosis. In contrast, DHA is oxidized into resolvins and neuroprotectins that have antioxidant, anti-inflammatory, and antiapoptotic effects.

EPA is another fatty acid that belongs to n-3 family. It is metabolized to 3-series PGs, TXs, and 5-series LTs by cyclooxygenases and 5-lipoxygenases. Metabolism of EPA also generates E series resolvins. In contrast, DHA is not a substrate for cyclooxygenase. DHA generates RvD series resolvins. In brain tissue, EPA competes with ARA for cyclooxygenases and lipoxygenases and decrease the generation of ARA-derived lipid mediators.

REFERENCES

[1]   Bazan, N. G. Brain Pathol. **2005**, 15, 159.
[2]   Bazan, N.G. Curr. Opin. Clin. Nutr. Metab. Care. **2007**, 10, 136.
[3]   Bazan, N.G. J. Lipid Res. **2009**, 50, S400.
[4]   Phillis, J.; Horrocks, L.A.; Farooqui, A.A. Brain Res. Rev. **2006**, 52, 201.
[5]   Farooqui, A.A.; Horrocks L.A.. Glycerophospholipids in Brain. Springer, New York **2007**.
[6]   Farooqui, A.A. Hot Topics in Neural membrane Lipidology. Springer, New York **2009**.
[7]   Simopoulos, A. P. Biomed. Pharmacother. **2002**, 56, 365; Simopoulos, A.P. Exp. Biol. Med. (Maywood). **2008**, 233, 674.
[8]   Farooqui, A.A. Beneficial Effects of Fish Oil on Human Brain, Springer, New York **2009**.
[9]   Serhan, C. N. Pharmacol. Ther. **2005**, 105, 7.
[10] Rapoport, S.I.. J Nutr. **2008**, 138, 2515.

[11]   Hirashima, Y.; Farooqui, A.A.; Mills, J.S.; Horrocks, L.A. J. Neurochem. **1992**, 59, 708.

[12]   Farooqui, A.A.; Yang, H.C.; Horrocks L.A.. Brain Res. Brain Res. Rev. **1995**, 21,152.

[13]   Hazen, S. L.; Gross, R. W. J. Biol. Chem. **1993**, 268, 9892.

[14]   Portilla, D.; Dai, G. J. Biol. Chem. **1996**, 271, 15451.

[15]   Garcia, M.C.; and Kim, H.Y. Brain Res. 1997, 768, 43.

[16]   Hong, S.; Gronert, K.; Devchand, P. R.; Moussignac, R. L.; Serhan C. N. J. Biol. Chem. **2003**, 278, 14677.

[17]   Marcheselli, V. L.; Hong, S.; Lukiw, W. J.; Tian, X. H.; Gronert, K.; Musto, A.; Hardy, M.; Gimenez, J. M.; Chiang, N.; Serhan C. N.; Bazan, N. G. J. Biol. Chem. **2003**, 278, 43807.

[18]   Serhan, C. N.; Arita, M.; Hong, S.; Gotlinger, K. Lipids **2004**, 39, 1125.

[19]   Wolfe, L.S.; Horrocks, L.A. Basic Neurochemistry (Siegel, G.J.; Agranoff, B.W.; Albers, R.W.; Molinoff P.B. eds), p. 475, Raven Press, New York **1994**.

[20]   Cimino, P.J.; Keene, C.D.; Breyer, R.M.; Montine, K.S.; Montine, T.J. Curr. Med. Chem. **2008**, 15,1863; Regan, J.W. Life Sci. **2003**, 74, 143; Sugimoto Y; Narumiya, S.J. Biol. Chem **2007**, 282, 11617.

[21]   Yacoubian, S.; Serhan, C.N. Nature Clinical Practice Rheumatology **2007**, 3, 570.

[22]   Decker, Y.; McBean, G.; Godson C. Am J Physiol Cell Physiol. **2009**, 296, C1420.

[23]   Serhan, C. N.; Levy B. Chem. Immunol. Allergy **2003**, 83,115.

[24]   Serhan, C. N. Curr. Opin. Clin. Nutr. Metab. Care **2005**, 8, 115.

[25]   Chiang, N.; Arita, M.; Serhan C. N. Prostaglandins Leukot. Essent. Fatty Acids **2005**, 73, 163.

[26]   Capra, V. Pharm. Rev. **2003**, 50, 1.

[27]   Wada, K.; Arita, M.; Nakajima, A.; Katayama, K.; Kudo, C.; Kamisaki, Y.; Serhan C. N. FASEB J. **2006**, 20:1785-1792.

[28]   Svensson, C.I.; Zattoni, M.; Serhan, C.N. J. Exp. Med. **2007**, 204, 245.

[29]   Serhan, C.N.; Chiang, N. Br. J. Pharmacol. **2008**, 153 Suppl 1, S200.

[30]   Mukherjee, P.K.; Marcheslli, V.L.; Barreiro, S.; Hu, J.; Bok, D.; Bazan N.G. Proc. Natl. Acad.Sci. **2007**, 104, 13152.

[31]   Bazan. N.G. Adv Exp Med Biol. **2008**, 613, 39.

[32]   Lukiw, W. J.; Cui, J. G.; Marcheselli, V. L.; Bodker, M.; Botkjaer, A.; Gotlinger, K.; Serhan C. N.; Bazan N. G. J. Clin. Invest. **2005**, 115, 2774.

[33]   Akbar, M.; Kim, H. Y. J. Neurochem. **2002**, 82, 655.

[34]   James, M.J.; Gibson, R.A.; Cleland, L.G. Am J Clin Nutr. **2000**, 71(1Suppl), 343S.

[35]   Calder, P. C. (2005) Biochem. Soc. Trans. 2005, 33, **423.**

[36]   Arita, M.; Oh, S. F.; Chonan, T.; Hong, S.; Elangovan, S.; Sun, Y. P.; Uddin, J.; Petasis, N. A.; Serhan, C. N. J. Biol. Chem. **2006**, 281, 22847.

[37]   Arita, M.; Ohira, T.; Sun, Y.P,; Elangovan, S.; Chiang, N.; Serhan, C.N. J. Immunol. **2007**, 178, 3912.

[38]   Hong, S.; Porter, T.F.; Lu, Y.; Oh, S.F.; Pillai, P.S.; Serhan, C.N. (2008). J. Immunol. **2008**, 180:3512.

[39]   Tjonahen, E.; Oh, S.F.; Siegelman, J.; Elangovan, S.; Percarpio, K.B.; Hong, S.; Arita, M.; Serhan, C.N. Chem Biol. **2006**, 13: 1193.

[40]   Zhao, Y.**;** Joshi-Barve, S.; Barve, S.; Chen L.H. J. Am. Coll. Nutr. **2004**, 23, 71.

[41]   Gilgun-Sherki, Y.; Melamed, E.; Offen, D. Curr. Pharmaceut. Design. **2006**, 12, 3509.

[42]   Wang, J. Y.; Wen, L. L.; Huang, Y. N.; Chen, Y. T.; Ku M. C. Curr. Pharmaceut. Design **2006**, 12, 3521.

[43]   Farooqui, A.A.; Horrocks, L.A.; Farooqui, T. J. Neurochem. **2007,** 101, 577.

CHAPTER 2

Neurochemical Aspects of Inflammation in Brain

Akhlaq A. Farooqui*

Department of Molecular and Cellular Biochemistry, The Ohio State University, Columbus, Ohio 43221,USA

Abstract: Neuroinflammation is a host defense mechanism involved in restoration of normal neural cell function. Most major CNS diseases are characterized by neuroinflammation. Following brain injury, microglial cells are activated and initiate a rapid response, which involves cell migration, proliferation, release of cytokines/chemokines and trophic and/or toxic effects. Cytokines/chemokines mediate the stimulation of phospholipases A_2 and cyclooxygenases. Stimulation of these enzymes results in breakdown of membrane glycerophospholipids with release of arachidonic acid and docosahexaenoic acid. Oxidation of arachidonic acid produces pro-inflammatory eicosanoids. Lyso-glycerophospholipids, the other products of reactions catalyzed by phospholipase A_2, are converted to pro-inflammatory platelet-activating factor. Eicosanoids and platelet activating factor intensify neuroinflammation. Lipoxin, an oxidized product of arachidonic acid through 5-lipoxygenase, is involved in resolution of inflammation and is anti-inflammatory. Docosahexaenoic acid is metabolized to resolvins and neuroprotectins. These lipid mediators retard the generation of eicosanoids. Levels of eicosanoids are markedly increased in many neurological disorders. Docosahexaenoic acid and its oxidized products retard neuroinflammation by inhibiting transcription factor NF-κB, preventing cytokine secretion, blocking the synthesis of eicosanoids, and modulating leukocyte trafficking. Depending on its timing and magnitude in brain tissue, inflammation serves multiple purposes. It is involved in protection of uninjured neurons and removal of degenerating neuronal debris and also in assisting repair and recovery processes. The dietary ratio of arachidonic acid to docosahexaenoic acid modulates inflammation in acute neural trauma and neurodegenerative diseases. Increase in dietary intake of docosahexaenoic acid offers the possibility of counterbalancing the harmful effects of high levels of arachidonic acid-derived pro-inflammatory eicosanoids.

Keywords: Neuroinflammation; pro-inflammatory; anti-inflammatory; arachidonic acid; docosa- hexaenoic acid; phospholipase A_2; cytokines; chemokines; ARA/DHA ratio

INTRODUCTION

Neuroinflammation is a protective mechanism that separates the injured brain tissue from uninjured area, destroys affected cells, and repairs the extracellular matrix [1]. Without a strong inflammatory response, brain tissue would be very vulnerable to acute neural trauma, neurodegenerative diseases, and microbial, viral, and prion infections.

Inflammatory response involves the participation of different cellular types of the immune system (macrophages, mast cells, T and B lymphocytes, dendritic cells) and resident cells of the CNS (microglia, astrocytes, neurons), adhesion molecules, cytokines and chemokines among other proteic components. During neuroinflammation, chemotaxis is an important process in the recruitment of non-neural cells into the brain tissue [2]. The lymphocyte recruitment implies the presence of chemokines and chemokine receptors, the expression of adhesion molecules, the interaction between lymphocytes and the blood brain barrier (BBB) endothelium, and finally their passage through the BBB to arrive at the site of inflammation. If this process is not controlled and prolonged, inflammation loses its protective function and can be the cause of damage. The main mediators of neuroinflammation are microglial cells. In the normal healthy brain, microglial cells have a ramified morphology and are called resting microglia. The resting microglia are activated during CNS injury and transformed into activated microglia characterized by amoeboid morphology. Activated microglial cells not only migrate rapidly to the injury site, but also engulf dead cells, and clear cellular debris. Although, the activation of microglia is a hallmark of brain pathology, but it remains controversial whether microglial cells have beneficial or detrimental functions in various neuropathological conditions. The chronic activation of microglia may induce and cause neuronal damage through the release of potentially cytotoxic molecules such as proinflammatory cytokines, reactive oxygen intermediates (ROS), proteinases and complement

*Address correspondence to: Department of Molecular and Cellular Biochemistry, The Ohio State University, Columbus, OH 43210, USA; Tel.: (614) 488-0361; Email: farooqui.1@osu.edu

Akhlaq A. Farooqui & Tahira Farooqui (Eds.)

proteins (see below) [1,2]. Information on signals and mechanisms that modulate microglial activation following CNS injury is beginning to be published. Thus, the inflammatory cascade is an attempt by the brain to eliminate the challenge imposed by the injury or infection, clear the system of the dead and damaged neurons, and rescue the normal functioning of this vital organ [1,2]. At the same time inflammation promotes neurogenesis. Inflammatory factors released during mild acute neuroinflammation usually stimulate neurogenesis; where as the factors released by uncontrolled inflammatory response block neurogenesis [1, 2, 3, 4].

Inflammatory response also involves recruitment of polymorphonuclear leukocytes (PMN) from the blood stream into brain tissue. PMN migration, a co-ordinated multistep process, is facilitated not only by chemotaxis, PMN adhesion to endothelial cells in the area of inflammation, but also by diapedesis, and migration of endothelial monolayer into the interstitium [2]. PMN eliminate invading antigens by phagocytosis and release free radicals and lytic enzymes into phagolysosomes. This is followed by a process called resolution, a turning off mechanism by neural cells to limit tissue injury. Acute inflammation normally resolves spontaneously, but the mechanism associated with this process remains elusive [3].

Two types of neuroinflammation, acute and chronic, occur in brain tissue. Acute neuroinflammation develops rapidly and accompanied by pain, whereas chronic inflammation develops slowly. Chronic neuroinflammation is accompanied by pain threshold, which body does not feel. As a result, the immune system continues to assault the brain and chronic inflammation lingers for years, ultimately reaching the threshold of detection [4]. Thus in brain, major hallmarks of neuroinflammation include activation and transformation of microglial cells into phagocytic cells, and to a lesser extent, induction of reactive astrocytosis. It remains controversial whether activation of microglial cells has beneficial or detrimental functions in neuroinflammatory reaction. The chronic activation of microglia may cause neuronal damage through the release of potentially cytotoxic molecules such as pro-inflammatory cytokines, reactive oxygen species, nitric oxide, proteinases and complement proteins. Very little is known about molecular mechanisms and internal and external factors that control and modulate the dynamics of acute and chronic neuroinflammation.

Table 1: Cytokines, chemokines, and other factors associated with the development of inflammation

Cytokines/chemokines/factors	Nature	Effect	Reference
TNF-α	Proinflammatory	Increased	[10, 77, 79]
IL-1β	Proinflammatory	Increased	[10, 77, 79]
IL-6	Proinflammatory	Increased	[10, 77, 79]
IL-17	Proinflammatory	Increased	[10, 77, 79]
IL-15	Proinflammatory	Increased	[10, 77, 79]
IL-18	Proinflammatory	Increased	[10, 77, 79]
IL-4	Antiinflammatory	Increased	[10, 77, 79]
IL-10	Antiinflammatory	Increased	[10, 77, 79]
IL-13	Antiinflammatory	Increased	[10, 77, 79]
TGF-α	Proinflammatory	Increased	[10, 77, 79]
Chemokine MIP-1α	Proinflammatory	Increased	[10, 77, 79]
Chemokine MCP-1	Proinflammatory	Increased	[10, 77, 79]
Chemokine MDC	Proinflammatory	Increased	[10, 77, 79]
ICAM-1	Proinflammatory	Increased	[10, 77, 79]
VCAM-1	Proinflammatory	Increased	[10, 77, 79]
E-selectin	-	Increased	[10, 77, 79]

It is becoming increasingly evident that neuroinflammation involves the interplay not only among microglia, astrocytes, neurons, PMN and endothelial cells, but also interactions (interplay) among various lipid mediators that originate from enzymic and non-enzymic degradation of neural membrane glycerophospholipids sphingolipids and cholesterol. In addition, receptors, like toll like receptors (TLR) and transcription factors such as peroxisome proliferator-activated receptor (PPAR) and NF-κB also play an important role in modulation of neuroinflammation. The purpose of this commentry is not only to describe neurochemical events associated with neuroinflammation, but also discuss the modulation of neuroinflammation by dietary fatty acids.

PARTICIPATION OF GLIAL CELLS IN NEUROINFLAMMATION

Microglials constitute 20% of the total glial cell population [5]. They can sense changes in the periphery and respond quickly to pathogenic stimuli in order to protect the brain. Microglial cells and astrocytes secrete a variety of immune system modulators including complement proteins, adhesion molecules, proinflammatory cytokines, which participate and modulate the dynamics of neuroinflammation [6-9]. A number of mechanisms including the expression and activation of secretory phospholipase A_2 (sPLA$_2$), cytosolic phospholipase A_2 (cPLA$_2$), cyclooxygenase-2 (COX-2), and lipoxygenases (LOX), that release free arachidonic acid (ARA) and convert it to its oxidized products (proinflammatory prostaglandin, PGE$_2$ and leukotriene (leukotriene B4), formation of lysoglycerophospholipids, generation of platelet-activating factor (PAF), and reactive oxygen species (ROS) [10], propagating and intensifying neuroinflammation. Furthermore, expression of inducible NOS (iNOS) generates high levels of nitric oxide (micromolar range), which reacts with superoxide anion to form peroxynitrite (ONOO$^-$). This metabolite has a half-life of approximately 1-2 seconds, thus it breaks down into hydroxyl radical and other toxic products, which promote and intensify inflammatory response. In addition, microglia, astrocytes, neurons, endothelial cells, and oligodendrocytes also express and secrete complement proteins, which play important roles in modulation induction and maintenence of neuroinflammation [11].

CYTOKINE AND NEUROINFLAMMATION

Cytokines are major effectors of the neuroinflammation. They are heterogeneous group of proteins and polypeptides involved in regulation of cell–cell interactions both in normal and pathological situations. Cytokines mediate cellular intercommunication through autocrine, paracrine, or endocrine mechanisms [12]. Their actions involve a complex network linked to feedback loops and protein kinase and PLA$_2$ cascades. Cytokines induce their effects by binding to specific membrane associated receptors that are composed of an extracellular ligand binding region (membrane spanning region) and an intracellular region that is activated by binding with cytokines, and facilitates the delivery of a signal to the nucleus [13]. Cytokines play an important role in neural cell response to infection and brain injury [14, 15]. Physiological levels of cytokines are needed for neural cell survival, but imbalanced secretion of cytokines is detrimental to neurons [13]. Tumor necrosis factor-α (TNF-α) and interleukin-1β (IL-1β) are major cytokines that are upregulated in the brain tissue after neural trauma and infection (Table **1**). In addition, interleukin-1 alpha (IL-1α), interleukin-3 (IL-3; colony-stimulating factor-1), interleukin-6 (IL-6), and tumor and growth factors (TGF- α and β), are also secreted by both microglia and astrocytes during neuroinflammation. IL-6, an anti-inflammatory cytokine is involved in the recovery process [6-9]. Synthesis and modulation of cytokine secretion creates an autoregulatory feedback loop involved in cytokine cascade [16]. In traumatized brain, astrocyte and microglial cells also secrete neurotrophic factors such as neurotrophin-3 and brain-derived neurotrophic factor (BDNF) which promote neuronal survival [1]. Furthermore, TNF-α, IL-1, and IFN-γ, are also involved in immunosuppressive functions. Their subsequent expression following neuroinflammation is associated with neurorepair and recovery processes [1]. Accumulating evidence suggests that actions of cytokines are dependent on the synergistic or antagonistic activities of various cytokines through a complex network that not only involves their feedback loops, but also modulation of many enzymes associated with production of lipid mediators [13, 16].

CONTRIBUTION OF NEURAL CELL RECEPTORS IN NEUROINFLAMMATION

Glutamate, dopamine, serotonin, histamine, retinoic acid, toll-like, and bradykinin receptors contribute to neuroinflammation (Table **2**). These receptors are linked to phospholipase A_2 (PLA$_2$), and phospholipase (PLC), and downstream cyclooxygenases, lipoxygenases, epoxygenases, sphingomyelinases and cholesterol hydroxylase (cytochrome P450 oxygenases) (Table **3**) through G protein dependent and G protein independent mechanisms [10]. Agonist-mediated hyper-stimulation of glutamate, biogenic amine, retinoic acid, and bradykinin receptors results in stimulation of isoforms of PLA$_2$ and PLC, breakdown of neural membrane glycerophospholipids and release of free arachidonic (ARA) or docosahexaenoic acids (DHA), which are oxidized by cyclooxygenase-2, 15-lipoxygenase, and epoxygenases. The oxidized products of ARA and DHA act as bioactive lipid mediators (Table **4**). These mediators are component of signal transduction network, which facilitates the transfer of signal (information) from plasma membrane to the nucleus to induce a biological response at the gene level.

Generation of cytokines also initiates signaling through Toll-like receptors (TLRs), which are transmembrane proteins that recognize pathogen and host-derived molecules (cell wall components, proteins, nucleic acids, and synthetic chemical compounds) released from pathogens or injured tissues and cells. They are critical components of the innate immune system. These receptors are expressed on microglial cells and dendritic cells. Recently, progress has been made in understanding the regulation of the innate immune system and it is shown that TLRs are closely associated with neuroinflammation-mediated signaling in brain [17]. These inflammatory responses may be initiated by the stimulation of toll-like receptors (TLRs). Thus, intracerebroventricular inoculation of TLR7 and/or TLR8 agonists to newborn mice results in induction of glial activation, up-regulation of IFN- γ,

TNF-α, CCL2, and CXCL10, and neuroinflammation in the CNS [18]. However, these responses are only of short duration compare to responses mediated by the TLR4 agonist LPS. Furthermore, some of the TLR7 and/or TLR8 agonists differ with each other in their ability to activate glial cells as evidenced by their ability to induce cytokine and chemokine expression both *in vivo* and *in vitro* [18]. Thus, TLR7 stimulation can induce neuroinflammatory responses in the brain, but individual TLR7 agonists may differ in their ability to stimulate cells of the CNS [18].

ENZYMIC LIPID MEDIATORS OF ARA AND DHA METABOLISM IN NEUROINFLAMMATION

As stated above, glycerophospholipids are transformed into lipid mediators, which are lipophilic molecules that facilitate signal transduction processes associated with molecular and cellular events in the brain. In glycerophospholipid sn-2 position of glycerol moiety is mainly esterified with (ARA and DHA) respectively [19, 20]. ARA and DHA are released by the action of cPLA$_2$ and PlsEtn-PLA$_2$ on phosphatidylcholine (PtdCho) and ethanolamine plasmalogen (PlsEtn), respectively. The action of cyclooxygenase converts ARA into eicosanoids, which include prostaglandins (PG), leukotrienes (LT), and hydroxyeicosatetraenoic acids (HETE) (Fig. 1).

Eicosanoids are closely associated with cell proliferation, differentiation, oxidative stress and neuroinflammation. Another, ARA-derived mediator is lipoxin. Eicosanoids induce proinflammatory effects, whereas lipoxins, a group of trihydroxytetraene eicosanoids, involved in the resolution of acute inflammation by modulating key steps in leukocyte trafficking and preventing neutrophil-mediated acute tissue injury produce anti-inflammatrory effects (Table 4) [19]. Lysophospholipid, the other product of PLA$_2$ catalyzed reaction either reacylated through acylation/ deacylation cycle into native glycerophospholipids or converted into platelet activating factor (PAF;1-O-alkyl-2-acetyl-*sn*-glycerophos- phocholine), which is a pro-inflammatory lipid mediator. It exerts its inflammatory effects by activating the PAF receptors that consequently activate leukocytes, stimulate platelet aggregation, and induce the release of cytokines and expression of cell adhesion molecules [21]. During the inflammatory process, PAF activates leukocytes tethered to the blood vessel wall via specific adhesion molecules expressed by endothelial cells. The physiological activity of PAF is not limited to its pro-inflammatory function. PAF is also involved in a variety of other settings including allergic reactions, brain function, and circulatory system disturbances such as atherosclerosis [21].

Table 2: Receptors associated with the induction of neuroinflammation.

Receptor	Reference
Glutamate receptor	[10, 19]
Dopamine receptor	[10, 19]
Serotonin receptor	[10, 19]
Histamine receptor	[10, 19]
Retinoic acid receptor	[38]
Cytokine receptors	[13, 14]
Toll like receptor	[18]
PPARα receptor	[80]

Among lysophospholipids, lyso-PtdCho is a chemoattractant that induces the expression of growth factors and adhesion molecules in endothelial cells. It also activates white blood cells. This activation increases their ability to permeate the endothelium. Lyso-PtdSer triggers the secretion of histamine by mast cells [22]. All these processes contribute to induction and maintenance of inflammatory reaction and apoptotic cell death. In addition, two endogenous ARA-containing molecules, 2-arachidonylglycerol (2-AG) and arachidonylethanolamide (anandamide) are also generated from the non-oxidative metabolism of ARA (Fig. **3**). They regulate microglial cell migration toward dying neurons during apoptotic cell death. Endocannabinoids also mediate anti-inflammatory, anti-apoptotic, and neuroproprotective properties [19]. Collective evidence suggests that endocannabinoids act as retrograde messengers that, by inhibiting neurotransmitter release via presynaptic CB_1 cannabinoid receptors, regulate the functionality of many synapses. In addition, the endocannabinoid system participates in the control of neuron survival. Neuronal survival through CB_1 receptors involves not only phosphatidylinositol 3-kinase /Akt and extracellular signal-regulated kinase pathways, but also the inhibition of glutamatergic neurotransmission. Through the involvement of above signaling mechanisms, CB_1 receptor activation protects neurons from acute neural trauma, neuroinflammatory conditions, and neurodegenerative diseases [10, 19].

Table 3: Enzymes associated with the development of inflammation.

Enzyme	Effect	Reference
sPhospholipase A_2	Increased	[10, 19]
Cyclooxygenase-2	Increased	[8, 10, 19]
Inducible nitric oxidase	Increased	[8, 10, 19]
NADPH oxidase	Increased	[78]
Matrix metalloproteinase	Increased	[77]
Caspase-3	Increased	[10, 19]
Sphingomyelinase	Increased	[10, 19]

Docosanoids produce antioxidant, anti-inflammatory and anti-apoptotic effects in the brain tissue. Docosanoids not only antagonize the effects of eicosanoids, but also modulate leukocyte trafficking and downregulate the expression of cytokines in glial cells. Eicosanoids and docosanoids act through their specific cysLT1 receptors that are located on neural cell surface [10, 19]. Thus, eicosanoids act through at least four types of eicosanoid receptors (EP, DP, FP, TP, IP) [24]. These receptors have a protein with seven hydrophobic transmembrane segments, and evoke cellular responses by distinct intracellular mechanisms [10, 19, 24]. Leukotrienes exert their effects through three types of leukotriene receptors such as (LTD4), cysLT2 (LTC4) and hydroleukotriene BLT (LTB4) receptors. These leukotrienes are potent inflammatory mediators, which also induce brain blood barrier (BBB) disruption and brain edema. At the injury site, PGE_2 is involved in modulating the immune response whilst its pro-inflammatory signaling is associated with vascular and microglial cell activation [10, 19]. Some prostaglandins, PGE_1, PGE_2 and PGF2 are inflammatory, whereas others are anti-inflammatory, for example PGD_2 and 15-deoxy-$\Delta^{12,14}$-prostaglandin- J_2 [10,19].

Table 4: Inflammatory properties of enzymically-derived lipid mediators of n-6 and n-3 fatty acids.

Lipid mediator	Property	Reference
Prostaglandins	Proinflammatory	[19, 24]
Leukotrienes	Proinflammatory	[19, 24]
Thromboxanes	Proinflammatory	[19, 24]
Lipoxins	Antiinflammatory	[19, 24]
PAF	Proinflammatory	[19, 24]
Lysophospholipids	Proinflammatory	[19, 24]
Docosatrienes	Proinflammatory	[19, 24]
Resolvins	Antiinflammatory	[19, 24]
2-AG	Antiinflammatory	[19, 24]
Anandimide	Antiinflammatory	[19, 24]

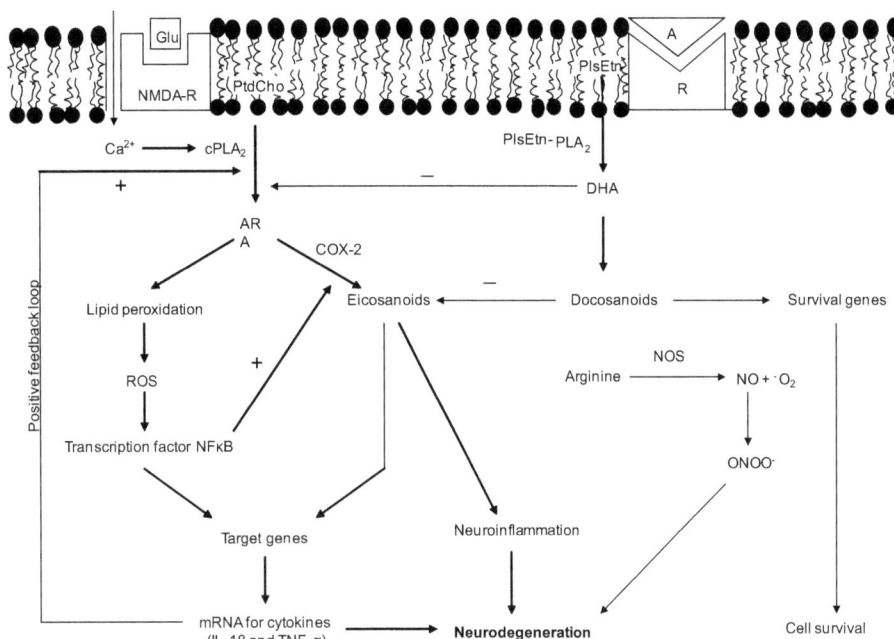

Figure 1: Generation of arachidonic acid-derived eicosanoids from glycerophospholipid and their contribution in the induction of neuroinlammation. Docosahexaenoic acid-derived lipid mediators, docosanoids produce anti-inflammatory effects and promote neural cell survival. N-Methyl-D-aspartate receptor (NMDA-R); kainic acid receptor (R); kainic acid (A); phosphatidylcholine (PtdCho); cytosolic phospholipase A_2 (cPLA$_2$); plasmalogen-selective phospholipase A_2 (PlsEtn-PLA$_2$); cyclooxygenase-2 (COX-2); nitric oxide synthase (NOS); peroxynitrite (ONOO$^-$); nuclear factor-kappa B (NF-κB); interleukin-1beta (IL-1β); tumor necrosis factor-α (TNF-α).

During inflammatory reactions, eicosanoids not only initiate inflammatory responses, but also mediate resolution. There are two phases in inflammatory responses: one at the onset for the generation of pro-inflammatory eicosanoids and the other at resolution for the synthesis of pro-resolving eicosanoids [25]. The first phase involves the stimulation of isoforms of PLA$_2$ and release of ARA. This is followed by the generation of PGE$_2$, LTB$_4$, and PAF through COX-2, LOX, and acetyl-CoA acetyltransferase reactions, respectively. The second phase of neuroinflammation involves the generation of lipoxins, and the pro-resolving prostaglandin, PGD$_2$ [25]. High levels of eicosanoids contribute to the development of cytotoxicity, vasogenic brain edema, and neuronal damage mediated by the participation of isoforms of PLA$_2$, PLC, PKC and cytokines and NF-κB (Figs. **1** and **2**) [10, 19, 24].

Lipoxins, a group of recently discovered ARA-derived lipid mediators, exert anti-inflammatory and proresolution effects on neural and non-neural cells. Lipoxins mediate resolution by interacting with their receptors (ALX and LXA receptors) [26, 27, 28]. They mediate a number of processes, including regression of pro-inflammatory cytokine production, inhibition of cell proliferation, and stimulation of phagocytosis of apoptotic leukocytes by macrophages [10, 19, 27, 28,]. Lipoxin A$_4$ (LXA$_4$) is one of the principal lipoxin synthesized by neural and non-neural cells. It serves as a "stop signal" that regulates key steps in leukocyte trafficking and prevents neutrophil-mediated tissue injury [27, 28]. LXA$_4$ receptor, which is coupled to G protein, has been identified in astrocytes and microglia. This suggests that astrocytes and microglial cells may be a target for lipoxin-mediated processes in the brain. Quantitative RT-PCR studies indicate that LXA$_4$ significantly inhibit the IL-β-mediated stimulation of IL-8 and ICAM-1 expression in astrocytes and microglia [29]. Furthermore, LXA$_4$ reduces the expression of IL-1β-mediated IL-8 protein levels. LXA4 retards IL-1β-mediated degradation of I-κBα, and the activation of an NF-κB regulated reporter gene construct. Collective evidence suggests that LXA$_4$ mediates anti-inflammatory effects in 1321N1 astrocytoma cells at least in part via an NF-κB-dependent mechanism [29].

The specific receptors for docosanoids are found in neural and non-neural tissues. These receptors include resolvin D receptors (resoDR1), resolvin E receptor (resoER1). Studies on isolation and characterization of these metabolites by lipidomics are in progress [10, 19].

MOLECULAR MECHANISMS ASSOCIATED WITH NEUROINFLAMMATION

The expression of genes involved in the inflammatory response is controlled transcriptionally and posttranscriptionally. The released cytokines act through their receptors causing activation of cascades of protein kinases and the pathway leading to activation of the transcription factor, nuclear factor kappa B (NF-κB) (Fig.2). In microglial cells NF-κB is present in the cytoplasm in an inhibitory form attached to its inhibitory protein, IκB. NF-κB activity is tightly controlled by the IκB kinase complex, consisting of IκB kinases IκKα, IκKβ, and IκKγ. IκKβ is essential for the inflammatory cytokine-mediated activation of NF-κB [30]. Upon stimulation IκB is rapidly phosphorylated, ubiquinated, and then degraded by proteasomes resulting in the release and subsequent nuclear translocation of active NF-κB [19, 20]. In the nucleus NF-κB induces the transcription of many genes associated with inflammatory and immune responses (Fig. 2). These include genes for enzymes, secretory phospholipase A_2 (sPLA$_2$), COX-2, inducible nitric oxide synthase (iNOS), NADPH oxidase and matrix metalloproteinases (MMPs), cytokines, such as TNF-α, IL-1β,

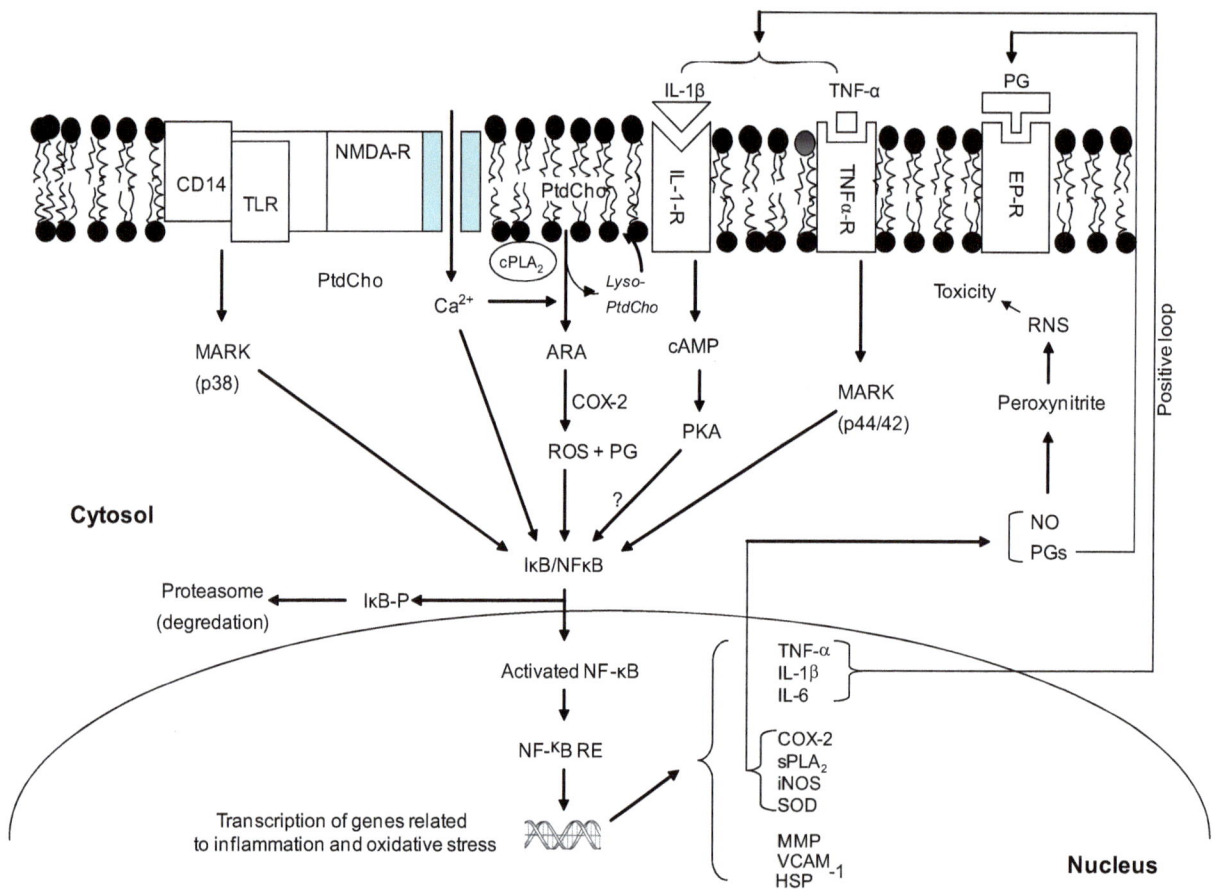

Figure 2: Hypothetical diagram showing the involvement of ROS in NF-κB-mediated gene transcription. N-Methyl-D- aspartate receptor (NMDA-R); cytosolic phospholipase A_2 (cPLA$_2$); lysophosphatidylcholine (lyso-PtdCho); arachidonic acid (ARA); interleukin-1β; cyclic adenosine monophosphate (cAMP); protein kinase A (PKA); tumor necrosis factor-α (TNF-α); TNF-α-receptor (TNF-α-R); interleukin-1β (IL-1β); IL-1β-receptor (IL-1β-R); interleukin-6 (IL-6); mitogen-activated protein kinase (MARK); nitric oxide (NO); prostaglandins (PGs); prostaglandin receptors (EP-R); nuclear factor-κB (NF-κB); NFκB, nuclear factor κB; nuclear factor κB-response element (NFκB-RE); inhibitory subunit of NFκB (IκB,); cyclooxygenase-2 (COX-2); inducible nitric oxide synthase (iNOS); secretory phospholipase A_2 (sPLA$_2$); superoxide dismutase (SOD); matrix metalloproteinase (MMP); vascular adhesion molecule-1 (VCAM-1); heat shock protein (HSP); TLR, toll-like receptor; and CD14 (LPS receptor), a 55kDa glycosylphosphatidylinositol anchored surface myeloid glycoprotein expressed on microglial cells for phagocytosis. Positive sign (+) indicates stimulation (proinflammatory) and negative sign (-) indicates inhibition (anti-inflammatory).

IL-6, and other proteins including, intracellular adhesion molecule-1 (ICAM-1), vascular adhesion molecule-1 (VCAM-1), E-selectin (Table 1). NF-κB is also stimulated by polyunsaturated fatty acids, which are generated by the action of isoforms of PLA$_2$. This action of NF-κB is inhibited by N-acetylcysteine as well as vitamin E [31], suggesting the involvement of reactive oxygen species (ROS) during NF-κB-mediated processes. Generation of ROS and NF-κB activation involves participation of NADPH oxidase. This process is an important component of the innate immune response against toxic agents and is closely associated with the cellular response to a variety of physiological and pathological signals [32, 33, 34, 35]. It is proposed that stimulation of cPLA$_2$ is followed by NF-κB recruitment to the plasma membranes where it interacts with NADPH oxidase [36]. The interaction between NF-κB and cPLA$_2$ provides the molecular basis for ARA release by cPLA$_2$ and generation of reactive species to activate the NADPH oxidase [36]. The ability of cPLA$_2$ to modulate superoxide production and generation of eicosanoid indicates its importance in inflammatory processes. Meanwhile, endothelial cells lining of local cerebral blood vessels is stimulated to produce adhesion molecules, causing the migration of peripheral circulating leukocytes into the compromised brain tissue, an event that amplifies inflammatory signaling cascades. A second group of transcription factors called peroxisome proliferator-activated receptors (PPARs) has also been implicated in neuroinflammation [37].

Activation of PPAR isoforms elicits anti-inflammatory activities in neural cells. Long-chain polyunsaturated fatty acids, eicosanoids, and oxidized phospholipids are endogenous ligands for PPAR-γ. In the presence of the peroxisome proliferator-activated receptor response element (PPAR-RE), PPAR heteromerizes with retinoid X-receptors (RXR), recruits the co-activator containing histone acetylase activity, and subsequently facilitates gene expression [38]. Mice deficient in PPAR have a prolonged response to inflammatory stimuli. PPAR ligands, in particular those of PPARα and PPARγ, inhibit the activation of inflammatory gene expression and negatively interfere with pro-inflammatory transcription factor signaling pathways in vascular and inflammatory cells [39]. Activation of Ca^{2+}-dependent constitutive nitric oxide synthases (NOS), endothelial NOS (eNOS) and neuronal NOS (nNOS) maintains basal levels of nitric oxide (NO). During neuroinflammation in induction of Ca^{2+}-independent iNOS in astrocytes and microglial cells along with upregulation in expression of adhesion molecules, cytokines and chemokines and receptors result in generation of high levels of NO, which reacts with superoxide anion and generates peroxynitrite (Fig. **1**). This metabolite is a major cytotoxic agent responsible for neuroinflammation-mediated neural cell death [40]. Cytokines and chemokines participate in the interaction among neurogliovascular cells and play an important role in the induction and maintenance of inflammation in brain [41].

NON-ENZYMIC METABOLITES OF ARA, EPA, AND DHA IN NEUROINFLAMMATION

Isoprostanes (IsoPs) are PG-like mediators formed non-enzymically by free radical-catalyzed peroxidation of esterified AA *in vivo* [42, 43] (Fig. **4**). They differ from PGs. In IsoPs the side chains are *cis* to the cyclopentane ring, whereas in PGs they have the *trans* orientation. The mechanism by which IsoPs are formed is analogous to the formation of PGs by COX enzymes [44]. Unlike PGs, the formation of IsoPs *in situ* initially takes place at the esterified AA on the glycerophospholipid molecule [45]. IsoPs are subsequently released in free form by the action of PLA$_2$ [45].

IsoPs are very potent vasoconstrictors in brain microvasculature. F$_2$-IsoP exerts its action in vascular beds by facilitating binding between endothelial cells and monocytes [45]. Binding between endothelial cells and monocytes is the key initial event in atherogenesis-related inflammation. Isoprostane-mediated monocyte adhesion is VCAM-1 independent but involves protein kinase A and mitogen-activated protein kinase kinase 1. F$_2$-IsoP also modulates the p38 MAPK pathway during monocyte adhesion [46]. All these processes relate to inflammation and atherosclerosis. Collectively, these studies suggest that IsoPs are not only feedback regulators related to neuroinflammation, but also to vasoconstriction, mitogenesis, and monocyte adhesion [46].

DHA undergoes non-enzymic oxidation. Compounds generated by this process are neuroprostanes (NP) (Fig. **4**) [43, 47, 48]. Similarly, non-enzymic oxidation of eicosapentaenoic acid (EPA) results in formation of F$_3$ isoprostane [49]. Non-enzymic oxidation of DHA also produces neuroketals (NK) [50]. Like IsoK, NK are very reactive. They form not only lactam and Schiff base adducts, but also generate lysine adducts suggesting that these metabolites may be involved in protein-protein cross-linking in brain tissue under oxidative stress. These metabolites may have neurochemical effects that intensify both neuroinflammation and oxidative stress in acute neural trauma and neurodegenerative diseases [19, 20, 47].

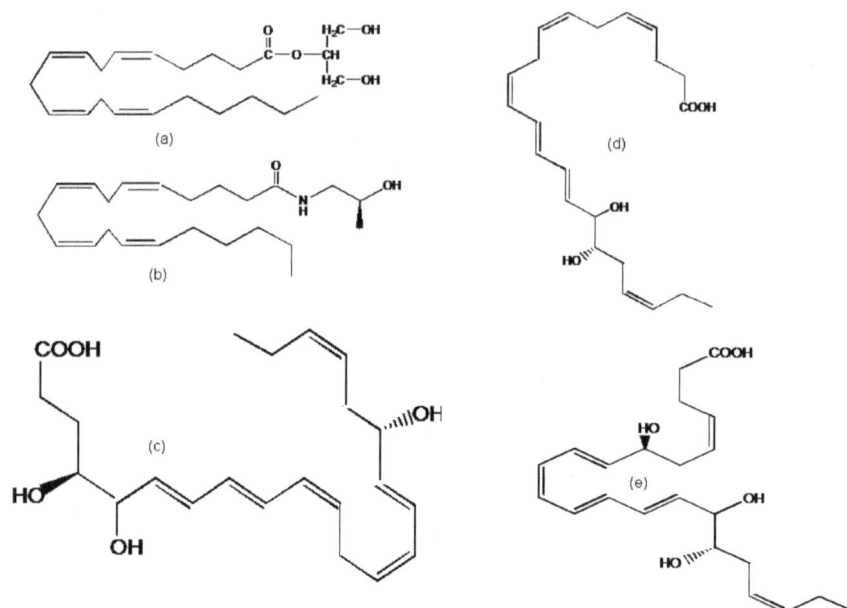

Figure 3: Chemical structures of arachidonic acid-derived lipid mediators. 2-Arachidonyl glycerol (a); anandamide (b); 16, 17S-docosatriene (c); 4S5, 17S-resolvin (d); and 7,16, 17S-resolvin (e).

MODULATION OF NEUROINFLAMMATION BY DIETARY FATTY ACIDS

Brain contains large amounts of ARA and DHA compared to other tissues. Despite of enrichment in the brain, ARA and DHA cannot be synthesized *de novo* by mammals; they, or their precursors, must be ingested from dietary sources and transported to the brain [10, 23, 51, 52]. The present day western diet has a ratio of ARA to DHA is about 15-20: 1. The Paleolithic diet on which human beings have evolved, and lived for most of their existence, has a ratio of ARA to DHA of 1:1. Changes in eating habits and agriculture development within the past 100 to 200 years and consumption of corn oil, sunflower oil, and safflower oil caused these changes in the ARA to DHA ratio. The decreased consumption of DHA-enriched foods and increased consumption of n-6 enriched vegetable oils is responsible for the 15-20:1 ARA: DHA ratio [53]. The richest source of DHA is fish oil. The consumption of DHA has numerous beneficial effects on the health of the human brain [51]. The beneficial effects of DHA may be due not only to its effect on the physicochemical properties of neural membranes, but also to modulation of neurotransmission [19], gene expression [54, 55, 56], enzyme, ion channel, receptor activities, and immunity [57]. EPA and DHA resemble each other in many biochemical effects including the decrease in production of the key immunoregulatory cytokines, IL-10, TNF-α and INFγ, and in prevention of lipopolysaccharide (LPS) toxicity [58, 59, 60]. They differ from each other in expression of specific genes and in many biochemical and physicochemical effects [60]. For example, EPA is hypotriglyceridemic and hypocholesterolemic, and DHA has no effect on plasma triglycerides [61]. DHA is less effective than EPA in inhibiting vascular smooth muscle proliferation. DHA is a more potent inhibitor than EPA of lymphocyte adhesion to endothelial cells [61]. Furthermore, DHA blocks voltage-activated sodium channels whilst EPA has no effect on membrane excitability and sodium channels in hippocampal neurons [62]. Similarly, DHA modulates certain voltage-gated K^+ channels in Chinese hamster ovary cells, whereas EPA has no effect on K^+ channels. DHA and EPA reduce chronic inflammation by attenuating NF-κB, in turn modulating the expression of proinflammatory cytokines including TNF-α and IL-1α and β. The intake of DHA and EPA reduces the synthesis of eicosanoids derived from ARA [63]. How DHA and EPA decrease the activation of NF-κB is not clear at present. However, these fatty acids may decrease the phosphorylation of IκB, thereby modulating the availability of NF-κB. This process can modulate the expression of the pro-inflammatory genes for COX-2, intracellular adhesion molecule-1 (ICAM-1), vascular adehe- sion molecule-1 (VCAM-1), E-selectin, TNF-α, IL-β, IL-6, nitric oxide synthase, and matrix metalloproteinases (MMP) [56]. These genes control the availability of lipid mediators such as PGs, LTs, and TXs, which not only modulate the intensity and duration of immune responses, but are also involved in neuroinflammation and pain.

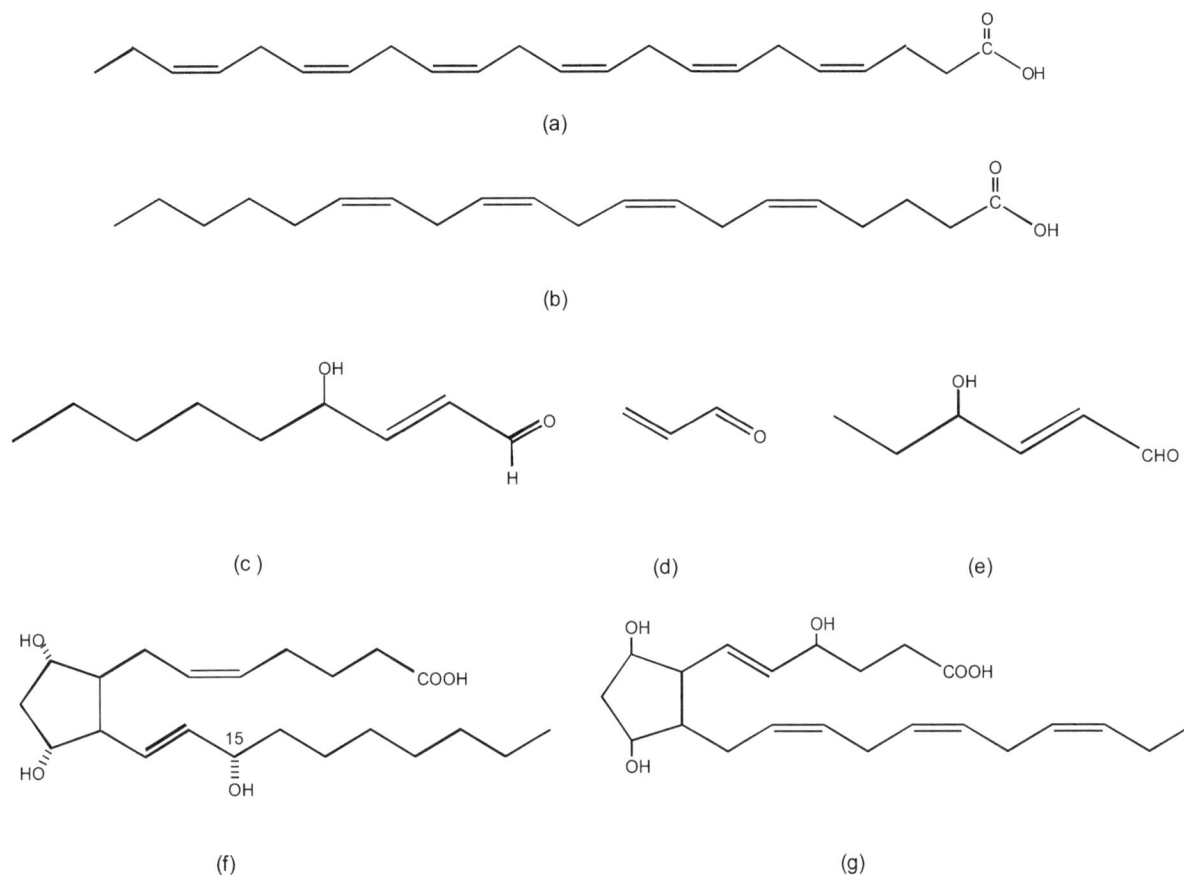

Figure 4: Chemical structures of non-enzymically derived lipid mediators of arachidonic acid and docosahexaenoic acid. Docosahexaenoic acid (a); arachidonic acid (b); 4-hydroxynonenal (4-HNE) (c); acrolein (d); 4-hydroxy-2-hexenal (4-HHE) (e); isoprostane (f); neuroprostane (g).

Neurochemically, enrichment of DHA and EPA in the diet competitively inhibits the oxygenation of ARA by cyclooxygenase thus suppressing the production of pro-inflammatory eicosanoids and pro-inflammatory cytokines [64]. The action of cyclooxygenases on EPA generates the 3-series prostaglandins and thromboxanes and the 5-series leukotrienes. These metabolites have different biological properties than the corresponding analogs produced by the metabolism of ARA. For example, TXA_3 is less active than TXA_2 in aggregating platelets and constricting blood vessels [64]. In contrast, the metabolism of EPA and DHA produces resolvins and neuroprotectins [65, 66]. These metabolites not only antagonize the effects of ARA-generated metabolites but also display potent actions on leukocyte trafficking as well as on glial cell functions by down-regulating expression of cytokines. Thus, resolvins and neuroprotectins inhibit both interleukin 1-β-mediated NF-κB activation and cyclooxygenase activation, indicating that resolvins and neuroprotectins not only counteract leukocyte-mediated injury, but also downregulate pro-inflammatory gene induction [65, 66, 67].

Collective evidence suggests that feeding EPA and DHA produces numerous immune responses including a decrease in lymphocyte proliferation, suppression of pro-inflammatory cytokines production, reduced gene expression of COX-2, and reduction in natural killer cell activity. These observations suggest that interactions among EPA, DHA, cytokines, eicosanoids, resolvins and neuroprotectins are quite complex and may be associated with beneficial effects of fish oil ingestion on inflammation and immune function [68]. DHA also targets TLR [69], cannabinoid receptors [70] and PPARγ-mediated signaling [71]. Modulation of these receptors by different dietary fatty acids may contribute to the regulation of acute and chronic inflammatory processes. Thus, a moderate intake of ARA and its precursors and the appropriate ratio between ARA and DHA may play an important role in physiologic functioning of the immune system and in modulation of inflammation in brain tissue. Very little is known about the optimal ARA to DHA ratio for the immunologic response against pathogens that can be effective in treating neuroinflammation. Studies on this topic are complicated by interactions between fatty acids and other nutrients such as vitamin E that are needed for normal immune function and response in mammalian tissues. The consumption of increased amounts of EPA and DHA results in a partial replacement of the ARA in cell membranes by EPA and DHA. This can lead to decreased production of ARA-derived mediators. This alone is a potentially beneficial anti-inflammatory effect of EPA and DHA [64].

Studies on the uptake and distribution of DHA into different glycerophospholipid classes indicate that the maximum incorporation of DHA occurs in ethanolamine plasmalogens, PlsEtn (Rapoport, 1999), which are unique glycerophospholipids with a vinyl ether group at the *sn*-1 position and ARA or DHA at the *sn*-2 position of the glycerol moiety. Myelin possesses the highest proportion of PlsEtn [21]. PlsEtn protect biomembranes against free radical attack [21].

In biomembranes, transition metal ions (copper and iron) initiate lipid peroxidation by generating peroxyl and alkoxyl radicals from the decomposition of lipid hydroperoxides [72]. Plasmalogen containing liposomes have a strong ability to chelate transition metal ions and thereby prevent the formation of peroxyl and alkoxyl radicals [73]. In contrast to the above view, studies based on the effect of menadione, an intracellular reactive oxygen species generator, on plasmalogen-deficient fibroblasts [74], and lactic acid production in astrocytic cultures suggest that plasmalogens do not play a major role in the protection of cells against superoxide anion radicals and lactic acid-induced oxidative stress [75]. Thus, more studies are required on this controversial topic.

The incorporation of DHA into ethanolamine plasmalogens may stabilize neural membranes. These PlsEtn may replace lost molecules. The losses may be due to neuroinflammatory stimulation of PlsEtn-selective PLA_2 or to oxidation of the vinyl ether group after exhaustion of other antioxidants. DHA also stimulates the synthesis of the peroxisomal enzymes needed for the synthesis of PlsEtn.

In summery, DHA and EPA act exclusively on neural cell membranes, in which they are esterified in glycerophospholipid molecules. DHA contribute to pre- and postnatal brain development, whereas EPA influences behavior and mood. Both DHA and EPA generate neuroprotective lipid mediators, which play important role in neuroprotective signaling. Higher intake of these fatty acids protects brain from stroke, neurodegenerative diseases, and neuropsychiatric disorders through anti-inflammatory activity. Recognition that inflammation may be closely involved in the pathogenesis of neuropsychiatric diseases is an important development [76, 81]. Patients with major depression show an increase in peripheral blood inflammatory biomarkers, including inflammatory cytokines, which have been shown to access the brain and interact with virtually every pathophysiologic domain known to be involved in depression, including neurotransmitter metabolism, neuroendocrine function, and neural plasticity [76, 81]. Indeed, activation of inflammatory pathways within the brain not only contributes to a decrease in neurotrophic support, but alterations in glutamate release/reuptake, as well as oxidative stress, leading to excitotoxicity and loss of glial elements. This is consistent with neuropathological features of depressive disorders.

CONCLUSIONS

Neuroinflammation is an active defensive process against diverse insults, metabolic and traumatic injuries, neurodegenerative diseases, and infection. Inflammation in the brain is driven by the activation of resident microglia, astrocytes and infiltrating peripheral macrophages, which release a plethora of anti- and pro-inflammatory cytokines, chemokines, neurotransmitters and ROS. It neutralizes toxic agents and blocks their detrimental effects. Although, neuroinflammation serves as a neuroprotective mechanism associated with repair and recovery, but persistent neuroinflammation causes brain damage. Most of the inflammatory reactions are initiated, maintained, and modulated by cytokines/chemokines and eicosanoids from microglial cells, astrocytes, macrophages, and endothelial cells in response to metabolic, traumatic and neurodegenerative insults. Cytokines propagate inflammation through the activation of phospholipases A_2, cyclooxygenases, and lipoxygenases, and generation of prostaglandins, leukotrienes, thromboxanes, and anti-inflammatory lipoxins [24]. The activation of phospholipases A_2 isoform (PlsEtn-PLA_2) by cytokines also generates DHA that is subsequently metabolized to resolvins and neuroprotectins. These lipid mediators are anti-inflammatory and are associated with resolution of inflammatory process. Collective evidence suggests that ARA and DHA play important roles in pro-and anti-inflammatory processes. Therefore, their ratio in the diet can modulate neuroinflammation. The ancient human diet had a ratio of ARA/DHA of 1:1, but modern diets contain a ARA/DHA ratio of 15:1 [53]. This dramatic increase in the ARA/DHA ratio has resulted in high levels of AA in neural membranes. Excess ARA generates high levels of prostaglandins, leukotrienes, and thromboxanes resulting in neuroinflammation. Based on the recent literature, the dietary intake of food rich in DHA can decrease or prevent inflammatory processes in brain tissue and can be beneficial for the neuroinflammation associated with acute neural trauma and neurodegenerative diseases.

REFERENCES

[1]   Correale, J; Villa, A. *J. Neurol.* **2004**, 251, 1304.

[2]   Diamond, P.; McGinty, A.; Sugrue, D.; Brady H. R.; Godson. C. *Clin. Chem. Lab Med.* **1999**, 37, 293.

[3]   Serhan, C. N.; Savill, J. *Nature Immunol.* **2005** 6, 1191.

[4]   Wood, P. L. *Neuroinflammation: Mechanisms and Management*, Humana Press, Totowa, New Jersey **1998**, pp. 1-59.

[5]   Kettenmann, H.; Ransom B. R. *Neuroglia*, Oxford University Press, New York, **2005,** pp. 1-250.

[6]   Wu, S.; M., Patel, D. D.; Pizzo S. V. *J. Immunol.* **1998**, 161, 4356.

[7]   Hays, S. J. *Curr. Pharm. Des* **1998**, 4, 335.

[8]   Minghetti, L.; Ajmone-Cat, M. A.; De Berardinis, M. A.; De Simone R. (2005) *Brain Res. Rev.* **2005**, 48, 251.

[9]   Sun, D.; Newman, T. A.; Perry, V. H.;Weller R. O. *Neuropathol. Appl. Neurobiol.* **2004**, 30, 374.

[10]   Farooqui, A.A. Hot Topics in Neural membrane Lipidology, Springer, New York, **2009,** pp. 1-382.

[11]   Hosokawa, M.; Klegeris, A.; Maguire, J.; McGeer, P. L. *Glia* **2003**, 42, 417.

[12]   Wilson, C. J.; Finch, C. E.; Cohen, H. J. *J. Am. Geriatr. Soc.* **2002**, 50, 2041.

[13]   Rothwell, N. J. (1999) *J. Physiol. (London)* **1999**, 514, 3.

[14]   Allan, S. M.; Rothwell, N. J. *Philos. Trans. R. Soc. Lond B Biol. Sci.* **2003**, 358, 1669.

[15]   Lucas, S. M.; Rothwell, N. J.; Gibson R. M. *Br. J. Pharmacol.* **2006**, 147 Suppl 1, S232.

[16]   Xing, Z.; Gauldie, J.; Cox, G.; Baumann, H.; Jordana, M.; Lei, X. F.; Achong M. K. *J. Clin. Invest* **1998**, 101, 311.

[17]   Lee, J. Y.; Plakidas, A.; Lee, W. H.; Heikkinen, A.; Chanmugam, P.; Bray, G.; Hwang, D. H. *J. Lipid Res.* **2003**, 44, 479.

[18]   Butchi, N.B.; Pourciau, S.; Du M.; Morgan T.W.; Peterson K.E. *J Immunol.* **2008**, 180, 7604.

[19]   Farooqui, A.A.; Horrocks, L.A. Glycerophospholipids in Brain: Phospholipases A$_2$ in Neurological Disorders, Springer, New York **2007**, pp. 1-376.

[20]   Farooqui, A. A.; Horrocks L. A.; Farooqui T. *J. Neurochem.* **2007**, 101, 577.

[21]   Farooqui, A.A. Metabolism and Function of Bioactive Ether Lipids in the Brain, Springer, New York **2008,** pp. 1-252.

[22]   Lloret, S.; Moreno, J. J. *J. Cell Physiol* **1995**, 165, 89.

[23]   Farooqui, A.A. Beneficial Effects of Fish Oil on Human Brain, Springer, New York **2009**.

[24]   Phillis, J.W.; Horrocks L.A.; Farooqui AA. Brain Res. Rev. **2006**, 52, 201.

[25]   Gilroy, D. W.; Newson, J.; Sawmynaden, P. A.; Willoughby, D. A.; Croxtall, J. D. *FASEB J.* **2004**, 18, 489.

[26]   Serhan, C. N.; Levy, B. *Chem. Immunol. Allergy* **2003**, 83, 115.

[27]   Serhan, C. N. *Prostaglandins Leukot. Essent. Fatty Acids* **2005**, 73, 141.

[28]   Chiang, N.; Arita, M.; Serhan C. N. *Prostaglandins Leukot. Essent. Fatty Acids* **2005**, 73, 163.

[29]   Decker, Y.; McBean, G.; Godson, C. *Am J Physiol Cell Physiol.* **2009**, 296, C1420.

[30]   Yamamoto, Y.; Gaynor, R. B. *Trends Biochem. Sci.* **2004**, 29, 72.

[31]   Mazière, C.; Conte, M. A.; Degonville, J.; Ali D.; Mazière, J. C. *Biochem. Biophys. Res. Commun.* **1999**, 265, 116.

[32]   Zhang, X.; Dong, F.; Ren, J.; Driscoll, M. J.; Culver, B. *Exp Neurol.* **2005**, 191, 318.

[33]   Rubin, B. B.; Downey, G. P.; Koh, A.; Degousee, N.; Ghomashchi, F.; Nallan, L.; Stefanski, E.; Harkin, D. W.; Sun, C. X.; Smart, B. P.; Lindsay, T. F.; Cherepanov, V.; Vachon, E.; Kelvin, D.; Sadilek, M.; Brown, G. E.; Yaffe, M. B.; Plumb, J.; Grinstein, S.; Glogauer, M.; Gelb, M. H. *J. Biol. Chem.* **2005**, 280, 7519.

[34]   Anrather, J.; Racchumi, G.; Iadecola, C. *J. Biol. Chem.* **2006**, 281, 5657-5667.

[35]   Frey, R. S.; Gao, X.; Javaid, K.; Siddiqui, S. S.; Rahman, A.; Malik, A. B. *J. Biol. Chem.* **2006**, 281, 16128.

[36]   Shmelzer, Z.; Haddad, N.; Admon, E.; Pessach, I.; Leto, T. L.; Eitan-Hazan, Z.; Hershfinkel, M.; Levy, R. *J. Cell Biol.* **2003**, 162, 683.

[37]   Drew, P. D.; Storer, P. D.; Xu, J. H.; Chavis, J. A. *Brain Res. Rev.* 2005, **48,** 322.

[38]   Farooqui, A. A.; Antony, P.; Ong, W. Y.; Horrocks, L. A.; Freysz, L. *Brain Res. Rev.* **2004**, 45, 179.

[39]   Moraes, L. A.; Piqueras, L.; Bishop-Bailey, D. *Pharmacol. Ther.* **2006**, 110, 371.

[40]   Kim, J. Y.; Lee, K. H.; Lee, B. K.; Ro J. Y. (2005) *Int. Arch. Allergy Immunol.* **2005**, 137, 104.

[41]   Minami, M.; Katayama, T.; Satoh, M. *J. Pharmacol. Sci.* **2006**, 100, 461.

[42]   Basu, S. *Free Radical Res.* **2004**, 38, 105.

[43]   Greco, A.; Minghetti, L. *Curr. Neurovasc. Res.* **2004**, 1, 341.

[44] Morrow, J. D. *Curr. Pharmaceut. Design* **2006**, 12, 895.

[45] Fam, S. S.; Morrow J. D. *Curr. Medicinal Chem.* **2003**, 10, 1723.

[46] Cracowski, J. L. *Chem. Phys. Lipids* **2004**, 128, 75-83.

[47] Roberts, L. J., II; Fessel, J. P.; Davies S. S. *Brain Path.* **2005**, 15, 143.

[48] Yin, H. Y.; Musiek, E. S.; Gao, L.; Porter, N. A.; Morrow J. D. *J. Biol. Chem.* **2005**, 280, 26600.

[49] Nourooz-Zadeh, J.; Halliwell, B.; Änggård E. E. *Biochem. Biophys. Res. Commun.* **1997**, 236, 467.

[50] Bernoud-Hubac, N.; Davies, S. S.; Boutaud, O.; Montine, T. J.; Roberts L. J., II *J. Biol. Chem.* **2001**, 276, 30964.

[51] Horrocks, L. A.; Farooqui A. A. *Prostaglandins Leukot. Essent. Fatty Acids* 2004, 70, 361.

[52] Marszalek, J. R.; Lodish H. F. *Annu. Rev. Cell Dev. Biol.* **2005**, 21, 633-657.

[53] Weylandt, K. H.; Kang J. X. *Lancet* **2005**, 366, 618.

[54] Farkas, T.; Kitajka, K.; Fodor, E.; Csengeri, I.; Lahdes, E.; Yeo, Y. K.; Krasznai Z.; Halver J. E. *Proc. Natl. Acad. Sci. USA* **2000**, 97, 6362.

[55] Puskás, L. G.; Kitajka, K.; Nyakas, C.; Barcelo-Coblijn, G.; Farkas T. *Proc. Natl. Acad. Sci. USA* **2003**, 100, 1580.

[56] De Caterina, R.; Massaro, M. *J. Membr. Biol.* **2005**, 206, 103.

[57] Yehuda, S.; Rabinovitz, S.; Carasso, R. L.; Mostofsky, D. I. *Neurobiol. Aging* **2002**, 23, 843.

[58] Lonergan, P. E.; Martin, D. S. D.; Horrobin, D. F.; Lynch, M. A. *J. Neurochem.* **2004**, 91, 20.

[59] Zhao, Y.; Joshi-Barve, S.; Barve, S.; Chen L. H. *J. Am. Coll. Nutr.* **2004**, 23, 71.

[60] Verlengia, R.; Gorjão, R.; Kanunfre, C. C.; Bordin, S.; Martins de Lima, T.; Fernandes Martins, E.; Newsholme, P.; Curi R. *Lipids* **2004**, 39, 857-864.

[61] Hashimoto, M.; Hossain, M. S.; Yamasaki, H.; Yazawa, K.; Masumura, S. *Lipids* **1999**, 34, 1297.

[62] Xiao, Y. F.; Li, X. Y. *Brain Res.* **1999**, 846, 112.

[63] Mills, S. C.; Windsor, A. C.; Knight S. C. *Clin. Exp. Immunol.* **2005**, 142, 216.

[64] Calder, F. C. *Biochem. Soc. Trans.* **2005**, 33, 423.

[65] Hong S., Gronert K., Devchand P. R., Moussignac R. L., and Serhan C. N. *J. Biol. Chem.* **2003**, 278, 14677.

[66] Marcheselli, V. L.; Hong, S.; Lukiw, W. J.; Tian, X. H.; Gronert, K.; Musto, A.; Hardy, M.; Gimenez, J. M.; Chiang, N.; Serhan, C. N.; Bazan N. G. *J. Biol. Chem.* **2003**, 278, 43807.

[67] Bazan, N. G. *Mol. Neurobiol.* **2005**, 31, 219.

[68] Song, C. in *Phospholipid Spectrum Disorders in Psychiatry and Neurology* Peet M., Glen L., and Horrobin D. F., eds., Marius Press, Carnforth, Lancashire, **2003**, pp. 415-422.

[69] Lee, J. Y.; Plakidas, A.; Lee W. H.; Heikkinen, A.; Chanmugam, P.; Bray, G.; Hwang D. H. *J. Lipid Res.* **2003**, 44, 479.

[70] Watanabe, S.; Doshi, M.; Hamazaki, T. *Prostaglandins Leukot. Essent. Fatty Acids* **2003**, 69, 51.

[71] Shiraki, T.; Kamiya, N.; Shiki, S.; Kodama, T. *J. Biol. Chem.* **2005**, 280, 14145.

[72] Murphy, R. C. *Chem. Res. Toxicol.* **2001**, 14, 463.

[73] Sindelar, P. J.; Guan, Z. Z.; Dallner, G.; Ernster L. *Free Radic. Biol. Med.* **1999**, 26, 318.

[74] Jansen, G. A.; Wanders R. J. A. *J. Inherited Metab. Dis.* **1997**, 20, 85.

[75] Fauconneau B., Stadelmann-Ingrand S., Favrelière S., Baudouin J., Renaud L., Piriou A., and Tallineau C. *Arch. Toxicol.* **2001**, 74, 695.

[76] Miller, A.H.; Maletic, V.; Raison, C.L. Biol. Psychitry 2009, an 14 Epub ahead of print.

[77] Block, M.L.; Hong, J.S. *Prog Neurobiol* **2005**, 76,77.

[78] Sun, G.Y.; Horrocks L.A.; Farooqui, AA. J Neurochem. **2007**, 103, 1.

[79] Aloisi, F. Glia **2001**, 36, 165.

[80] Beck, S.; Lambeau, G.; Scholz-Pedretti, K.; Gelb, M. H.; Janssen, M. J. W.; Edwards, S. H.; Wilton, D. C.; Pfeilschifte, J.; Kaszkin, M. *J. Biol. Chem.* **2003**, 278, 29799.

[81] Raison, C.L.; Capuron, L.; Miller, A.H. Trends Immunol. 2006, 27.

CHAPTER 3

Suitable Biomarkers of Oxidative Stress in Neurodegenerative Diseases

Fabio Coppedè[a] and Lucia Migliore[b,*]

[a] *Department of Neuroscience, University of Pisa, Via Roma 67, 56126 Pisa, Italy and* [b]*Department of Human and Environmental Sciences, University of Pisa, Via S. Giuseppe 22, 56126 Pisa, Italy*

Abstract: Free radicals have been implicated and considered as associated risk factors for a variety of human disorders including neurodegenerative diseases, although it is not yet clear whether oxidative stress acts as a causative agent of neuronal degeneration. In human tissues, a condition of oxidative stress can be revealed through searching for specific biomarkers of oxidative damage to lipids, proteins and nucleic acids. Markers of oxidative damage to lipids include 4-hydroxynonenal, malondialdehyde, lipid hydroperoxides and isoprostanes, and thiobarbituric acid–reactive substances. Protein carbonyls and protein nitration are common markers of oxidative damage to proteins. The levels of 8-hydroxyguanine, or alternatively 8-hydroxy-2'- deoxyguanosine, can be measured in brain specimens, blood and urines, and are commonly used as a marker of oxidative DNA damage. Besides DNA a growing body of evidence indicates that also RNA undergoes oxidative damage. Studies have been performed also on markers of antioxidant defence such as oxidative modifications of plasma proteins, the activity of enzymes of the antioxidant defence (superoxide dismutase and catalase), the levels and the activity of proteins involved in the repair of oxidative DNA damage and the state of components of blood glutathione system. Significant biological changes related to a condition of oxidative stress have been found not only in brain tissues but also in biological fluids such as urine, peripheral blood or cerebrospinal fluid, and in peripheral tissues such as blood cells and fibroblasts of individuals affected by Alzheimer's disease, Mild Cognitive Impairment, Parkinson's disease and Amyotrophic Lateral Sclerosis.

Keywords: Oxidative stress; neurodegenerative diseases; free radicals; reactive oxygen species; 4-hydroxynonenal; malondialdehyde; F2-isoprostanes

INTRODUCTION

Neurodegenerative diseases include complex multifactorial pathologies, monogenic disorders and disorders characterized by inherited, sporadic and transmissible forms. Alzheimer's diseases (AD), Parkinson's disease (PD) and Amyotrophic Lateral Sclerosis (ALS) represent three major neurodegenerative diseases, affecting several million people worldwide. They are defined as complex multifactorial disorders since both familial and sporadic forms are known.

Familial forms represent only a minority of the cases (ranging from 5 to 10% of the total), whereas the vast majority of AD, PD and ALS occurs as sporadic forms, likely resulting from complex interactions between genetic and environmental factors superimposed on slow, sustained neuronal dysfunction due to aging. Studies in families have led to the discovery of several causative genes responsible for the familial forms; in parallel large-scale studies in populations have contributed to identify genetic and environmental influences on risk and progression of the sporadic forms [1].

Oxidative stress describes a condition in which cellular antioxidant defences are insufficient to keep the levels of reactive oxygen species (ROS) below a toxic threshold. Increasing evidence suggests crucial implications for oxidative stress in several steps of the pathogenesis of many neurodegenerative diseases, and the current opinion is that it could have a causative role, instead of being an epiphenomenon of the pathological processes. Significant biological changes related to a condition of oxidative stress have been found not only in brain tissues but also in peripheral tissues of individuals affected by AD, Mild cognitive impairment (MCI), which is a pre-dementia phase preceding AD, by PD and ALS. Overall, there is substantial evidence indicating that a condition of increased oxidative damage to lipids, proteins and nucleic acids characterizes neurodegenerative diseases [2]. Chemical

*Address correspondence to: Department of Human and Environmental Sciences, University of Pisa Via S. Giuseppe 22, 56126 Pisa, Italy; Tel: +390502211029; Fax: +390502211034; E-mail: l.migliore@geog.unipi.it

Akhlaq A. Farooqui & Tahira Farooqui (Eds.)

species containing one or more unpaired electrons are denoted as free radicals. The most common cellular free radicals are hydroxyl radical, superoxide radical, and nitric oxide (NO). Free radicals and related molecules are classified as ROS for the ability of unpaired electrons to act as electron acceptors from other molecules, leading to their oxidation. A wide variety of ROS are produced in the course of the normal metabolism in biological systems since they have several important physiological functions, but their accumulation beyond the needs of the cell can potentially damage lipids, proteins, and nucleic acids. Mitochondrial respiration is the major source of ROS in human cells, since during the transport of high-energy electrons through the electron transport chain (ETC) the high-energy electrons can react with O_2 to form superoxide. It is estimated that up to 2% of the O_2 consumed by healthy mitochondria is converted to superoxide. NO is a gas present in most of the cells of the body where it acts as a vasorelaxant agent, as a neurotransmitter, or participates in the immune response. However NO is thermodynamically unstable and, under pathologic conditions involving the production of ROS, undergoes oxidative/reductive reactions, producing reactive nitrogen species (RNS) leading to protein oxidation, nitrosylation, and nitration [3].

In human tissues, a condition of oxidative stress can be revealed through searching for specific biomarkers of oxidative damage to lipids, proteins and nucleic acids. The brain is particularly vulnerable to oxidative damage due to its relative lack of antioxidant capacity, high concentration of unsaturated fatty acids, and high consumption rate of oxygen. Oxidative attack to membrane unsaturated fatty acids results in the formation of lipid hydroperoxides and isoprostanes, and highly reactive aldehydes and alkenals such as malondialdehyde (MDA) and 4-hydroxynonenal (4-HNE). The most general indicator and by far the most commonly used marker of protein oxidation is protein carbonyl content. Protein nitration and 4-HNE-protein adducts represent other indicators of oxidative attack on proteins. Among DNA bases, guanine is the most susceptible to oxidative attack; therefore the measure of the content of 8-hydroxyguanine (8-OHG) is the most commonly used indicator of oxidative DNA damage. Concerning neurodegenerative diseases, markers of oxidative damage can be searched in postmortem brain tissues as well as in biological fluids of living individuals, including blood, urine and cerebrospinal fluid (CSF), or in peripheral tissues, such as lymphocytes and fibroblasts. Alternatively biological fluids from model systems can also be used as markers for oxidative stress [4]. Table **1** lists several examples of biomarkers of oxidative dam-age detected in different tissues of individuals affected by AD, MCI, PD or ALS.

Table 1: Some examples of biomarkers of oxidative damage detected in neurodegenerative diseases.

Disease	Biomarker of	Biomarker	Tissue	Result	Reference
AD	Oxidative damage to lipids	4-HNE	Brain CSF	Increased levels	[5]
PD	Oxidative damage to lipids	4-HNE	Brain	Increased levels	[5]
ALS	Oxidative damage to lipids	4-HNE	Spinal cord motor neurons CSF	Increased levels	[5]
PD	Oxidative damage to lipids	MDA	Substantia nigra, Blood CSF	Increased levels	[8]
AD	Oxidative damage to lipids	TBARS	Brain	Increased levels	[7]
PD	Oxidative damage to lipids	TBARS	Plasma	Increased levels	[9]
AD MCI	Oxidative damage to lipids	F2-isoprostanes	Brain CSF Plasma	Increased levels Increased in AD No difference	[11] [11] [12]
PD	Oxidative damage to lipids	F2-isoprostanes	Brain Plasma Urine	No difference	[14] [13] [13]
ALS	Oxidative damage to lipids	F2-isoprostanes	Urine	Increased levels	[15]
AD	Oxidative damage to proteins	HNE-protein adducts	Brain	Increased levels	[16]
AD MCI	Oxidative damage to proteins	Protein carbonyls	Brain	Increased levels	[16]
EAD	Oxidative damage to proteins	HNE-protein adducts	Brain	Increased levels	[17]
AD	Oxidative damage to proteins	HNE-protein adducts	Brain	Increased levels	[16]
AD MCI	Oxidative damage to proteins	Protein nitration	Brain	Increased levels	[18]

PD	Oxidative damage to proteins	Protein nitration	Brain	Increased levels	[19]
ALS	Oxidative damage to proteins	Protein carbonyls	Spinal cord	Increased levels	[23]
ALS	Oxidative damage to proteins	Protein nitration	CSF	Increased levels	[25]
AD	Oxidative DNA damage	8-OHG 8-OHA 5-OHC 5-OHU	Brain regions	Increased levels	[27,28]
MCI	Oxidative DNA damage	8-OHG 8-OHA 4,6-diamino-5-formamidopyri-midine	Brain regions	Increased levels	[29]
AD	Oxidative DNA damage	8-OHdG	Ventricular CSF, Lymphocytes	Increased levels	[30]
AD MCI	Oxidative DNA damage	Oxidized bases (Comet assay)	Lymphocytes	Increased levels	[31,32]
PD	Oxidative DNA damage	8-OHG	Substantia nigra	Increased levels	[33,34]
PD	Oxidative DNA damage	8-OHG 8-OHdG	Serum CSF	Increased levels	[37]
PD	Oxidative DNA damage	DNA SSB Oxidised purines	Blood lymphocytes	Increased levels	[38]
ALS	Oxidative DNA damage	8-OHdG	Motor cortex Spinal cord Plasma, Urine CSF	Increased levels	[39, 40]
AD	Oxidative RNA damage	8-OHG 8-OHdG	Brain	Increased levels	[44,45,47]
AD	Oxidative RNA damage	Oxidized rRNA	Brain	Increased levels	[46]
MCI	Oxidative RNA damage	8-OHG	Brain	Increased levels	[49]
PD	Oxidative RNA damage	8-OHG	Substantia nigra	Increased levels	[34]
ALS	Oxidative RNA damage	Oxidized mRNA	Motor neurons	Increased levels	[51]
AD MCI	Antioxidant defence	GSH	Plasma	Decreased levels	[55,56]
PD	Antioxidant defence	GSH	Substantia nigra	Decreased levels	[58]
AD	Antioxidant defence	GSH-Px CAT	Erytrocyte	Decreased activity	[57]
PD	Antioxidant defence	GST	Plasma	Decreased levels	[64]
ALS	Antioxidant defence	GSH-Px SOD1	Erytrocyte	Decreased activity	[65,69]
AD MCI	Antioxidant defence DNA repair	OGG1	Brain	Decreased activity	[74-77]
PD	Antioxidant defence DNA repair	OGG1	Substantia nigra	Increased levels	[80]
ALS	Antioxidant defence DNA repair	OGG1	Spinal motor neurons	Decreased activity	[86]

BIOMARKERS OF OXIDATIVE DAMAGE TO LIPIDS

The oxidative degradation of lipids, known as lipid peroxidation, is the process during which ROS "steal" electrons from the lipids in cell membranes, resulting in the formation of fatty acid radicals (Fig. **1**). Fatty acid radicals are unstable molecules, and soon react with molecular oxygen forming a peroxyl-fatty acid radical, which in turn reacts with another free fatty acid producing a different fatty acid radical and lipid peroxide. This process proceeds by a free radical chain reaction mechanism. 4-HNE (Fig. **1**) is the major alkenal formed during peroxidation of arachidonic and linoleic acids and has been widely used as a biomarker for oxidative damage in tissues. Particularly, 4-HNE was shown to react with proteins to form stable adducts that can be used as markers of oxidative stress-induced cellular damage [5]. The presence of 4-HNE is increased in brain tissue and cerebrospinal fluid (CSF) of

AD patients, and in spinal cord and CSF of ALS patients. Immunohistochemical studies show presence of 4-HNE in neurofibrilary tangles and in senile plaques in AD, in the cytoplasm of the residual motor neurons in sporadic ALS, in Lewy bodies in neocortical and brain stem neurons in Parkinson's disease (PD) and in diffuse Lewy bodies disease (DLBD). Thus, increased levels of 4-HNE in neurodegenerative disorders and immunohistochemical distribution of 4-HNE in brain tissues indicate patho- physiological contribution of oxidative stress in these diseases, suggesting a role for 4-HNE in formation of abnormal filament deposits [5]. Other markers of oxidative damage to lipids include MDA, lipid hydroperoxides and isoprostanes, and thiobarbituric acid reactive substances (TBARS); MDA (Fig. **1**) is a reactive aldehyde resulting from ROS degradation of polyunsaturated lipids. This compound forms covalent protein adducts and reacts with deoxyadenosine and deoxyguanosine in DNA, forming DNA adducts. MDA can be measured by high-performance liquid chromatography (HPLC) and by the thiobarbituric acid (TBA) test. Indeed, MDA and other TBARS condense with two equivalents of TBA to give a fluorescent red derivative that can be assayed spectrophotometrically [6]. Several studies failed to find differences in MDA and lipid hydroperoxides levels between AD and control brains, on the contrary TBARS levels have been found increased in several regions of AD brains in different experimental conditions [7]. Increased MDA levels have been observed in the substantia nigra, in blood and in the CSF of PD patients [8]. Increased TBARS have been observed in plasma of PD patients and in animal models of the disease [9]. Conflicting results have been obtained when evaluating MDA and TBARS levels in tissues of both ALS patients and ALS animal models; however, from arachidonic acid (Fig. **1**), are especially useful as *in vivo* biomarkers of lipid peroxidation. F2-isoprostanes are increased in brain regions of AD and MCI individuals, as well as in the CSF of patients with early AD. On the contrary, quantification of F2-isoprostanes in plasma and urine of AD patients has produced conflicting data [11]. Although, plasma F2-isoprostane has been proposed as a diagnostic marker for AD and MCI, a recent study on nondemented control individuals and patients with AD, MCI, or PD showed that mean plasma levels of F2A isoprostane did not differ significantly between the four diagnostic groups. Moreover, within the MCI and AD groups, F2A isoprostane levels did not correlate with duration of memory impairment or with several investigators report increased MDA levels in ALS tissues [10]. F2-isoprostanes, one group of lipid peroxidation products derived cognitive test scores, suggesting that while CSF isoprostane levels are elevated in AD, plasma isoprostane measures were neither sensitive nor specific for the clinical diagnosis of MCI or AD [12]. Plasma and urine isoprostane levels have been measured in plasma and urines of PD patients and PD patients with dementia (PDD), with the result that isoprostane do not seem to be substantially elevated in PD and are not associated with severity of cognitive impairment in PDD [13]. Moreover, F2-isoprostanes were not increased in the substantia nigra of PD patients and dementia with Lewy body disease [14]. Urinary isoprostane levels were found to be elevated in sporadic ALS [15].

BIOMARKERS OF OXIDATIVE DAMAGE TO PROTEINS

Concerning protein oxidation in AD brain, oxidative attack on proteins results in the formation of protein carbonyls (Fig. **2**), often with loss of functionality of the parent molecule, and protein carbonyls are a common finding in brain samples from AD and MCI individuals. Moreover, in AD pathogenesis there is also evidence for lipid peroxidation-derived protein modifications, such as 4-HNE-modified proteins (Fig. **2**) [16]. Butterfierld and Sultana have summarized their findings of oxidatively modified proteins obtained using a redox proteomics approach in AD and MCI brain tissues and in animal models of the disease [16]. Their analyses indicated several oxidatively-modified proteins in AD brains that can be classified as those dealing with energy metabolism, glutamate reuptake, activity of the proteasome, neurotransmission, cellular defence system, long term potentiation involved in formation of memory, and others involved in the maintenance of neuronal structure and functions. Particularly, one of the enzymes specifically modified by oxidation in AD brains is ubiquitin carboxyl terminal hydrolase L-1 (UCHL-1), resulting in a decreased activity of the proteasome and impaired degradation of oxidized, aggregated and misfolded proteins. The analysis of MCI brains revealed that protein oxidation is significantly increased in the hippocampus of MCI subjects when compared to controls. Among oxidatively modified proteins in MCI hippocampus there are alpha-enolase, glutamine synthetase, pyruvate kinase M2 and peptidyl-prolyl cis/trans isomerase. The interactome of these proteins revealed functional involvement in several processes, including energy metabolism, synaptic plasticity and mitogenesis/ proliferation. Therefore, alterations due to oxidative inactivation of these proteins could lead to the progression from MCI to AD [16]. Early Alzheimer's disease (EAD) is the intermediary stage between mild cognitive impairment (MCI) and late-stage AD. The symptoms of EAD mirror the disease advancement between the two phases since EAD is a stage following MCI and characterized by full blown dementia. Proteomics was recently utilized to investigate 4-HNE-bound brain proteins in EAD compared to those in control. In total, six proteins were found to be excessively covalently bound by HNE in EAD inferior parietal lobule compared to age-related control brain. These proteins play roles in antioxidant defence (manganese superoxide dismutase), neuronal communication and neurite outgrowth (dihydropyriminidase-related protein 2), and energy metabolism (alpha-enolase, malate dehydrogenase, triosephosphate isomerase, and F1 ATPase, alpha subunit). This study showed that there is an overlap of brain proteins in EAD with previously identified oxidatively modified proteins in MCI and late-stage AD. The results are consistent with the hypothesis that oxidative stress, in particular lipid peroxidation, is an early event in the progression of AD [17]. Protein nitration (Fig. **2**) also is an important oxidative modification observed in MCI and AD brains, and proteomic analysis identified nitrated proteins in both MCI and AD, including proteins involved in glucose metabolism. A proteomics approach to identify nitrated brain proteins in the inferior parietal lobule from four subjects with EAD revealed that eight proteins were found to be significantly nitrated in EAD: peroxyredoxin 2, triose phosphate isomerase, glutamate dehydrogenase, neuropolypeptide h3, phosphoglycerate mutase1, H(+)-transporting ATPase, alpha-enolase, and fructose-1,6-bisphosphate aldolase. Many of these proteins were also nitrated in MCI and late stage AD, suggesting that nitrated proteins might have a role in stages of AD [18].

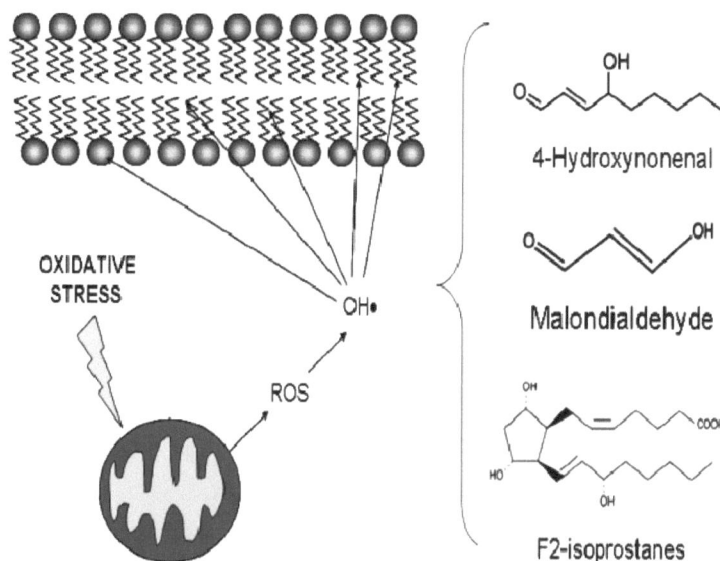

Figure 1: Some examples of lipid peroxidation products commonly used as biomarkers of oxidative stress. Reactive oxygen species (such as hydroxyl radical = •OH), mainly produced during mitochondrial respiration, attack fatty acids in membrane phospholipids resulting in the formation, among others, of 4-Hydroxynonenal (4-HNE), Malondialdehyde (MDA) and F2-isoprostanes

Extensive evidence has accumulated from post-mortem brain tissues suggesting that oxidative and nitrative stress is an integral factor in neurodegeneration associated with PD. Danielson and Andersen have recently reviewed oxidative and nitrative protein modifications in PD [19]; these include proteins that are also mutated in familial forms of the disease (α-synuclein, parkin, Dj-1, and PINK1) and proteins or complexes, such as tyrosine hydroxylase (the initial enzyme in the biosynthesis of dopamine), the mitochondrial complex I and the proteasome, whose inactivation results in increased oxidative stress [19]. Lipoxidative damage of aldolase A, enolase 1, and glyceraldehyde dehydrogenase (GAPDH) was found in the frontal cortex in the majority of cases of PD, and dementia with Lewy bodies (DLB). These three enzymes participate in glycolysis and energy metabolism in the adult human brain, and. their oxidative modifications might partly account for impaired metabolism and function of the frontal lobe in PD [20].

The identification in 1993 of superoxide dismutase-1 (SOD1) mutations as the cause of 20% of familial amyotrophic lateral sclerosis cases, led to a substantial amount of research into the mechanisms of SOD1-mediated toxicity. Recent exp- eriments have demonstrated that oxidation of wild-type SOD1 leads to its misfolding, causing it to gain many of the same toxic properties as mutant SOD1. It has been therefore suggested that oxidized/misfolded SOD1 could lead to ALS even in individuals who do not carry a SOD1 mutation [21]. Transgenic mice overexpressing the mutant human *SOD1* gene showed nitrated and oxidized proteins in the motor cortex, the cerebellar cortex and nucleus of hypoglossal nerves (regions related with movement). Significantly elevated protein nitration and nitric oxide synthesis

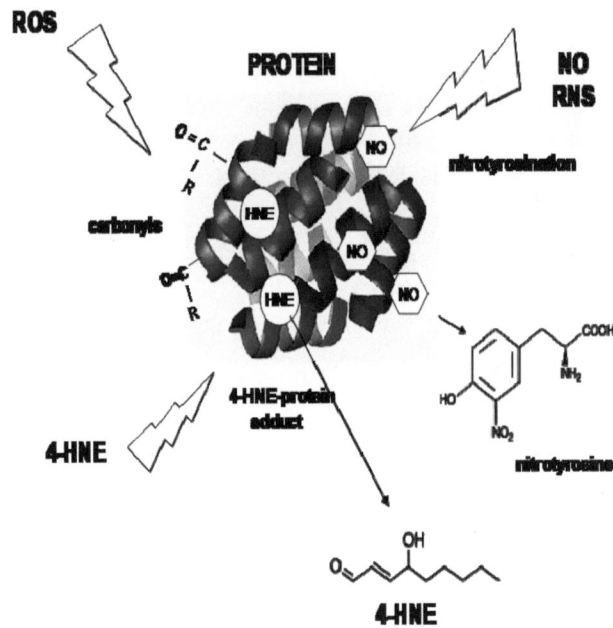

Figure 2: Schematic representation of oxidative attack to proteins. Reactive Oxygen Species (ROS) attack to proteins leads to the oxidation of amino acid residues, forming protein carbonyls. Nitric oxide (NO) and its oxidative derivatives (Reactive Nitrogren Species = RNS) lead to the oxidation of tyrosine residues on proteins, resulting in the formation of nitrotyrosine. In addition, lipid derivatives, such as 4-hydroxynonenal (4-HNE) can react with proteins to form adducts that can be analyzed. Protein carbonyls, protein nitrotyrosination and 4-HNE-protein adducts are commonly used biomarkers of protein oxidation.

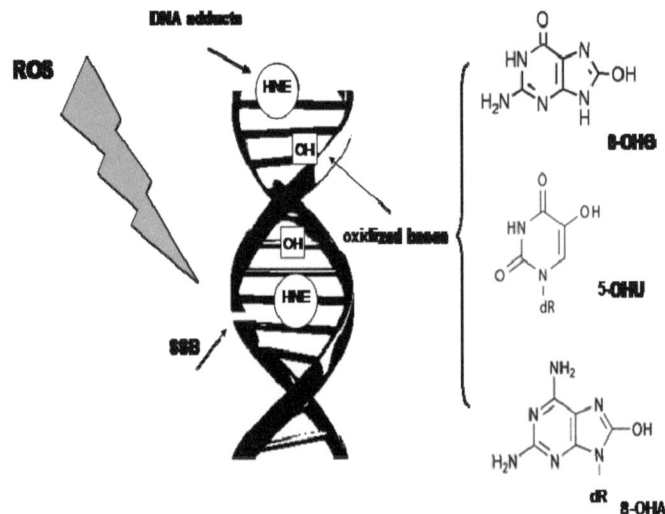

Figure 3: Schematic representation of oxidative attack to the DNA. Reactive oxygen species (ROS) are able to induce DNA single strand breaks (SSB) or lead to the generation of at least 20 different oxidized base adducts, the most frequent being 8-hydroxyguanine (8-OHG). Other oxidized bases commonly used as biomarkers of oxidative DNA damage are 8-hydroxyadenine (8-OHA) and 5-hydroxyuracil (5-OHU). 4-Hydroxynonenal (4-HNE), resulting from lipid peroxidation of membrane phospholipids can lead to DNA-HNE adducts.

were also observed in brain tissues and CSF of mutant *SOD1* mice. This study correlated mutation of the *SOD1* gene to increased nitric oxide, nitration and oxidation of proteins in ALS [22]. Four proteins in the spinal cord of G93A-SOD1 transgenic mice have higher specific carbonyl levels compared to those of non-transgenic mice. These proteins are SOD1, translationally controlled tumor protein (TCTP), ubiquitin carboxyl- terminal hydrolase-L1, and alphaB-crystallin, suggesting that their oxidative modification may play an important role in the neurodegeneration of ALS [23]. A study performed in 8 ALS patients and 5 controls showed that the protein carbonyl content of the ALS spinal cords significantly increased in all examined cases. In most ALS patients, proteins with 125 kDa, 70 kDa and 36kDa were highly oxidized. The 70-kDa protein was identified immunochemically to be neurofilament 68 [24]. Increased levels of nitrated proteins have been observed in the CSF of sALS patients [25]. Proteomic analysis of HNE-modified proteins in G93A-SOD1 transgenic mice showed a significant increase of HNE-adducts to α-enolase, heatshock protein and DRP-2 in the spinal cord tissue. These proteins are key enzymes involved in energy metabolism, stress response and neuronal development and repair, respectively, and their oxidative modification supports a pivotal role for lipid peroxidation in the degenerative process [26].

BIOMARKERS OF OXIDATIVE DAMAGE TO NUCLEIC ACIDS

Nucleic acids (nuclear DNA, mito- chondrial DNA, and RNA) are one of the several cellular macromolecules damaged by ROS, particularly by the hydroxyl radical. Oxidative DNA damage may play an important role in neurodegenerative diseases such as AD, PD and ALS, since modification of DNA bases can lead to mutation and altered protein synthesis. ROS attack on DNA bases can lead to the generation of at least 20 different oxidized base adducts, the most frequent being 8-hydroxyguanine (8-OHG) (Fig.3). Increased levels of 8-OHG have been observed in both nuclear DNA (nDNA) and mitochondrial DNA (mtDNA) isolated from brain regions of AD subjects compared with controls [27]. Subsequent studies reported increased levels of 8-OHG, 8-hydroxyadenine (8-OHA), 5-hydroxycitosine (5-OHC) and 5-hydroxyuracile (5-OHU), which derives from the degradation of cytosine), in several brain regions of AD individuals, and it was observed that mtDNA has approximately 10-fold higher levels of oxidized bases than nDNA, that guanine is the most vulnerable base to DNA damage, and that multiple oxidized bases are significantly higher in AD brain specimens in comparison with controls [28]. The quantification of multiple oxidized bases in both nDNA and mtDNA in several postmortem brain regions of MCI subjects revealed increased levels of 8-OHG and 4,6-diamino-5-formamidopyrimidine in both nDNA and mtDNA, and increased levels of 8-OHA in nDNA, suggesting that oxidative DNA damage is an early event in AD and not a secondary phenomenon [29]. Increased levels of 8-hydroxy-2'-deoxyguanosine (8-OHdG) were observed in the ventricular CSF of AD subjects, and in lymphocytes of AD individuals, besides a significantly lower level of plasma antioxidants [30]. Increased oxidative DNA damage was observed by applying a modified version of the comet assay which allows the detection of oxidised purines and pyrimidines in lymphocytes of AD patients [31]. Subsequently, by means of the same assay, we observed that the amounts of both primary DNA damage and oxidized bases were significantly higher in AD and MCI patients, respect to controls, giving a further indication that oxidative DNA damage is an earlier event in AD pathogenesis, which is detectable also in peripheral tissues of MCI subjects [32].

Increased oxidative DNA damage, measured as the content of 8-OHG, was observed in the substantia nigra of PD patients [33]. Subsequent studies in PD patients revealed that cytoplasmic 8-OHG immunoreactivity was intense in neurons of the substantia nigra, and present to a lesser extent in neurons of the nucleus raphe dorsalis and oculomotor nucleus, and occasionally in glia. These results were consistent with the hypothesis that ROS largely attack both RNA and mtDNA in PD brains [34]. Though rare at any particular site, multiple somatic mtDNA point mutations induced by oxidative damage or by other mechanisms may accumulate with age in the brain and thus could play a role in aging and neurodegenerative diseases. However, the study of levels of somatic point mtDNA mutations in the substantia nigra revealed no significant differences between PD patients and age-matched controls [35]. On the contrary, it was observed an accumulation of high levels of mtDNA deletions in substantia nigra neurons in aging and PD, suggesting that somatic mtDNA deletions might be one of the contributing factors leading to mitochondrial dysfunction, human aging, and neurodegeneration in PD [36]. 8-OHdG and 8-OHG levels were measured in the serum and CSF of PD patients. Compared to age-matched controls, the mean levels of serum and CSF 8-OHdG/8-OHG were significantly higher in PD individuals [37]. We determined peripheral markers for oxidative DNA damage in PD by testing for spontaneous and induced chromosomal damage, DNA strand breaks, and oxidized bases both in peripheral blood and cultured lymphocytes. Compared to healthy controls, PD patients showed higher frequencies of chromosomal damage (micronuclei) and a significant increase in the levels of single strand breaks (SSB). Significant differences were also obtained in the distribution of oxidised purine bases between the two groups [38].

Nuclear DNA 8-OHdG levels were increased in sporadic ALS motor cortex and in both sporadic and familial ALS spinal cord [39]. Increased 8-OHdG levels were also observed in plasma, urine, and CSF of ALS individuals [40].

Oxidative damage to mitochondrial DNA, measured as the content of 8-OHdG, was observed in spinal motor neurons of transgenic ALS mice, bearing SOD1 mutations. Increased 8-OHdG levels were also observed in the nuclear DNA from the spinal cord, frontal cortex and striatum from G93A SOD1 transgenic mice [41]. It has been demonstrated that the mutant G93A SOD1 lacks its ability to enter the nucleus and protect the DNA against oxidative damage. Indeed the comet assay analysis revealed increased oxidative DNA damage in both mice and cell cultures bearing the mutant SOD1 enzyme [42]. In a recent study, increased 8-OHdG levels have been correlated with increased levels of oxidized forms of coenzyme Q10 (CoQ10) [43].

In 1999 Nunomura *et al.* first demonstrated RNA oxidation in AD vulnerable neurons, measured as the content of 8-OHG and 8-OHdG [44]. Subsequently, it has been demonstrated that RNA oxidation is not random but highly selective. Importantly, many identified oxidized mRNA species have been implicated in the pathogenesis of AD, including p21ras, mitogen-activated protein kinase (MAPK) kinase 1, α-enolase 1, carbonyl reductase, apolipoprotein D, transferrin, phosphotriesterase protein, glutamate dehydrogenase, calpain, and SOD1 [45]. It was then demonstrated that also ribosomal RNA (rRNA) in AD hippocampal neurons is oxidized by bound redox-active iron [46]. It has been established that 30-70% of the mRNAs isolated from AD frontal cortices is oxidized [47]. Additional studies indicate that declines in protein synthesis and ribosome function occur in the earliest stage of AD, suggesting a possible role for rRNA oxidation as a potential mediator of decreased protein synthesis in AD [48]. Oxidatively modi- fied RNA has been recently detected also in MCI hippocampal regions, further suggesting that RNA oxidation contributes to the early stages of AD [49].

PD has been associated with oxidative damage to cytoplasmic DNA and RNA in substantia nigra neurons, by means of 8-OHG immunoreactivity [34]. Mutations in DJ-1 lead to a monogenic form of early onset recessive parkinsonism, but it has been also observed that DJ-1 is more oxidized in cortex from cases of sporadic PD compared to controls [50]. It has been also demonstrated that mRNA oxidation is a common feature in ALS patients as well as in many different transgenic mice expressing familial ALS-linked mutant SOD1. In mutant SOD1 mice, increased mRNA oxidation primarily occurs in the motor neurons and oligodendrocytes of the spinal cord at an early, pre-symptomatic stage [51]. Overall, there is accumulating evidence obtained from studies on either human samples or experimental models suggesting that oxidative RNA damage is a feature in vulnerable neurons at early-stage of AD, PD, dementia with Lewy bodies and ALS, and indicating that RNA oxidation actively contributes to the onset or the development of these disorders [52].

BIOMARKERS OF ANTIOXIDANT DEFENCE

The cell possesses several enzymatic and non enzymatic antioxidant defence mechanisms to counteract the daily production of oxidative reactive species. These include, among others, antioxidant molecules such as vitamins and glutathione (GSH), antioxidant enzymes (superoxide dismutase, catalase, glutathione S-transferases, glutathione peroxidase) and proteins participating in the repair of oxidative DNA damage. Glutathione is required for many critical cell processes, but plays a particularly important role in the maintenance and regulation of the thiol-redox status of the cell. Increased oxidation of red blood cells GSH was observed in AD patients, and this was correlated with the cognitive status of the patients [53]. Measurements of GSH levels in AD brains have yielded contradictory results [54]. Plasma GSH was decreased in AD patients, compared with the control group, and was associated with a significant increase in oxidative stress markers (i.e., glutathione disulfide (GSSG), 4-HNE, protein carbonyl content, and nitrotyrosine) [55]. A recent analysis of protein carbonyls and erythrocyte glutathione system plasma levels showed an increase in protein modification, a decrease in GSH levels and GSH/GSSG ratio in AD and MCI patients compared to age-matched control subjects [56]. Kharrazi *et al.* examined whether antioxidant defence mechanism exacerbates the risk of AD in individual carrying the *APOE*-ε4 allele. They determined the enzymatic activities of the erythrocyte SOD1, glutathione peroxidase (GSH-Px), catalase (CAT) and serum level of total antioxidant status (TAS), showing that the TAS level and the activities of enzymatic antioxidants CAT and GSH-Px were significantly lower and the SOD1 activity was significantly higher in AD patients compared to controls. Moreover, AD patients with the *APOE*-ε4 allele had significantly lower serum TAS concentration and lower erythrocytes GSH-Px and CAT activities, but significantly higher erythrocytes SOD1 activity than the non-*APOE*- ε4 carriers [57].

One of the most robust and significant alterations in the antioxidant defence in PD is a decrease in GSH concentrations, confirmed by multiple studies indicating that the substantia nigra is abnormally depleted of GSH in

PD patients, and several lines of evidence indicate that GSH depletion might be pivotal in PD and other neurodegenerative diseases [58]. Environmental agents associated over the years with the risk of PD, including pesticides and metals, have in common the capacity to challenge the antioxidant status of the substantia nigra and deplete its GSH content [59]. Glutathione-S–transferases (GSTs) are a family of dimeric proteins whose primary function is to use GSH to detoxify electrophiles capable to bind to DNA (such as 4-HNE), and the toxic effects of pesticides can be attenuated through a GSTs mediated metabolism. An association between *GSTP1* polymor- phisms and PD was described in a pesticide-exposed population [60]. Both the *GSTT1* and the *GSTM1* null genotypes have been associated with PD risk [61]. Studies on *Drosophila* models of PD with mutations in the *parkin* gene indicate that overespression of GSTS1 in dopaminergic neurons suppresses neurodegeneration [62]; moreover a role in modulating age at onset of both AD and PD has been proposed for polymorphisms in the *GSTO1* and *GSTO2* genes [63]. We observed a decreased GST activity in plasma of PD patients, consistent with a reduced GSTA4 activity but not due to *GSTA4* mutations [64].

The activity of GSH-Px as well as the activities of other antioxidative enzymes: SOD1, CAT, glutathione reductase (GR) in erythrocytes, as well as the activity of plasma GST, and the plasma content of vitamins E and C were evaluated in sporadic ALS patients. The results revealed significantly decreased activity of both GSH-Px and SOD1 in sALS patients compared with the control group [65]. Increase in oxidized NO products and reduction in oxidized glutathione were observed in CSF from patients with sporadic ALS [66]. GST activity assayed with 1-chloro-2,4-dinitrobenzene (substr- ate for transferase activity) and cumene peroxide (substrate for peroxidase activity) was significantly decreased in peripheral blood mononuclear cells of ALS patients, as well as the GSTP expression was reduced on both mRNA and protein level. The mean peroxidase activity was however significantly increased in CSF and serum of ALS patients [67]. The activity of SOD1, CAT, GSH-Px and GR was determined in erythrocytes from sporadic ALS patients, familial ALS patients with *SOD1* mutations, asymptomatic carriers with *SOD1* mutations, and control subjects. The *SOD1* gene mutation decreased SOD1 and GSH-Px activity. The disease also contributed to decreased SOD1 activity in sporadic ALS patients in comparison with the control group. The disease also influenced CAT and GR activity. CAT activity was decreased in both sporadic ALS and familial ALS patients. In all three patient groups, GR activity was higher than in the control group [68]. It has been observed that GSH depletion induces motor neuron degeneration both in cell cultures and in ALS mice models [69]. A recent study on motor neuronal cell line (NSC-34) was performed to investigate whether wild-type and familial amyotrophic lateral sclerosis-linked G93A mutant SOD1 (G93A-SOD1) modified the GSH pool and glutamate cysteine ligase (GCL), the rate-limiting enzyme for GSH synthesis. Mutant SOD1 induced an adaptive process involving the upregulation of GSH synthesis, even at very low expression levels. However, cells with a high level of G93A-SOD1 cultured for 10 weeks show- ed GSH depletion and a decrease in expression of the modulatory subunit of GCL. Cells with a low level of G93A-SOD1 maintained higher GSH levels and GCL activity, showing that the exposure time and the level of the mutant protein modulate GSH synthesis. Therefore, fail- ure of the regulation of the GSH pathway caused by G93A-SOD1 may contribute to motor neuron vulnerability [70].

We have discussed in the previous section (biomarkers of oxidative damage to proteins) that several proteins can be oxidatively modified in neurodegenerative diseases, resulting in reduced enzymatic activity, among them there are proteins of the antioxidant defence pathway. In this context, it is of great interest that SOD1 has been identified to be specifically modified by oxidative stress in AD, PD and ALS, indicating overlapping pathways of neurodegeneration [71].

Accumulating evidences suggest that impairments of the DNA repair ability, particularly of the DNA base excision repair pathway that specifically repairs oxidative DNA lesions, might be relevant to neurodegeneration [72]. We have recently reviewed the complex interaction between oxidative DNA damage and DNA repair in AD [73]. In summary, several lines of evidence point to the fact that oxidative damage, and particularly oxidative DNA damage, is one of the earliest detectable events in AD pathogenesis being observable even in MCI subjects prior to the deposition of amyloid plaques and the manifestation of dementia. Therefore, a few years ago, it was suggested that the BER process, which is the major repair pathway of oxidative DNA lesions, might be impaired in AD individuals (for a review see ref [73]).

The analysis of AD brain specimens for the activity of 8-oxoguanine DNA glycosylase (OGG1), which is the DNA glycosylase that specifically removes 8-OHG from the DNA, revealed that it was decreased in nuclear extracts from

AD hippocampal and parahippocampal gyri, superior and middle temporal gyri, and inferior parietal lobule [74]. Decreased levels of the mitochondrial OGG1 were observed in AD orbitofrontal gyrus and entorhinal cortex [75]. More recently, significant BER deficiencies, due to decreased activities of uracil DNA glycosylase (UNG), OGG1 and DNA polymerase β (Pol β), were observed in brain specimens obtained from 10 sporadic AD patients compared with the activities measured in samples obtained from 10 age-matched controls. Moreover, the authors detected the BER impairment in brains of 9 amnestic MCI patients, where it correlated with the severity of the disease [76]. Subsequently, OGG1 protein and incision activities were measured in nuclear and mitochondrial fractions from frontal, temporal, and parietal lobes from 8 MCI individuals, 7 late onset AD patients, and 6 age-matched normal control subjects. OGG1 activity resulted significantly decreased in nuclear specimens from frontal, temporal, and parietal lobes in MCI and AD and in mitochondria from AD frontal and temporal lobes and MCI temporal lobe. Nuclear OGG1 protein was significantly decreased in AD frontal lobe and MCI and AD parietal lobe. No differences in mitochondrial OGG1 protein levels were found. Overall, these results suggest that a decreased OGG1 activity occurs early in the progression of AD, and may contribute to elevated 8-OHG levels observed in AD and MCI brains [77]. OGG1 mutations, each of them resulting in impaired OGG1 activity, were identified and characterized in AD individuals, but not in matched controls [78]. A common Ser326Cys polymorphism is known in the *OGG1* gene, resulting in decreased protein activity. We have investigated the possible contribution of the *OGG1* Ser326Cys polymorphism to the risk of sporadic AD, observing no difference in the distribution of the mutant allele and genotypes between AD subjects and healthy matched controls, and concluding that this polymorphism does not contribute to the risk for sporadic forms of the disease [79]. Increased OGG1 levels have been found in the substantia nigra of PD patients [80]. Manganese exposure is considered to be one of the putative environmental PD risk factors, and both wild-type and OGG1 knockout mice were exposed to manganese from conception to postnatal day 30 to study its effects on nigrostriatal neurons [81]; in both groups exposure to manganese led to alterations in the neurochemistry of the nigrostriatal system. After exposure, dopamine was reduced in the caudate of OGG1 knockout mice, a loss that was paralleled by an increase in the dopamine index of turnover. In addition, the reduction of dopamine in caudate putamen correlated with the accumulation of 8-OHG in midbrain [81].

Several studies suggest impaired BER activity in ALS. The frontal cortical levels and the activity of the apurinic/apyrimidinic endonuclease 1 (APE1, the enzyme that removes the sugar-phosphate moiety, after the removal of an oxidized base by a specific DNA glycosylase) were found to be significantly reduced in sporadic ALS patients than in controls [82]. Others observed an increased APE1 activity in ALS motor neurons [83]. Missense mutations in the gene encoding APE1 were found in ALS patients [84]. The investigation of the common *APE1* Asp148Glu polymorphism as sporadic ALS risk factor yielded conflicting results [85].

CONCLUDING REMARKS

The use of several biomarkers of oxidative damage has led to a challenge of the opinion that oxidative stress could be only an epiphenomenon in neurodegenerative diseases, likely arising as a consequence of the neuro- degenerative process. Several studies performed in MCI individuals, in pre-symptomatic animal models of neurodegenerative diseases as well as in pre-symptomatic individuals bearing neurodegenerative-causative mutations, have revealed that oxidative damage is one of the earliest detectable events in the neurodegenerative process, often arising before the onset of several disease symptoms, and likely contributing to the progression of the disease. However, the question of whether or not oxidative stress is a cause rather than a consequence of neuro- degeneration is still largely debated. It is plausible that both hypotheses are correct since oxidative damage could act either in the initial stages of the neurodegenerative process as well as being triggered and increase by the compromising of intracellular pathways occurring within the progression of the disease [2]. For instance, at the time oxidative damage was observed in AD, it was supposed that amyloid β (Aβ) aggregates were the main source of oxidative stress; however, recent evidence suggests that oxidative stress is one of the earliest events in AD, and drives Aβ production. Moreover, it seems that Aβ peptides might be produced to function as scavengers of reactive oxidative species. Only with the persistence of oxidative stress, the production of Aβ peptides overcomes their cellular turnover, so that they start to aggregate and their anti-oxidant function evolve into pro-oxidant, ultimately leading to neuronal death [2,88]. Several studies are still required to further address the role of oxidative damage in neurodegenerative diseases.

Findings on the involvement of oxidative stress as a primary event in AD and in many neurodegenerative diseases, have led to the idea that antioxidant-based therapies or specific diet consumptions could be of help in aging and neurodegenerative diseases. In general the potential therapeutical treatments are currently divided in two different categories: vitamin and non-vitamins antioxidants, this last mainly including the phytochemicals. Many vitamins directly scavenge ROS and in parallel can upregulate the antioxidant capacity of oxidative defence system of the body. Among the antioxidant treatments, those using vitamins, Vitamin E, Vitamin E analogs, and Vitamin C have been evaluated since several years. Oral supplementation of vitamin C (ascorbic acid) and vitamin E (D-alfa-tocopherol acetate) alone and in combination have been shown to decrease oxidative DNA damage in animal studies in vivo, in vitro, and in situ. Results of a prospective observational study (n = 4740) suggested that the combined use of vitamin E 400 IU daily and vitamin C 500 mg daily for at least 3 years was associated with the reduction of AD prevalence and incidence [89]. However subsequent meta-analyses reached the conclusions that in the absence of prospective, randomized, controlled clinical trials documenting benefits that outweigh recently documented morbidity and mortality risks, vitamin E supplements should not be recommended for primary or secondary prevention of AD [90]. In 2005, over 187 studies on specific antioxidants in the prevention of AD were evaluated [91]. Among non-vitamin agents that show promise in helping prevent AD there are: aged garlic extract, curcumin, melatonin, resveratrol, Ginkgo biloba extract, green tea. Indeed so many intervention trials with antioxidants are presently considered inconclusive in demons- trating benefits in humans. The main observation supporting these findings is that the administration of antioxidants in subjects who already had extensive pathology may be too late. Moreover the doses used may not be the optimal. There is some evidence that lower doses and/or mixtures of antioxidants might have more benefit than higher doses of single agents [92]. Although epidemiological data on diet and AD have been conflicting [93] higher adherence to the Mediterranean Diet (rich of fruits, vegetables, legumes, cereals and fish) is associated with a trend for reduced risk for developing MCI and with reduced risk for MCI conversion to AD [94].

Therapeutic interventions for the prodromal stages of dementia are currently being investigated with a view to delaying if not preventing disease onset. Uncertainty as to whether cognitive disorder in a given individual will progress towards dementia and adverse drug side effects has led to hesitancy to promote preventive pharmacotherapies. Antioxidant therapies may thus provide a low-risk alternative, targeting very early biological changes. There is increasing evidence in neurodegenerative diseases, and particularly in AD, that epigenetic modifications of the promoter of disease-related genes might be another mechanism relevant to disease pathogenesis [95]. Recently Zawia *et al.* [96] proposed a model linking epigenetic modifications, oxida- tive DNA damage, DNA repair and AD. One of the best known epigenetic events is the modification of the methylation pattern of CpG dinucleo- tides in the promoter region of specific genes. The authors observed that environmental influences occurring during brain development, such as exposure to metals, inhibit DNA-methyltransferases, thus resulting in hypomethylation of the promoters of genes associated with AD, such as that encoding the amyloid β precursor protein (APP). This early life imprint was sustained and triggered later in life to increase the levels of APP and Aβ. Increased Aβ production led to increased oxidative stress and oxidative DNA damage. Whereas AD-related genes were overexpressed late in life, others were repressed, suggesting that the early life perturbations resulted in hypo-methylation of some genes as well as hypermethylation of others. However, hypermethylated genes are more suscep- tible to Aβ-induced oxidative DNA damage because methylation of cytosines at CpG sites restricts repair of an adjacent 8-OHG. Therefore, according to these Authors, although the conditions leading to early life hypo- or hypermethylation of specific genes are not yet fully understood, these changes can have an impact on gene expression and imprint susceptibility to oxidative DNA damage in the aged brain [96]. This scenario opens a new promising field of research that is the study of the relationship between oxidative damage and epigenetic modifications in neuro- degenerative diseases.

REFERENCES

[1] Williamson, J.; Goldman, J.; Marder, K.S.;*Neurologist.* **2009**,15, 80; Lesage, S.; Brice, A.; *Hum. Mol. Genet.* **2009**, 18, R48; Valdmanis, P.N.; Daoud, H.;Dion, P.A.; Rouleau, G.A. *Curr Neurol Neurosci Rep.* **2009**, 198.

[2] Mancuso, M.; Coppedè, F.; Migliore, L.; Siciliano, G.; Murri, L. *J.Alzheimers Dis.* **2006**, 10, 59; Moreira, P.I.; Nunomura, A.; Honda, K.; Aliev, G.; Casadenus, G.; Zhu, X.; Smith, M.A.; Perry, G. in: *Oxidative Stress and Neurodegenerative Disorders* Qureshi, S. and Parvez, H. (Eds.), Elsevier, Amsterdam, 2007, pp. 451–466. Migliore, L; Coppedè, F. *Mutat Res.* **2009**, 674, 73.

[3] Chabrier, P.E.; Demerle-Pallardy, C.; Auguet, M. *Cell. Mol. Life Sci.* **1999**,1029. Skulachev, V.P. *Mol. Aspects Med.* **1999**, 139. Schulz, J.B.; Lindenau, J.; Seyfried, J.; Dichgans, J. *Eur. J. Biochem.* **2000**, 267, 4904.

[4]   Migliore, L; Fontana, I.; Colognato, R.; Coppedè, F.; Siciliano, G.; Murri, L. Searching for the role and the most suitable biomarkers of oxidative stress in Alzheimer's disease and in other neurodegenerative diseases. *Neurobiol Aging* **2005**, 26,587.

[5]   Yoritaka, A.; Hattori, N.; Uchida, K.; Tanaka, M.; Stadtman, E.R.; Mizuno, Y. *Proc Natl Acad Sci U S A.* **1996**, 93, 2696. Sayre, L.M.; Zelasko, D.A., Harris, P.L.; Perry, G.; Salomon, R.G.; Smith, M.A. *J Neurochem.* **1997**, 68, 2092; Lovell, M.A.; Ehmann, W.D.; Mattson, M.P.; Markesbery, W.R. *Neurobiol Aging.* **1997**, 18, 457; Markesbery, W.R.; Lovell, M.A. *Neurobiol Aging.* **1998**, 19, 33; Pedersen, W.A.; Fu, W.; Keller, J.N.; Markesbery, W.R.; Appel, S.; Smith, R.G.; Kasarskis, E.; Mattson, M.P. *Ann Neurol.* **1998**, 44,819.Smith, R.G.; Henry, Y.K.; Mattson, M.P.; Appel, S.H. *Ann Neurol.* **1998**, 44, 696. Castellani, R.J.; Perry, G.; Siedlak, S.L.; Nunomura, A.; Shimohama, S.; Zhang, J.; Montine, T.; Sayre, L.M.; Smith, M,A. *Neurosci Lett.* **2002**, 319, 25. Zarkovic, K. *Mol Aspects Med.* **2003**, 24, 293;

[6]   Bermejo, P.; Gómez-Serranillos, P.; Santos, J.; Pastor, E.; Gil, P.; Martín-Aragón, S. Determination of malonaldehyde in Alzheimer's disease: a comparative study of high-performance liquid chromatography and thiobarbituric acid test. *Gerontology* **1997**, 43, 218.

[7]   Mariani, E.; Polidori, M.C.; Cherubini A.; Mecocci, P. Oxidative stress in brain aging, neurodegenerative and vascular diseases: an overview. *J Chromatogr B Analyt Technol Biomed Life Sci.* **2005**, 827, 65.

[8]   Dexter, D.T.; Carter, C.J.;Wells, F.R.; Javoy-Agid, F.; Agid, Y.; Lees, A.; Jenner, P.; Marsden, C.D. *J Neurochem.* **1989**, 52, 381; Kalra, J.; Rajput, A.H.; Mantha, S.V.; Chaudhary, A.K.; Prasad, K. *Mol Cell Biochem.* **1992**,112, 181; Ilic, T.V.; Jovanovic, M.; Jovicic, A.; Tomovic, M. *Funct Neurol.* **1999**, 14, 141; Sharma, A.; Kaur, P.; Kumar, B.; Prabhakar, S.; Gill, K.D. *Parkinsonism Relat Disord.* **2008**, 14, 52; Chen, C.M.; Liu, J.L.; Wu, Y.R.; Chen, Y.C.; Cheng, H.S.; Cheng, M.L.; Chiu, D.T. *Neurobiol Dis.* **2009**, 33, 429.

[9]   Marzatico, F.; Café, C.; Taborelli, M.; Benzi, G. *Neurochem Res.* **1993**, 18, 1101; Younes-Mhenni, S.; Frih-Ayed, M.; Kerkeni, A.; Bost, M.; Chazot, G. *Eur Neurol.* **2007**, 58, 78.

[10]   Hall, E.D.; Oostveen, J.A.; Andrus, P.K.; Anderson, D.K.; Thomas, C.E. *J Neurosci Methods.* **1997**, 76, 115; Ferrante, R.J.; Browne, S.E.; Shinobu, L.A.; Bowling, A.C.; Baik, M.J.; MacGarvey, U.; Kowall, N.W.; Brown, R.H. Jr.; Beal, M.F. *J Neurochem.* **1997**, 69, 2064. Bonnefont-Rousselot, D.; Lacomblez, L.; Jaudon, M.; Lepage, S.; Salachas, F.; Bensimon, G.; Bizard, C.; Doppler, V.; Delattre, J.; Meininger, V. *J Neurol Sci.* **2000**, 178, 57. Fang, L.; Huber-Abel, F.; Teuchert, M.; Hendrich, C.; Dorst, J.; Schattauer, D.; Zettlmeissel, H.; Wlaschek, M.; Scharffetter-Kochanek, K.; Tumani H.; Ludolph, A.C.; Brettschneider, J. *J Neurol Sci.* **2009**, in press.

[11]   Montine, T.J.; Montine, K.S.; McMahan, W.; Markesbery, W.R.; Quinn, J.F.; Morrow, J.D. *Antioxid Redox Signal.* **2005**, 7, 269. Markesbery, W.R.; Kryscio, R.J.; Lovell, M.A.; Morrow, J.D. *Ann Neurol.* **2005**, 58, 730.

[12]   Irizarry, M.C.; Yao, Y.; Hyman, B.T.; Growdon, J.H.; Praticò, D. Plasma F2A isoprostane levels in Alzheimer's and Parkinson's disease. *Neurodegener Dis.* **2007**, 4, 403.

[13]   Connolly, J.; Siderowf, A.; Clark, C.M.; Mu, D.; Praticò, D. F2 isoprostane levels in plasma and urine do not support increased lipid peroxidation in cognitively impaired Parkinson disease patients. *Cogn Behav Neurol.* **2008**, 21, 83.

[14]   Fessel, J.P.; Hulette, C.; Powell, S.; Roberts, L.J. 2nd.; Zhang, J. Isofurans, but not F2-isoprostanes, are increased in the substantia nigra of patients with Parkinson's disease and with dementia with Lewy body disease. *J Neurochem.* **2003**, 85, 645.

[15]   Mitsumoto, H.; Santella, R.M.; Liu, X.; Bogdanov, M.; Zipprich, J; Wu, H.C.; Mahata, J.; Kilty, M.; Bednarz, K.; Bell, D.; Gordon, P.H.; Hornig, M.; Mehrazin, M.; Naini, A.; Flint Beal, M.; Factor-Litvak, P. Oxidative stress biomarkers in sporadic ALS. *Amyotroph Lateral Scler.* **2008**, 9, 177.

[16]   Butterfield, D.A.; Sultana, R. Redox proteomics identification of oxidatively modified brain proteins in Alzheimer's disease and mild cognitive impairment: insights into the progression of this dementing disorder. *J Alzheimers Dis.* **2007**, 12, 61.

[17]   Reed, T.T.; Pierce, W.M.; Markesbery, W.R.; Butterfield, D.A. Proteomic identification of HNE-bound proteins in early Alzheimer disease: Insights into the role of lipid peroxidation in the progression of AD. *Brain Res.* **2009**, 1274, 66.

[18]   Reed, T.T.; Pierce, W.M. Jr.; Turner, D.M.; Markesbery, W.R.; Butterfield, D.A. Proteomic identification of nitrated brain proteins in early Alzheimer's disease inferior parietal lobule. *J Cell Mol Med.* **2008**, in press.

[19]   Danielson, S.R.; Andersen, J.K; Oxidative and nitrative protein modifications in Parkinson's disease. *Free Radic Biol Med.* **2008**, 44, 1787.

[20]   Gómez, A.; Ferrer, I. Increased oxidation of certain glycolysis and energy metabolism enzymes in the frontal cortex in Lewy body diseases. *J Neurosci Res.* **2009**, 87, 1002.

[21]   Kabashi, E.; Valdmanis, P.N.; Dion, P.; Rouleau, G.A. Oxidized/misfolded superoxide dismutase-1: the cause of all amyotrophic lateral sclerosis? *Ann Neurol.* **2007**,

[22]   Liu, D.; Bao, F.; Wen, J; Liu, J. Mutation of superoxide dismutase elevates reactive species: comparison of nitration and oxidation of proteins in different brain regions of transgenic mice with amyotrophic lateral sclerosis. *Neuroscience.* **2007**, 146, 255.

[23]   Poon, H.F.; Hensley, K.; Thongboonkerd, V.; Merchant, M.L.; Lynn, B.C.; Pierce, W.M.; Klein, J.B.; Calabrese, V.; Butterfield, D.A. Redox proteomics analysis of oxidatively modified proteins in G93A-SOD1 transgenic mice--a model of familial amyotrophic lateral sclerosis. *Free Radic Biol Med.* **2005**, 39, 453.

[24]   Niebrój-Dobosz, I.; Dziewulska, D.; Kwieciński, H. Oxidative damage to proteins in the spinal cord in amyotrophic lateral sclerosis (ALS). *Folia Neuropathol.* **2004**, 42,151.

[25]   Tohgi, H.; Abe, T.; Yamazaki, K.; Murata, T.; Ishizaki, E.; Isobe, C. Remarkable increase in cerebrospinal fluid 3-nitrotyrosine in patients with sporadic amyotrophic lateral sclerosis. *Ann Neurol* **1999**, 46, 129.

[26]   Perluigi, M.; Fai Poon, H.; Hensley, K.; Pierce, W.M.; Klein, J.B.;Calabrese, V.; De Marco, C.; Butterfield, D.A. Proteomic analysis of 4-hydroxy-2-nonenal-modified proteins in G93A-SOD1 transgenic mice–a model of familial amyotrophic lateral sclerosis. *Free Radic Biol Med* **2005**, 38, 960.

[27]   Mecocci, P.; MacGarvey, U.; Beal, M.F.; Oxidative damage to mitochondrial DNA is increased in Alzheimer's disease. *Ann. Neurol.* **1994**, 36, 747.

[28]   Lyras, L.; Cairns, N.J.; Jenner, A.; Jenner, P.; Halliwell, B. *J Neurochem.* **1997**, 68, 2061; Gabbita, S.P.; Lovell, M.A.; Markesbery, W.R. *J Neurochem.* **1998**, 71, 2034. Wang, J.; Xiong, S.; Xie, C.; Markesbery, W.R.; Lovell, M.A. *J. Neurochem.* **2005**, 93, 953.

[29]   Wang, J; Markesbery, W.R.; Lovell, M.A. Increased oxidative damage in nuclear and mitochondrial DNA in mild cognitive impairment. *J. Neurochem.* **2006**, 96, 825.

[30]   Lovell, M.A.; Gabbita, S.P.; Markesbery, W.R. *J. Neurochem.* **1999**, 72, 771. Mecocci, P.; Polidori, M.C.; Cherubini, A.; Ingegni, T.; Mattioli, P.; Catani, M.; Rinaldi, P.; Cecchetti, R.; Stahl, W.; Senin, U.; Beal, M.F. *Arch Neurol* **2002**, 59, 794.

[31]   Kadioglu, E., Sardas, S.; Aslan, S.; Isik, E.; Esat Karakaya, A. Detection of oxidative DNA damage in lymphocytes of patients with Alzheimer's disease. *Biomarkers* **2004**, 9, 203.

[32]   Migliore, L.; Fontana, I.; Trippi, F.; Colognato, R.; Coppedè, F.; Tognoni, G.; Nucciarone, B.; Siciliano, G. Oxidative DNA damage in peripheral leukocytes of mild cognitive impairment and AD patients. *Neurobiol Aging,* **2005**, 26, 567.

[33]   Alam, Z.I.; Jenner, A.; Daniel, S.E.; Lees, A.J.; Cairns, N.; Marsden, C.D.; Jenner, P.; Halliwell, B. Oxidative DNA damage in the parkinsonian brain: an apparent selective increase in 8-hydroxyguanine levels in substantia nigra. *J Neurochem.* **1997**, 69, 1196.

[34]   Zhang, J.; Perry, G.; Smith, M.A.; Robertson, D.; Olson, S.J.; Graham, D.G.; Montine, T.J. Parkinson's disease is associated with oxidative damage to cytoplasmic DNA and RNA in substantia nigra neurons. *Am J Pathol.* **1999**, 154, 1423.

[35]   Simon, D.K.; Lin, M.T.; Zheng, L.; Liu, G.J.; Ahn, C.H.; Kim, L.M.; Mauck, W.M.; Twu, F.; Beal, M.F.; Johns, D.R. Somatic mitochondrial DNA mutations in cortex and substantia nigra in aging and Parkinson's disease. *Neurobiol Aging.* **2004**, 25, 71.

[36]   Bender, A; Krishnan, K.J.; Morris, C.M.; Taylor, G.A.; Reeve, A.K.; Perry, R.H.; Jaros, E.; Hersheson, J.S.; Betts, J.; Klopstock, T.; Taylor, R.W.; Turnbull, D.M. High levels of mitochondrial DNA deletions in substantia nigra neurons in aging and Parkinson disease. *Nat Genet.* **2006**, 38, 515.

[37]   Kikuchi, A.; Takeda, A; Onodera, H.; Kimpara, T.; Hisanaga, K.; Sato, N.; Nunomura, A.; Castellani, R.J.; Perry, G.; Smith, M.A.; Itoyama, Y. Systemic increase of oxidative nucleic acid damage in Parkinson's disease and multiple system atrophy. *Neurobiol Dis.* **2002**, 9, 244.

[38]   Migliore, L.; Scarpato, R.; Coppedè, F.; Petrozzi, L.; Bonuccelli, U.; Rodilla, V. *Int J Hyg Environ Health.* **2001**, 204, 61; Migliore, L.; Petrozzi, L.; Lucetti, C.; Gambaccini, G.; Bernardini, S.; Scarpato, R.; Trippi, F.; Barale, R.; Frenzilli, G.; Rodilla, V.; Bonuccelli, U. *Neurology.* **2002**, 58, 1809.

[39]   Ferrante, R.J.; Browne, S.E.; Shinobu, L.A.; Bowling, A.C.; Baik, M.J.; MacGarvey, U.; Kowall, N.W.; Brown, R.H. Jr.; Beal, M.F. Evidence of increased oxidative damage in both sporadic and familial amyotrophic lateral sclerosis. *J Neurochem.* **1997**, 69, 2064.

[40]   Bogdanov, M.; Brown, R.H.; Matson, W.; Smart, R.; Hayden, D.; O'Donnell, H.; Beal, M.F.; Cudkowicz, M. Increased oxidative damage to DNA in ALS patients. *Free Radic Biol Med.* **2000**, 29, 652.

[41]   Warita, H.; Hayashi, T.; Murakami, T.; Manabe, Y.; Abe, K. *Brain Res Mol Brain Res.* **2001**, 89, 147. Aguirre, N.; Beal, M.F.; Matson, W.R.; Bogdanov, M.B. *Free Radic Res.* **2005**, 39, 383.

[42]   Sau, D.; De Biasi, S.; Vitellaro-Zuccarello, L.; Riso, P.; Guarnieri, S.; Porrini, M.; Simeoni, S.; Crippa, V.; Onesto, E.; Palazzolo, I.; Rusmini, P.; Bolzoni, E.; Bendotti, C.; Poletti, A. Mutation of SOD1 in ALS: a gain of a loss of function. *Hum Mol Genet.* **2007**, 16, 1604.

[43]   Murata, T.; Ohtsuka, C.; Terayama, Y. Increased mitochondrial oxidative damage and oxidative DNA damage contributes to the neurodegenerative process in sporadic amyotrophic lateral sclerosis. *Free Radic Res.* **2008**, 42, 221.

[44]   Nunomura, A.; Perry, G.; Pappolla, M.A.; Wade, R.; Hirai, K.; Chiba, S.; Smith, M.A. RNA oxidation is a prominent feature of vulnerable neurons in Alzheimer's disease. *J Neurosci.* **1999**, 19, 1959.

[45]   Shan, X.; Tashiro, H.; Lin, C.L. The identification and characterization of oxidized RNAs in Alzheimer's disease. *J Neurosci.* **2003**, 23, 4913.

[46]   Honda, K.; Smith, M.A.; Zhu, X.; Baus, D.; Merrick, W.C.; Tartakoff, A.M.; Hattier, T.; Harris, P.L.; Siedlak, S.L.;, Fujioka, H.; Liu, Q.; Moreira, P.I.; Miller, F.P.; Nunomura, A.; Shimohama, S.; Perry, G. Ribosomal RNA in Alzheimer disease is oxidized by bound redox-active iron. *J Biol Chem.* **2005**, 280, 20978.

[47] Shan, X.; Lin, C.L. Quantification of oxidized RNAs in Alzheimer's disease. *Neurobiol Aging.* **2006**, 27, 657.

[48] Ding, Q.; Markesbery, W.R.; Cecarini, V.; Keller, J.N. Decreased RNA, and increased RNA oxidation, in ribosomes from early Alzheimer's disease. *Neurochem Res.* **2006**, 31, 705.

[49] Lovell, M.A.; Markesbery, W.R. Oxidatively modified RNA in mild cognitive impairment. *Neurobiol Dis.* **2008**, 29, 169.

[50] Blackinton, J.; Kumaran, R.; van der Brug, M.P.; Ahmad, R.; Olson, L.; Galter, D.; Lees, A.; Bandopadhyay, R.; Cookson, M.R. Post-transcriptional regulation of mRNA associated with DJ-1 in sporadic Parkinson disease. *Neurosci Lett.* **2009**, 452, 8.

[51] Chang, Y.; Kong, Q.; Shan, X.; Tian, G., Ilieva, H.; Cleveland, D.W.; Rothstein, J.D.; Borchelt, D.R.; Wong, P.C.; Lin, C.L. Messenger RNA oxidation occurs early in disease pathogenesis and promotes motor neuron degeneration in ALS. *PLoS One.* **2008**, 3, e2849.

[52] Nunomura, A.; Hofer, T.; Moreira, P.I.; Castellani, R.J.; Smith, M.A.; Perry, G. RNA oxidation in Alzheimer disease and related neurodegenerative disorders. *Acta Neuropathol.* **2009**, 118, 151.

[53] Vina, J.; Lloret, A.; Ortı´, R.; Alonso, D. Molecularbases of the treatment of Alzheimer's disease with antioxidants: prevention of oxidative stress. *Mol. Aspects Med.* **2004**, 25, 117.

[54] Ballatori, N.; Krance, S. M.; Notenboom, S.; Shi, S.; Tieu, K.; Hammond C.L. Glutathione dysregulation and the etiology and progression of human diseases *Biol. Chem.*, **2009**, 390, 191.

[55] Bermejo, P.; Martín-Aragón, S.; Benedí, J.; Susín, C.; Felici, E.; Gil, P.; Ribera, J.M.; Villar, A.M. Peripheral levels of glutathione and protein oxidation as markers in the development of Alzheimer's disease from Mild Cognitive Impairment. *Free Radic Res.* **2008**, 42, 162.

[56] Calabrese, V.; Sultana, R.; Scapagnini, G.; Guagliano, E.; Sapienza, M.; Bella, R.; Kanski, J.; Pennisi, G.; Mancuso, C.; Stella, A.M., Butterfield, D.A. Nitrosative stress, cellular stress response, and thiol homeostasis in patients with Alzheimer's disease. *Antioxid Redox Signal.* **2006**, 8, 1975.

[57] Kharrazi, H.; Vaisi-Raygani, A.; Rahimi, Z.; Tavilani, H.; Aminian, M.; Pourmotabbed, T.; Association between enzymatic and non-enzymatic antioxidant defense mechanism with apolipoprotein E genotypes in Alzheimer disease. *Clin Biochem.* **2008**, 41, 932.

[58] Riederer, P.; Sofic, E.; Rausch, W.D., Schmidt, B.; Reynolds, G.P.; Jellinger K.; Youdim, M.B. Transition metals,ferritin, glutathione, and ascorbic acid in parkinsonian brains. *J Neurochem* **1989**, 52, 515–520.

[59] Kidd, P.M. Parkinson's disease as multifactorial oxidative neurodegeneration: implications for integrative management. *Altern Med Rev* **2000**, 5, 502.

[60] Menegon, A.; Board, P.G.; Blackburn, A.G.; Mellick, G.D.; Le Couteur, D.G. Parkinson's disease, pesticides, and glutathione transferase polymorphisms. *Lancet* **1998**, 352, 1344.

[61] Perez-Pastene, C.; Graumann, R.; Díaz-Grez, F.; Miranda, M.; Venegas, P.; Godoy, O.T.; Layson, L.; Villagra, R.; Matamala, J.M.; Herrera, L.; Segura-Aguilar, J. *Neurosci Lett.* **2007**, 418, 181; Singh, M.; Khan, A.J.; Shah, P.P.; Shukla, R.; Khanna, V.K.; Parmar, D. *Mol Cell Biochem.* **2008**, 312, 131.

[62] Whitworth, A.J.; Theodore, D.A.; Greene, J.C.; Benes, H.P.; Wes, D.; Pallanck, L.J. Increased glutathione S-transferase activity rescues dopaminergic neuron loss in a Drosophila model of Parkinson's disease. *Proc Natl Acad Sci USA* **2005**, 102, 8024.

[63] Li, Y.J.; Scott, W.K.; Zhang, L.; Lin, P.I.; Oliveira, S.A.; Skelly, T.; Doraiswamy, M.P.; Welsh-Bohmer, K.A.; Martin, E.R.; Haines, J.L.; Pericak-Vance, M.A.; Vance, J.M. Revealing the role of glutathione S-transferase omega in age-at-onset of Alzheimer and Parkinson diseases. *Neurobiol Aging.* **2006**, 27, 1087.

[64] Coppedè, F.; Armani, C.; Bidia, D.D.; Petrozzi, L.; Bonuccelli, U.; Migliore, L. Molecular implications of the human glutathione transferase A-4 gene (hGSTA4) polymorphisms in neurodegenerative diseases. *Mutat Res.* **2005**, 579, 107.

[65] Apostolski, S.; Marinković, Z.; Nikolić, A.; Blagojević, D.; Spasić, M,B. Michelson AM. Glutathione peroxidase in amyotrophic lateral sclerosis: the effects of selenium supplementation. *J Environ Pathol Toxicol Oncol.* **1998**,17, 325.

[66] Tohgi, H.; Abe, T.; Yamazaki, K.; Murata, T.; Ishizaki, E.; Isobe, C. Increase in oxidized NO products and reduction in oxidized glutathione in cerebrospinal fluid from patients with sporadic form of amyotrophic lateral sclerosis. *Neurosci Lett.* **1999**, 260, 204-6.

[67] Kuźma, M.; Jamrozik, Z.; Barańczyk-Kuźma, A. Activity and expression of glutathione S-transferase pi in patients with amyotrophic lateral sclerosis. *Clin Chim Acta.* **2006**, 364, 217.

[68] Nikolić-Kokić, A.; Stević, Z.; Blagojević, D.; Davidović, B.; Jones, D.R.; Spasić, M.B. Alterations in anti-oxidative defence enzymes in erythrocytes from sporadic amyotrophic lateral sclerosis (SALS) and familial ALS patients. *Clin Chem Lab Med.* **2006**, 44, 589.

[69] Chi, L.; Ke, Y.; Luo, C.; Gozal, D.; Liu, R. Depletion of reduced glutathione enhances motor neuron degeneration *in vitro* and in vivo. *Neuroscience.* **2007**, 144, 991.

[70] Tartari, S.; D'Alessandro, G.; Babetto, E.; Rizzardini, M.; Conforti, L.; Cantoni, L. Adaptation to G93Asuperoxide dismutase 1 in a motor neuron cell line model of amyotrophic lateral sclerosis: the role of glutathione. *FEBS J.* **2009**, 276, 2861.

[71] Choi, J.; Rees, H.D.; Weintraub, S.T.; Levey, A.I.; Chin, L.S.; Li, L. Oxidative modifications and aggregation of Cu,Zn-superoxide dismutase associated with Alzheimer and Parkinson diseases. *J Biol Chem.* **2005**, 280, 11648.

[72] Martin, L.J. DNA damage and repair: relevance to mechanisms of neurodegeneration. *J Neuropathol Exp Neurol.* **2008**, 67, 377.

[73] Coppedè, F.; Migliore, L. DNA damage and repair in Alzheimer's disease. *Curr Alzheimer Res.* **2009**, 6, 36.

[74] Lovell, M.A.; Xie, C.; Markesbery, W.R. Decreased base excision repair and increased helicase activity in Alzheimer's disease brain. *Brain Res* **2000**, 855, 116.

[75] Iida, T.; Furuta, A.; Nishioka, K.; Nakabeppu, Y.; Iwaki, T. Expression of 8-oxoguanine DNA glycosylase is reduced and associated with neurofibrillary tangles in Alzheimer's disease brain. *Acta Neuropathol* **2002**, 103, 20.

[76] Weissman, L.; Jo, D.G.; Sørensen, M.M.; de Souza-Pinto, N.C.; Markesbery, W.R.; Mattson, M.P.; Bohr, V.A. Defective DNA base excision repair in brain from individuals with Alzheimer's disease and amnestic mild cognitive impairment. *Nucleic Acids Res* **2007**, 35, 5545.

[77] Shao, C.; Xiong, S.; Li, G.M.; Gu, L.; Mao, G.; Markesbery, W.R.; Lovell, M.A. Altered 8-oxoguanine glycosylase in mild cognitive impairment and late-stage Alzheimer's disease brain. *Free Radic Biol Med.* **2008**, 45, 813.

[78] Mao, G.; Pan, X.; Zhu, B.B.; Zhang, Y.; Yuan, F.; Huang, J.; Lovell, M.A.; Lee, M.P.; Markesbery, W.R.; Li, G.M.; Gu, L. Identification and characterization of OGG1 mutations in patients with Alzheimer's disease. *Nucleic Acids Res* **2007**, 35, 2759.

[79] Coppedè, F.; Mancuso, M.; Lo Gerfo, A.; Manca, M.L.; Petrozzi, L.; Migliore, L.; Siciliano, G.; Murri, L. A Ser326Cys polymorphism in the DNA repair gene hOGG1 is not associated with sporadic Alzheimer's disease. *Neurosci Lett* **2007**, 414, 282.

[80] Fukae, J.; Takanashi, M.; Kubo, S.; Nishioka, K.; Nakabeppu, Y.; Mori, H.; Mizuno,Y.; Hattori, N. Expression of 8-oxoguanine DNA glycosylase (OGG1) in Parkinson's disease and related neurodegenerative disorders. *Acta Neuropathol.* **2005**, 109, 256.

[81] Cardozo-Pelaez, F.; Cox, D.P.; Bolin, C. Lack of the DNA repair enzyme OGG1 sensitizes dopamine neurons to manganese toxicity during development. *Gene Expr.* **2005**, 12, 315.

[82] Kisby, G.E.; Milne, J.; Sweatt, C. Evidence of reduced DNA repair in amyotrophic lateral sclerosis brain tissue. *Neuroreport.* **1997**, 8, 1337.

[83] Shaikh, A.Y.; Martin, L.J. DNA base-excision repair enzyme apurinic/apyrimidinic endonuclease/redox factor-1 is increased and competent in the brain and spinal cord of individuals with amyotrophic lateral sclerosis. *Neuromolecular Med.* **2008**, 2, 47.

[84] Olkowski, Z.L. Mutant AP endonuclease in patients with amyotrophic lateral sclerosis. *Neuroreport* **1998**, 9, 239.

[85] Hayward, C.; Colville, S.; Swingler, R.J.; Brock, D.J. *Neurology* **1999**, 52, 1899. Tomkins, J.; Dempster, S.; Banner, S.J.; Cookson, M.R.; Shaw, P.J. *Neuroreport* **2000**, 11, 1695. Greenway, M.J.; Alexander, M.D.; Ennis, S.; Traynor, B.J.; Corr, B.; Frost, E.; Green, A.; Hardiman, O. *Neurology* **2004**, 63, 1936. Coppedè, F.; Gerfo, A.L.; Carlesi, C.; Piazza, S.; Mancuso, M.; Pasquali, L.; Murri, L.; Migliore, L.; Siciliano, G. *Neurobiol Aging.* **2008**, Epub May 13.

[86] Kikuchi, H.; Furuta, A.; Nishioka, K.; Suzuki, S.O.; Nakabeppu, Y.; Iwaki, T. Impairment of mitochondrial DNA repair enzymes against accumulation of 8-oxo-guanine in the spinal motor neurons of amyotrophic lateral sclerosis. *Acta Neuropathol* **2002**, 103, 408.

[87] Coppedè, F.; Mancuso, M.; Lo Gerfo, A.; Carlesi, C.; Piazza, S.; Rocchi, A.; Petrozzi, L.; Nesti, C.; Micheli, D.; Bacci, A.; Migliore, L.; Murri, L.; Siciliano, G. Association of the hOGG1 Ser326Cys polymorphism with sporadic amyotrophic lateral sclerosis. *Neurosci Lett.* **2007**, 420, 163.

[88] Yan, S.D.; Yan, S.F.; Chen, X.; Fu, J.; Chen, M.; Kuppusamy, P.; Smith, M.A.; Perry, G.; Godman, G.C.; Nawroth, P.; Zweier, J.L.; Stern, D. *Nat. Med.* **1995**, 1, 693. Tamagno, E.; Bardini, P.; Obbili, A.; Vitali, A.; Borghi, R.; Zaccheo, D.; Pronzato, M.A.; Danni, O.; Smith, M.A.; Perry, G.; Tabaton, M. *Neurobiol. Dis.* **2002**, 10, 279. Smith, M.A.; Casadesus, G.; Joseph, J.A.; Perry, G. *Free Radic. Biol. Med.* **2002**, 33, 1194. Zhu, X.; Su, B.; Wang, X.; Smith, M.A.; Perry, G. *Cell. Mol. Life Sci.* **2007**, 64, 2202.

[89] Zandi, P.P.; Anthony, J.C.; Khachaturian, A.S.; Stone, S.V.; Gustafson, D.; Tschanz, J.T.; Norton, M.C.; Welsh-Bohmer, K.A.; Breitner, J.C.; Cache County Study Group. Reduced risk of Alzheimer disease in users of antioxidant vitamin supplements: the Cache County Study. *Arch Neurol.* **2004**, 61, 82.

[90] Boothby, L.A.; Doering, P.L. Vitamin C and vitamin E for Alzheimer's disease. *Ann Pharmacother.* **2005**, 39, 2073.

[91] Frank, B.; Gupta, S.; A review of antioxidants and Alzheimer's disease. Ann Clin Psychiatry. **2005**, 17, 269.

[92] Halliwell, B.; The wanderings of a free radical. *Free Radic Biol Med.* **2009**, 46, 531.

[93] Luchsinger, J.A.; Noble, J.M.; Scarmeas, N. Diet and Alzheimer's Disease. *Curr Neurol Neurosci Rep.* **2007**, 7, 366.

[94] Scarmeas, N.; Stern, Y.; Mayeux, R.; Manly, J.; Schupf, N.; Luchsinger, J.A. Mediterranean Diet and Mild Cognitive Impairment. *Arch Neurol.* **2009**, 66, 216.

[95]   Migliore, L.; Coppedè, F. Genetics, environmental factors and the emerging role of epigenetics in neurodegenerative diseases. *Mutat Res.* **2008**, E-pub Oct 31.

[96]   Zawia, N.H.; Lahiri, D.K.; Cardozo-Pelaez, F. Epigenetics, oxidative stress, and Alzheimer disease. *Free Radic Biol Med.* **2009**, 46, 1241.

CHAPTER 4

Kainic Acid-Mediated Neural Cell Death in Brain: Interplay Among Glycerophospholipid-, Sphingolipid-, and Cholesterol-Derived Lipid Mediators

Wei-Yi Ong and Akhlaq A. Farooqui[*]

Department of Anatomy, National University of Singapore, Singapore 119260, and Department of Molecular and Cellular Biochemistry*, The Ohio State University, Columbus, OH, USA

Abstract: In addition to integral proteins, neural membranes are composed of glycerophospholipids, sphingolipids, and cholesterol, which provide membranes with structural integrity (suitable stability, fluidity, and permeability). Kainic acid (KA), an excitotoxin, markedly upregulates glycerophospholipids, sphingolipids, and cholesterol metabolism resulting not only in loss of essential lipids and inducing changes in neural membrane fluidity and permeability, but also in elevations in glycerophospholipid, sphingolipid, and cholesterol-derived lipid mediators. These processes result in depolarization, sustained increase in Ca^{2+} and stimulation of Ca^{2+}-dependent enzymes including PLA_2, PLC, nitric oxide synthase, calpains, and endonucleases. Sustained stimulation of these enzymes, generation, and interplay among glycerophospholipid-, sphingolipid-, and cholesterol-derived lipid mediators along with mitochondrial dysfunction, decrease in ATP levels, changes in redox status of neural cell may be responsible for neurodegeneration through apoptosis and necrosis in KA-mediated neurotoxicity. Other KA-mediated neurochemical changes include synaptic reorganization associated with recapitulation of hippocampal development and synaptogenesis following KA-induced seizures.

Keywords: Lipid mediators; kainic acid-mediated neurotoxicity; glycerophospholipids, sphingolipids, cholesterol; mitochondrial dysfunction

INTRODUCTION

Neural membranes are composed of glycerophospholipids, sphingolipids, cholesterol, and proteins. Glycerophospholipids and sphingolipids contribute to lipid asymmetry, whereas cholesterol and sphingolipids form lipid microdomains or lipid rafts, which float within the membrane and are united by certain groups of proteins [1]. Neural membrane glycerophospholipids include phosphatidylcholine; phosphatidylethanolamine (PtdEtn); phosphatidylserine (PtdSer); and phosphatidylinositol (PtdIns). PtdEtn and PtdSer are located in the cytofacial side of the inner monolayer, whereas PtdCho is localized monolayer, on the exofacial side of the outer monolayer. Among the glycerophospholipids, PtdEtn, plasmenyl ethanolamine (PlsEtn), and PtdSer contain high levels of docosahexaenoyl groups (22:6n-3) at the sn-2 position of the glycerol moiety, whereas phosphatidylcholine (PtdCho), phosphatidylinositol (PtdIns), and phosphatidic acid (PtdH) contain high levels of arachidonoyl groups (20:4n-6) [2, 3]. Thus, glycerophospholipids are storage depots for arachidonic acid (ARA) and docosahexaenoic acid (DHA). ARA-derived enzymic and non-enzymic lipid mediators include eicosanoids, lipoxins, 4-hydroxynonenal (4-HNE), isoprostane, isoketal, and isofuran, whereas DHA-derived enzymic and non-enzymic lipid mediators are docosanoids (docosatrienes, resolvins, and neurotrotectins), neuroprostane, neuroketal, and neurofuran [4, 5]. Ceramide (N-acyl-sphingosine) not only forms the backbone of sphingolipids, but a central molecule in sphingolipid metabolism. It resides and functions within lipid rafts. Ceramide-enriched rich rafts are stable and have been implicated in diverse biological processes, such as cell growth, differentiation, senescence and apoptosis. Ceramide is generated by *de novo* synthesis or in response to stress or agonists through the action of sphingomyelinases on sphingomyelin. Ceramide is either phosphorylated to ceramide-1-phosphate or hydrolyzed by ceramidases to yield sphingosine, which is phosphorylated by sphingosine kinase to yield sphingosine-1-kinase. Thus, ceramide-derived second metabolites include ceramide-1-phosphate, sphingosine, and sphingosine-1-phosphate [1]. Dysregulation of the relative levels of ceramide-derived lipid metabolites have been implicated in a wide array of neurodegenerative disorders and neural malformations [6, 7].

Cholesterol is an essential component of neural membranes. It plays an important role in glycerophospholipid organ

---

*Address correspondence to: Department of Molecular and Cellular Biochemistry, The Ohio State University, Columbus, OH 43210, USA; Tel.: (614) 488-0361; Email: farooqui.1@osu.edu

Akhlaq A. Farooqui & Tahira Farooqui (Eds.)

ization, dynamics, signal transduction, and sorting. It increases membrane thickness and decreases transbilayer permeability [8, 9]. In brain, most cholesterol is present in myelin and other cellular membranes, where it functions as a modulator of the lateral segregation of lipids into cholesterol-poor and cholesterol-rich domains. Cholesterol is metabolized into 24-hydroxycholesterol, 25-hydroxycholesterol, and 27-hydroxycholesterol by cytochrome P450-dependent oxygenases. Cholesterol is also oxidized to cholesterol oxides and converted into cholesterol ester via acyl-CoA:cholesterol acyltransferase [8, 9].

Kainic acid (KA) is a cyclic and nondegradable analog of glutamate. Its systemic administration in adult rats causes persistent seizures, memory loss and neuronal damage [10]. KA interacts with its receptors and produces selective degeneration of neurons, especially in the hippocampus after intraventricular or intracerebral injections [10, 11]. The role of KA receptors depends on their precise membrane and subcellular localization in presynaptic, extrasynaptic and postsynaptic domains. KA receptors are formed from the combination of five subunits, three of them having several splice variants. The subunits and splice variants show great divergence in their C-terminal cytoplasmic tail domains, which have been implicated in intracellular trafficking of homomeric and heteromeric receptors [12]. At a presynaptic level, KA receptors modulate short and long term synaptic plasticity while at the postsynaptic level, KA receptors allow the influx of Na^+ and the efflux of K^+. KA receptor mediated movement of Na^+ and K^+ ions is closely associated with depolarization and alterations in Ca^{2+} levels [13]. KA injections can also cause neuronal death through the activation of AMPA receptors, which may have different Ca^{2+} permeability depending on subunit composition [13].

During KA-induced neurotoxicity, increase in intracellular calcium activates calcium-dependent enzymes, including $cPLA_2$, nitric oxide synthase, calpains, sphingomyelinases, and cytochrome P450 oxidases generating glycerophospholipid-, sphingolipid-, and cholesterol derived lipid mediators [14, 15, 16, 17]. Under physiological conditions, ARA and its metabolites produced by the stimulation of KA-receptor at the postsynaptic level, cross the synaptic cleft to act at presynaptic KAR, and thus may act as retrograde messengers for the activation of protein kinase C isozymes [14, 15]. This process may be involved in the induction of mossy fiber long-term potentiation in the hippocampus. During KA-induced toxicity, elevated levels of ARA and its metabolites produce a variety of detrimental effects on neural membrane structures, activities of membrane enzymes, and neurotransmitter uptake systems [14, 15]. Furthermore the peroxidation of ARA results in generation of 4-HNE, an α, β unsaturated aldehyde that not only reacts with lysine, cysteine, and histidine residues in proteins but also binds to free amino acids and deoxyguanosine [15]. 4-HNE has been reported to cause a number of deleterious effects in cells including inhibition of DNA synthesis, disturbance in calcium homeostasis, and inhibition of mitochondrial respiration [14]. In addition to the above neurochemical changes, stimulation of ornithine decarboxylase, expression immediate early genes, growth factors and heat shock proteins have been shown to occur in KA-mediated neurotoxicity [14, 15, 16]. The purpose of this commentary is to discuss the importance of glycerophospholipid-, ceramide-, and cholesterol-derived lipid mediators interplay in kainic acid neurotoxicity.

KAINIC ACID-MEDIATED ALTERATIONS IN BRAIN GLYCEROPHOSPHOLIPIDS

Systemic administration of KA into adult rats stimulates rat brain $cPLA_2$ and $sPLA_2$ activities (Fig. **1a**) [18]. Increase in $cPLA_2$ immunoreactivity is observed in neurons at 1 and 3 days after KA injection [19]. KA injections also increase $cPLA_2$ immunoreactivity in astrocytes after 1, 2, 4, 11 weeks. Increased $cPLA_2$ activity in neurons in KA-mediated toxicity may be involved in neurodegeneration, whereas the elevation of $cPLA_2$ immunoreactivity in astrocytes is associated with gliosis (Fig. **2**) [19]. KA-mediated increase in $cPLA_2$ immunoreactivity is accompanied by a marked increase in $cPLA_2$ mRNA levels (Fig. **1b**) [20]. Quinacrine, a non-specific PLA_2 inhibitor, and CNQX retard this increase in $cPLA_2$ activity and $cPLA_2$ mRNA indicating that generation of ARA is a receptor-mediated process. In cortical cell cultures, KA also stimulates plasmalogen selective-PLA_2 ($PlsEtn-PLA_2$) activity in a dose- and time-dependent manner. Bromoenol lactone (BEL), a $PlsEtn-PLA_2$ inhibitor, and CNQX, a KA/AMPA antagonist, blocks this stimulation [21]. Increased ARA release through the stimulation of PLA_2 isoforms subsequently results in an increase in lipid peroxidation, generation of free radicals, accumulation of eicosanoids, lipid peroxides, 4-HNE, and isoprotanes and 4-HNE-modified proteins [4, 16]. Accumulation of the above metabolites is higher in brain tissue because of its high content of polyunsaturated fatty acids and high oxygen consumption for metabolism. In addition, eicosanoids promote inflammation and lipid peroxides act as proinflammatory agonists facilitating chronic inflammation mediated by inflammatory cytokines. Collective evidence suggests that isoforms of PLA_2 are stimulated by

KA and other glutamate analogs in neural cell culture systems as well as in brain. This stimulation of these isoforms decreases levels of PtdCho and PlsEtn, and increases degradation products of ARA [16]. Generation of high levels of ARA-derived lipid mediators and ROS along with mitochondrial dysfunction, decrease in ATP level, alteration in redox status along with changes in neural membrane permeability and loss of ion homeostasis results in neuronal injury in KA-mediated neurotoxicity [5, 16].

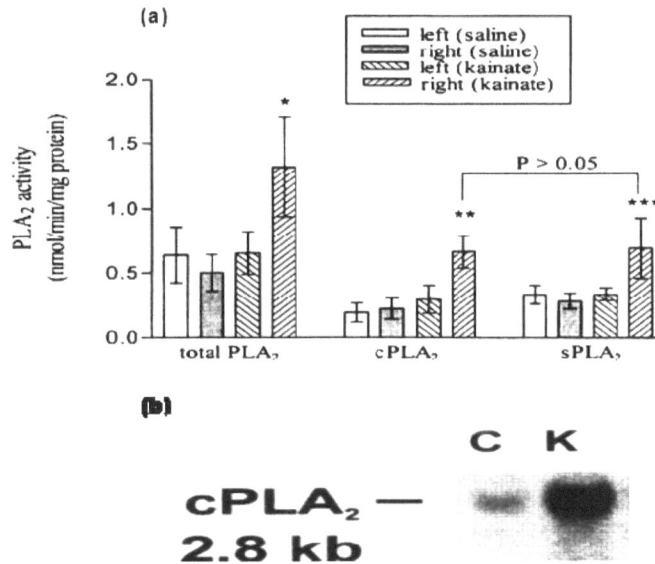

Figure 1: (a) cPLA$_2$ activity in the hippocampus of normal control rats and 3 days after kainate injections. An increase in activities of both cPLA$_2$ and sPLA$_2$ is observed in the kainate lesioned hippocampus. Reproduced with kind permission from Thwin *et al.*, 2003, Experimental Brain Research 150:427-433, Springer. Fig. **1**. (b) Northern blot analysis of cPLA$_2$ mRNA in the hippocampus of control rats, and 3 days after kainate injections. Increase in cPLA$_2$ mRNA expression is detected in the kainate lesioned hippocampus (K) compared to that PBS injected controls (C). Reproduced with kind permission from Ong et al., 2003, Experimental Brain Research 148:521-524, Springer.

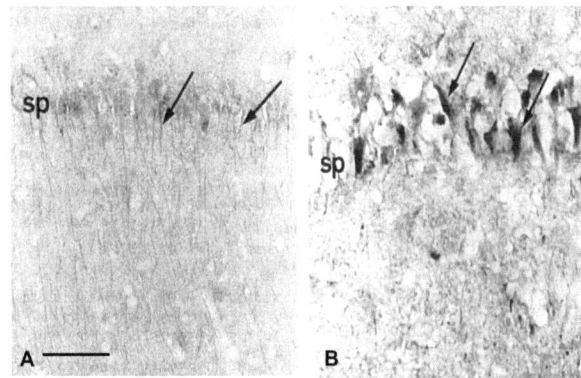

Figure 2: Immunohistochemical analysis of cPLA$_2$ in the hippocampus of normal control rats. (A) and 3 days after kainate injections (B). Light labeling to cPLA$_2$ is present in neurons of the normal hippocampus (arrows in A), whilst dense labeling is observed in affected pyramidal neurons, in the kainate lesioned hippocampus (arrows in B). sp: stratum pyramidale. Scale: A = 130 μm, B = 80 μm. Reproduced with kind permission from Sandhya et al., 1998, Brain Research 788:223-231, Elsevier.

KA treatment produces significant increase in iron levels in the hippocampus at 2 weeks (~42%), 4 weeks and (~76%), and 8 weeks (~88%) after injection, compared to controls (Fig. **3a**). This increase in iron is not due to the diffusion and redistribution of iron, but is caused by increased iron uptake into brain tissue. A significant increase in density ratios of divalent metal ion transporter-1 (DMT-1)/ β-actin bands is observed on Western blots in the 1-week, 1-month, and 2-months post-KA-injected hippocampus compared to uninjected and 1-day post-KA-injected hippocampus. The increase in DMT-1 protein is paralleled by an increase in DMT-1 immunoreactivity in astrocytes. Light staining for DMT-1 is observed in the uninjected, saline-injected, and 1-day post-KA-injected rat hippocampus. In contrast, an upregulation of DMT-1 is seen in reactive glial cells at 1 week, 1 month, and 2 months post-KA injection. Electron microscopic studies show that that the glial cells have morphological features of astrocytes [22, 23]. KA-mediated neurotoxicity also results in a progressive change in the oxidative state of iron in the lesioned areas. Light staining for iron is observed in the hippocampus of normal or saline-injected rats and 1-day post-kainate-injected rats.

There is increase in the Fe^{3+} form of iron in glial cells of the degenerating hippocampus at 1 week and 1 month post-KA injection, and increase in both Fe^{3+} and Fe^{2+} forms in these cells, at 2 months post-KA injection. A possible consequence of the accumulation of Fe^{2+} may be free radical generation and free radical-mediated damage in the KA-lesioned areas [22, 23]. Increase in iron levels in KA-mediated neurotoxicity is related with increase in isoprostanes, the non-enzymic metabolite and novel marker for oxidative stress (Fig. **3b**) [5].

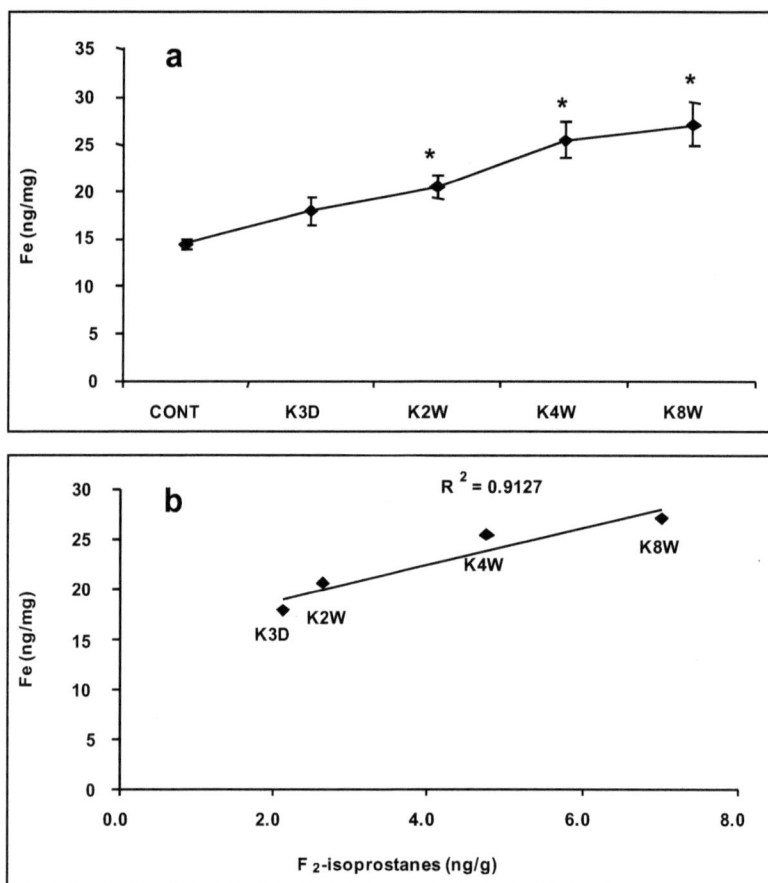

Figure 3a: Iron levels in the rat hippocampus. CONT, K3D, K2W, K4W and K8W indicate control, 3 days, 2 weeks, 4 weeks, and 8 weeks post-kainate injection. The mean and standard error are indicated. *Significant difference compared to control group ($P < 0.05$). Figure **3b**: Correlation between iron levels and F2-isoprostane levels in the kainate-injected rat hippocampus by linear correlation analysis. K3D, K2W, K4W and K8W indicate control, 3 days, 2 weeks, 4 weeks, and 8 weeks post-kainate injection. R2 = 0.9127

KA-MEDIATED ALTERATIONS IN SPHINGOLIPID METABOLISM

Intracerebroventricular injections of KA in rats produce significant increase in ceramide immunoreactivity and levels in the hippocampus, at 1 day and 3 days post-injection compared to controls. No increase in serine palmitoyltransferase (SPT) immunolabeling is seen in degenerating neurons, but occasional increase in SPT immunoreactivity is observed in glial cells in the KA injected rat brain in CA fields. This enzyme catalyzes the condensation of serine and palmitoyl CoA leading to the synthesis of keto-sphinganine, which is reduced and acylated by ceramide synthase to dihydroceramide. This metabolite is then desaturated to yield ceramide [6, 7].

Tandem mass spectrometric profiling of lipid extract from KA injected and control hippocampal tissues indicate major reduction in glycerophospholipids, and significant increase in NAPE (N-acylated ethanolamine, anadamide), and ceramide with different molecular species, including 16:0, 18:0, 20:0, 22:0 and 24:1 fatty acids in KA injected hippocampus (Fig. **4**) [24].

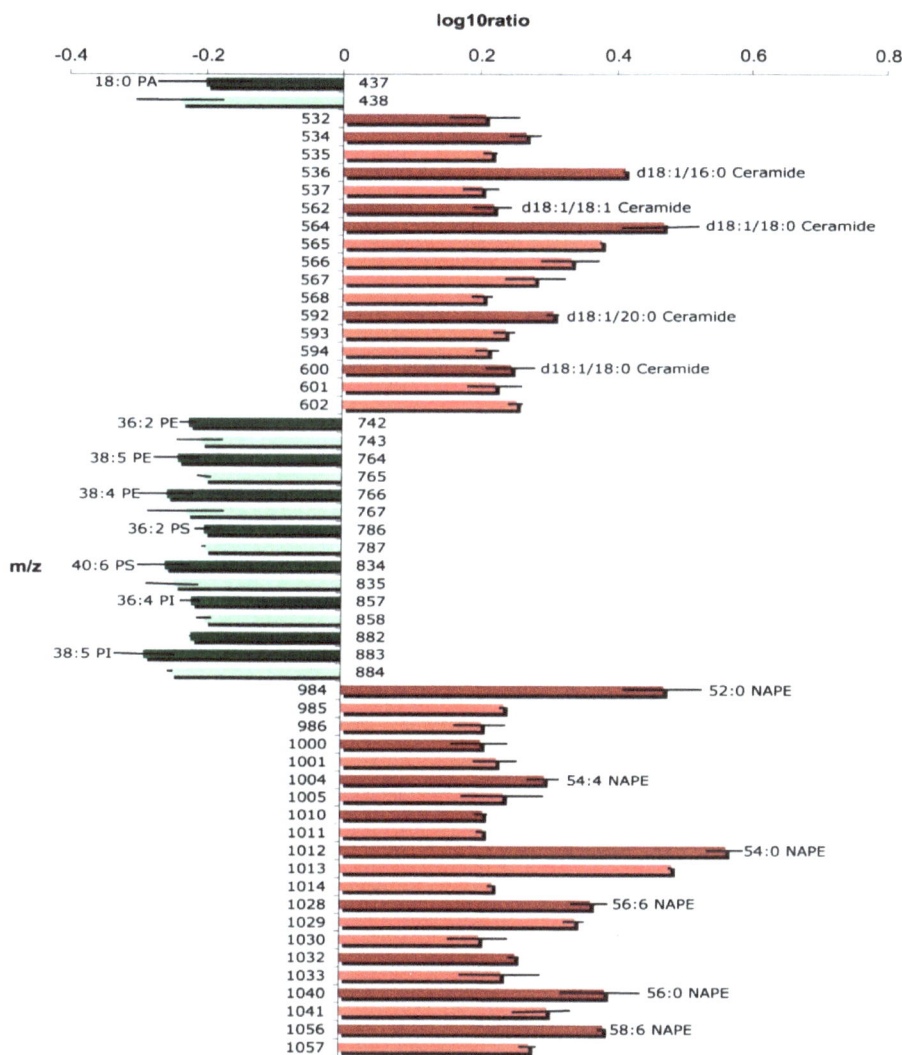

Figure 4: Lipidomic analysis of the hippocampus 3 days after kainate injections. The red or green bars indicate increases (red) or decreases (green) compared to controls. Increase in ceramide species (red bars at top of figure) and NAPE (cannabinoid receptor agonist, red bars at bottom of figure), but decreases in glycerophospholipids (green bars at middle of figure) is detected in the kainate lesioned hippocampus. Reproduced with kind permission from Guan *et al.*, 2006, FASEB J 20:1152-1161.

The increase in ceramide levels is associated with increased expression and activity of SPT, after 1 to 3 weeks intra-cerebroventricular KA injections [25]. Immunohistochemical analyses indicate baseline expression of SPT in neurons, and gradual increase in immunoreactivity in astrocytes post KA treatment (Fig. **5**). The expression of SPT in reactive astrocytes suggests that these cells may be involved in turnover of sphingolipids in kainate induced excitotoxic brain injury.

Treatment of organotypic rat hippocampal slices with KA increases ceramide levels with elevation in 16:0, 18:0, 20:0 molecular species, at one day after KA treatment. Addition of SPT inhibitors (myriocin, or L-cycloserine) after KA treatment not only reduces total ceramide levels, but also partially reduces cell death caused by KA in vitro. In contrast, treatment with sphinogomyelinase inhibitor (GW4869) has no effect. These observations indicate that a substantial portion of the ceramide is generated at 1 to 4 weeks after kainate treatment is due to the activity of SPT. The action of sphingomyelinases cannot, however, be excluded [25]. Based on GC/mass analysis, some decrease in the sphingomyelin content of the hippocampus may also account for the increase in ceramide [24, 25].

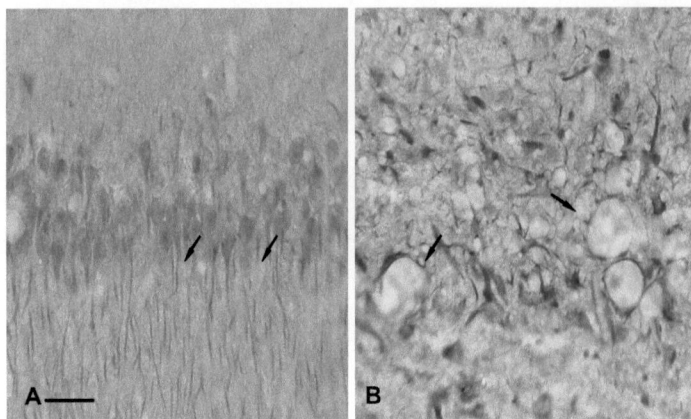

Figure 5: Immunohistochemical analysis of serine palmitoyltransferase (SPT) expression in the hippocampus of normal control rats (A) and 3 days after kainate injections (B). Light labeling to SPT is present in neurons of the normal hippocampus (arrows in A), whilst dense labeling is observed in astrocytes in the kainate lesioned hippocampus (arrows in B). Scale = 50 μm. Reproduced with kind permission from He *et al.*, 2007, Journal of Neuroscience Research 85:423-432, Wiley-Liss

KA treatment of rat hippocampal slices not only decreases neuronal MAP2 staining, but also increases LDH release into the culture media indicating onset of neurodegeneration following KA treatment. The addition of SPT inhibitors (L-cycloserine or myriocin) to hippocampal slices, after KA treatment, partially protects against the reduction in MAP2 staining and increase in LDH release [25]. In contrast to inhibition of ceramide biosynthesis by L-cycloserine or myriocin, the incubation of slices with an inhibitor of sphingomyelinase has no significant protective effect. Thus, inhibition of ceramide biosynthesis has a significant neuroprotective effect after KA-mediated neural cell injury. Ceramide has been reported to modulate the opening of the mitochondrial permeability transition pores (PTP), which disrupts the transmembrane potential, thus causing the release of cytochrome c and the generation of hydrogen peroxide. These molecules induce the release of APAF-1 and caspase-3 activation and subsequent poly-ADP-ribose polymerase (PARP) cleavage and DNA fragmentation during apoptotic cell death [1, 17]. Ceramide also induces changes in the expression of the Bcl-2 family of proteins by activating specific transcription factors such as NFκ-B and c-jun [17].

KA-MEDIATED ALTERATIONS IN CHOLESTEROL METABOLITES

Intraventricular injections of KA increase cholesterol staining with filipin in the CA1 region of rat hippocampus [26]. Hippocampal sections also show increase in cholesterol immunostaining in cell bodies and dendritic fields of neurons. No increase in cholesterol staining is seen in glial cells. The maximal increase in cholesterol (from 23.0 ± 15.8 to 98.6 ± 16.7μg/mg) is observed at 2 weeks post KA-injection [27]. Our recent studies indicate that increase in cholesterol level in KA-mediated neurodegeneration is possibly due to a disproportion between increased cholesterol and lack of export, and not because of increased biosynthesis of cholesterol (Kim, Jittiwat, Ong, Farooqui, and Jenner unpublished). There is significant increase in level of cholesterol oxidation products (COPs) or oxysterols, including 7-ketocholesterol in the degenerating brain after kainate-lesions [27] (Fig. **6**). KA treatment also increases cholesterol 24-hydroxylase immunoreactivity in glial cells of the affected CA fields. The cholesterol 24-hydroxylase positive glial cells show GFAP immunoreactivity, indicating that they are astrocytes [27].

Studies on rat hippocampal slices show significant increase in cholesterol and oxysterols after kainate treatment. Incubation of slices with kainate plus lovastatin results in significantly lower levels of oxysterols and reduced cell death, compared to those that have been treated with KA alone [27]. This observation indicates that lovastatin has neuroprotective effect on hippocampal neurons possibly through reduction in oxysterols. In hippocampal slices, glutathione protects neurons from oxysterol-mediated toxicity [26]. Collectively, these studies suggest that oxysterols may be important factors in aggravating oxidative damage to neurons after kainate induced excitotoxic brain injury [26, 27].

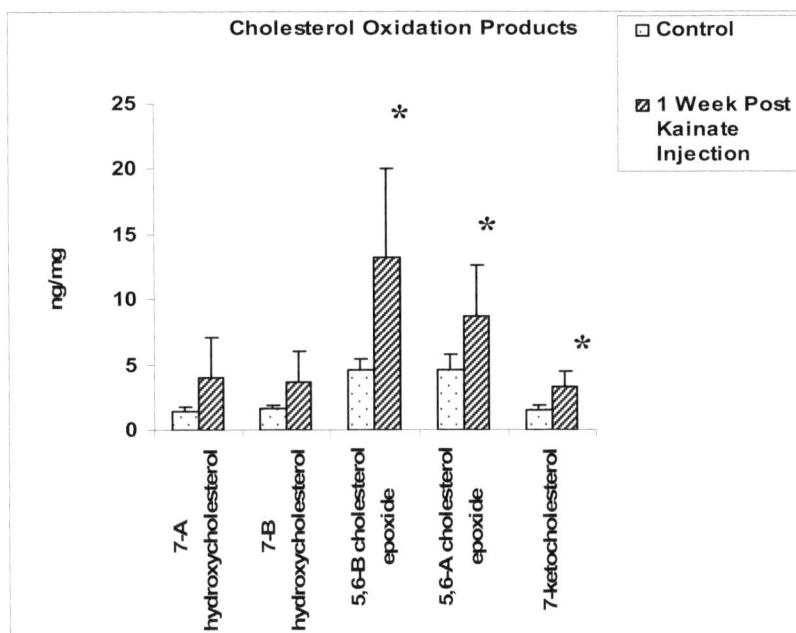

Figure 6: Oxysterol levels in the hippocampus of normal control rats, and at 1 day and 1 week after kainate injections. * Significant difference compared to control group (P < 0.05). (from Kim- JH *et al.* unpublished).

In brain, cholesterol is also converted into neurosteroids. This process involves cholesterol side-chain cleavage enzyme P450scc and the enzyme 3β-hydroxy-steroid dehydrogenase/δ5-δ4-isomerase (3β-HSD) [28]. Cholesterol side-chain cleavage enzyme P450scc is localized in neurons, and it converts cholesterol to pregnenolone. The latter is then metabolized to other hormones such as DHEA and estradiol. KA injections result into a loss of immunoreactivity to P450scc in pyramidal neurons, and induction of P450 immunoreactivity in astrocytes [29]. Western blots show a net decrease in the level of P450scc enzyme protein in the hippocampus of rats after KA injections indicating that in KA-mediated toxicity, cholesterol is converted to oxysterols rather than metabolized to neurosteroids [29]. Furthermore, since estradiol is commonly thought to be a neuroprotective steroid, the decrease in neurosteroids may contribute to neurodegeneration.

Collectively, the above studies indicate that KA-mediated neurotoxicity is accompanied by changes in the composition of neural membrane glycerophospholipids, sphingolipids, cholesterol and its oxidation products but not in neurosteroids. These alterations in membrane constituents are closely associated with KA-mediated neurodegeneration [5, 14, 16, 17, 24, 26].

INTERPLAY AMONG GLYCEROPHOSPHOLIPID, SPHINGOLIPID AND CHOLESTEROL-DERIVED LIPID MEDIATORS IN KA-MEDIATED NEUROTOXICITY

Optimal neural membrane functions are maintained by specific glycerophospholipid, sphingolipid and cholesterol composition [1]. Both external and internal stimuli can alter levels of neural cell glycerophospholipid, sphingolipid, and cholesterol levels by regulating enzyme activities associated with their metabolism. KA-mediated neurotoxicity enhances glycerophospholipid, sphingolipid, and cholesterol metabolism stimulating activities of PLA_2 isoforms, SPT, and cholesterol hydroxylases as well as the oxidative modification of arachidonic acid and cholesterol [16, 25, 26, 26]. This enhancement results in increased generation of glycerophospholipid, sphingolipid, and cholesterol-derived lipid mediators (Fig. **7**). KA-mediated alterations in levels of lipid mediators may disturb normal signal transduction homeostasis, and threaten neural cell survival due to increase in intensity of signal transduction associated with inflammation and oxidative stress in the brain tissue. Interactions among glycerophospholipid, sphingolipid and cholesterol-derived lipid mediators occur at several sites (Fig. **7**) [16].

Thus, in KA-mediated neurotoxicity, increased synthesis of ceramide through the stimulation of SPT or the action of sphingomyelinases induces a spontaneous association of ceramide molecules to form ceramide-enriched microdomains, which fuse to large ceramide-enriched membrane platforms. These large ceramide-enriched membrane platforms may be associated with a transmembrane signaling mechanism for a subset of cell surface receptors [30]. In neural membranes, both ceramide and its metabolites (ceramide-1-phosphate and sphingosine-1-phosphate) stimulate the generation of arachidonic acid and ROS through the stimulation of PLA_2 isoforms and arachidonic acid cascade [16]. Similarly, $cPLA_2$-derived arachidonic acid enhances the generation of ceramide via sphimgomyelinase stimulation [1, 5, 17]. Simultaneous intensification of these processes following KA-mediated toxicity not only indicate that glycerophospholipid and sphingolipid metabolism support each other for the generation of lipid mediators, but are also involved in an interplay (cross talk) that lowers the levels of essential glycerophospholipid species, but increases levels of ceramide species in neural membranes and bring about cellular demise [16, 24]. It should be noted that as with ceramide metabolizing enzymes (enzymes that mediate the release of ROS) are localized in membrane rafts, and the integrity of ceramide-enriched rafts may be required for continuous cellular ROS release during KA-mediated neurotoxicity (Fig. 7) [16, 31]. In a feed-forward mechanism, ROS stimulate acid sphingomyelinase and generate ceramide-enriched membrane platforms. This alters optimal sphingolipid to cholesterol ratio, and facilitates neural cell demise via KA-mediated alterations in glycerophospholipid and spingolipid-derived lipid mediators. Similarly, KA stimulates the hydrolysis of plasmalogens, the vinylether containing glycerophospholipids, by plasmalogen-selective-PLA_2 (PlsEtn-PLA_2) in dose- and time-dependent manner [21], and ceramide decreases the levels of plasmalogens by inhibiting PlsEtn-PLA_2 in rat brain slices [32]. The decrease in plasmalogen levels by ceramide can be blocked by quinacrine, ganglioside, and bromoenol lactone. These compounds inhibit PlsEtn-PLA_2 activity. Thus, it is likely that interplay between plasmalogen and sphingomyelin-derived lipid mediators may modulate inflammation and oxidative stress, processes that are closely associated with KA neurotoxicity [21, 32].

INTERPLAY BETWEEN CERAMIDE AND CHOLESTEROL METABOLISM IN KA-MEDIATED TOXICITY

Very little information is available on *in vivo* interactions between ceramide and cholesterol in brain [1, 17]. In an artificial membrane system, displacement of cholesterol by ceramide induces marked changes in molecular composition, liquid ordered properties, and function of lipid rafts [33, 34]. As stated above, in KA-mediated neurotoxicity cholesterol immunoreactivity and activities and immunoreactivities of cholesterol hydroxylases are markedly increased [27]. These enzymes convert cholesterol into 24-hydroxycholesterol, 25-hydroxycholesterol, and 27-hydroxychol-esterol. Except 24-hydroxycholesterol, 27- and 25-Hydroxycholesterol also undergo 7 α-hydroxylation with subsequent oxidation to 7 α-hydroxy-3-oxo-delta 4 steroids in neurons, astrocytes and Schwann cells [35]. The conversion of cholesterol into hydroxycholesterol is an important mechanism for the maintenance of brain cholesterol homeostasis [36].

Hydroxycholesterols induce cytotoxicity in neural and endothelial cells. They induce apoptotic cell death. In human neuroblastoma cells, SH-SYSY 24-hydroxycholesterol increases caspase-3 and decreases the number of viable cells [36]. Caspase-3 mediates apoptotic cell death by stimulating other caspases and cleaving a number of enzymes (protein kinase C, cytosolic phospholipase A_2, calcium-independent phospholipase A_2, phospholipase C), cytoskeletal proteins (α-spectrin, β-spectrin, actin, vimentin, Bcl-2 family of apoptosis related proteins), and DNA modulating enzymes, poly (ADP-ribose) polymerase [17, 36]. Although the molecular mechanism associated with hydroxyl- and ketocholesterol-mediated toxic effect is not fully understood, 7-ketocholesterol is known to trigger the stimulation of NADPH oxidase, generation of superoxide anions, loss of mitochondrial transmembrane potential ($\Delta\Psi$m), the release of cytochrome c, and activation of caspase-3. These processes are closely associated with apoptotic cell death (**Fig.7**) [37]. Furthermore, in retinal pigment epithelial cells oxysterols not only increase (2-4 fold) ROS production, but also enhances IL-8 gene expression and IL-8 protein secretion in the following decreasing order: 25-hydroxycholesterol > 24-hydroxycholesterol > 7-ketocholesterol [38]. Accumulating evidence suggests that oxysterols have neurotoxic effects on neural cell cultures and among them; 24-hydroxycholesterol can be used as a marker for neurodegeneration [39]. It should be noted that less than 1 % of the total excretion of 24-hydroxycholesterol is via the cerebrospinal fluid. This small fraction reflects neuronal damage and rate of neuronal loss rather than the total number of metabolically active neuronal cells. Patients with neurodegenerative disorders (Alzheimer and multiple sclerosis) show increased concentrations of 24S-hydroxycholesterol in cerebrospinal fluid, in parallel with decreased concentrations in the circulation [16].

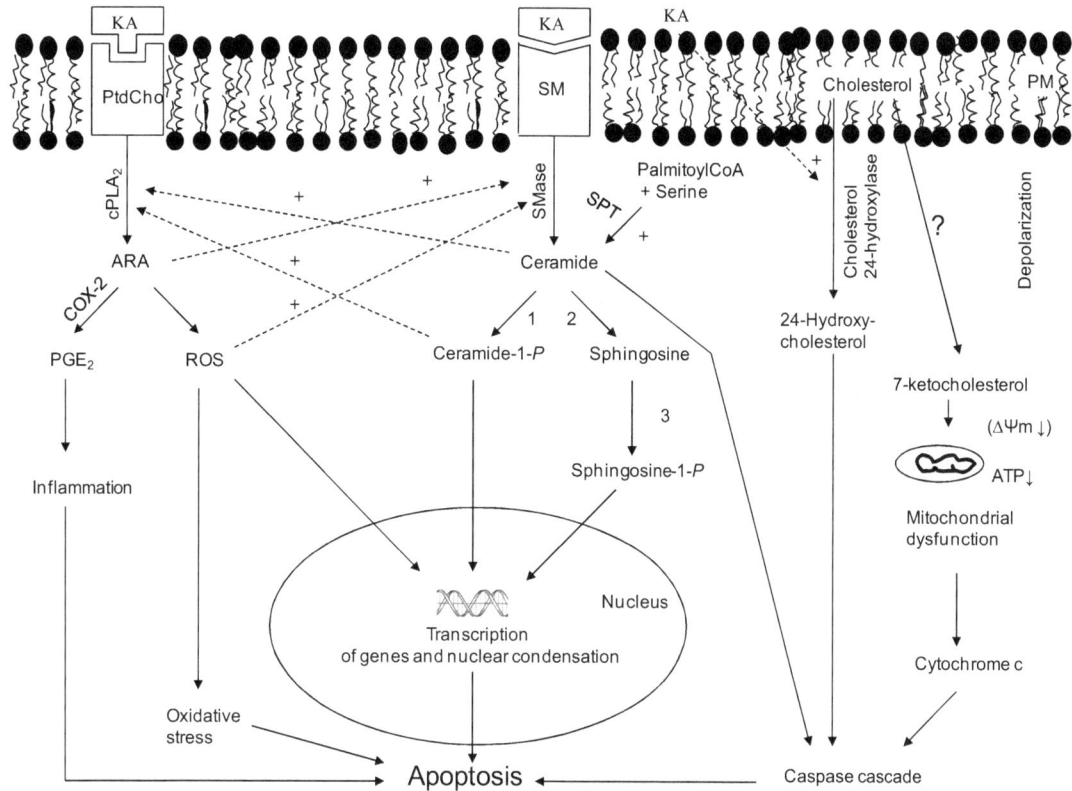

Figure 7: A hypothetical diagram showing interplay among glycerophospholipid-, sphingolipid-, and cholesterol-derived lipid mediators and their involvement in apoptotic cell death. Phosphatidylcholine (PtdCho); sphingomyelin (SM); cytosolic phospholipase A$_2$ (PLA$_2$); sphingomyelinase (SMase); (SPT); arachidonic acid (ARA); cyclooxygenase-2 (COX-2); reactive oxygen species (ROS); ceramide -1-kinase (1); ceramidase (2); sphingosine-1-kinase (3); ceramide-1-phosophate (ceramide-1-P); sphingosine-1-phosphate (sphingosine-1-P); and plasma membrane (PM).

INTERPLAY BETWEEN GLYCERO PHOS-PHOLIPID AND CHOLESTEROL METABOLISM IN KA-MEDIATED NEUROTOXICITY

It is well known that cholesterol and glycerophospholipids contents and distribution modulate the physicochemical and functional properties of cellular membranes [2]. In glycerophospholipid bilayers, cholesterol modulates many biophysical properties including ordering of the fatty acyl chains, condensing of the lipids in the bilayer plane, and promotion of the liquid-ordered phase [40]. These effects depend not only on the type of glycerophospholipids in the bilayer, but also on the nature of the underlying molecular interactions. Cholesterol has been shown to interact more favorably with sphingomyelin than with phosphatidylcholines. In neural membranes, cholesterol is not only associated with lipid rafts, but the ratio between cholesterol and glycerophospholipids as profound effects on a number of critical membrane functions and signal processes [1, 16]. KA-mediated changes in cholesterol may markedly affect neural membrane function by regulating fluidity and permeability [16]. The generation of high levels of oxy- and hydroxycholesterol in KA-mediated neurotoxicity may have serious consequences for neural cell survival. Oxysterols exert tight control over neural cell cholesterol trafficking by altering cholesterol influx/efflux [40]. Oxysterols also modulate Ca^{2+} signals and interact with lipid metabolites of glycerophospholipid and sphingolipid metabolism [5, 41]. 7-Ketocholesterol induces apoptosis through the production of superoxide anions [37]. All these processes may contribute to neural cell death in KA-mediated neurotoxicity.

CONCLUSIONS

KA is an agonist for a subtype of ionotropic glutamate receptor. KA administration not only increases production of ROS and induces mitochondrial dysfunction, but also results in apoptosis in neurons in many regions of the brain, particularly in the hippocampal subregions of CA1 and CA3. Systemic injection of KA to rats also results in activation of glial cells and inflammatory responses typically associated with neurodegenerative processes. In KA-mediated neurotoxicity, the stimulation of KA receptors activates PLA$_2$ with a rapid release of ARA. Neural membrane depolarization during KA-mediated neurotoxicity facilitates elevation in intracellular Ca^{2+}. Increase in Ca^{2+} not only stimulates lipolysis, proteolysis and increase the production 4-HNE, but also promotes depletion of ATP, and loss of glutathione. The processes also alter cellular redox, which plays an important role in KA-induced cell death. The pathological consequences of altered glycerophospholipid, sphingolipid, and cholesterol metabolism in KA-induced neurotoxicity result in elevation in lipid mediators, which contribute to apoptotic as well as necrotic cell death depending upon the intensity of oxidative stress and abnormality in mitochondrial function. Other neurochemical changes may be related to synaptic reorganization following KA-induced seizures and may be involved in recapitulation of hippocampal development and synaptogenesis.

REFERENCES

[1]   Farooqui, A.A.; Horrocks, L.A.; Farooqui, T. J. Neurosci. Res., **2006**, 85, 1834.
[2]   Farooqui, AA, Horrocks, LA.; Farooqui, T. Chem. Phys. Lipids, **2000**, 106, 1.
[3]   Tillman, T.S.; Cascio, M. Cell. Biochem. Biophys. **2003**, 38, 161.
[4]   Farooqui, A.A.; Horrocks, L.A. Neuroscientist **2006**, 12, 245.
[5]   Farooqui, A. A.; Horrocks, L. A. In Glycerophospholipids in the Brain: Springer, New York, **2007**.
[6]   Gómez-Muñoz, A. Biochim. Biophys. Acta. **2006**, 1758, 2049.
[7]   Smith, W.L.; Merrill, A.H, Jr. J. Biol. Chem. **2002**, 277, 25841.
[8]   Björkhem, I.; Lütjohann, D.; Diczfalusy, U.; Ståhle, L.; Ahlborg, G.; Wahren J. J. Lipid Res. **1998**, 39, 1594.
[9]   Pfrieger, F.W. Bioessays, **2003**, 25, 72.
[10]   Coyle, J. T. J. Neurochem., **1983**, 41.
[11]   Lerma, J. Neuron **1997**, 19, 1155.
[12]   Coussen, F. Neuroscience **2009**, 158, 25.
[13]   Choi, D. W. Neuron **1988**, 1, 628.
[14]   Farooqui, A. A.; Horrocks, L. A. Brain Res. Rev. **1991**, 16, 171.
[15]   Farooqui, A. A.; Ong, W. Y.; Horrocks, L. A. Curr. Drug Targets Cardiovasc. Haematol. Disord. **2004**, 4, 85.
[16]   Farooqui, A. A.; Ong, W. Y.; Horrocks, L. A. Neurochemical Aspects of Excitotoxicity, Springer, New York. **2008**, pp. 1-290
[17]   Farooqui, A.A. Hot Topics in Neural Membrane Lipidology, Springer, New York **2009**, pp. 1-36.
[18]   Thwin, M.M.; Ong, W.Y.; Fong, C.W.; Sato, K.; Kodama K.; Farooqui A.A., Gopalakrishnakone P. Exp. Brain Res. **2003**, 150, 427.
[19]   Sandhya, T.L.; Ong, W.Y.; Horrocks, L.A.; Farooqui, A.A. A light and electron microcopic study of cytoplasmic phospholipase A$_2$ and cyclooxygenase-2 in the hippocampus after kainate lesion. Brain Res. **1998**, 788, 223.
[20]   Ong, W.Y.; Lu, X.R.; Ong, B.K.; Horrocks, L.A.; Farooqui, A.A.; Lim, S.K. Exp. Brain Res. **2003**, 148, 521.
[21]   Farooqui, A. A.; Ong, W. Y.; Horrocks L. A. Peroxisomal Disorders and Regulation of Genes, Roels, F., Baes, M., de Bies, S. Eds., Kluwer Academic/Plenum Publishers: London. **2003**, pp. 335.
[22]   Wang, X.S.; Ong, W.Y.; Connor, J.R. Exp. Neurol. **2002**, 177, 193.
[23]   Wang, X.S.; Ong W.Y.; Connor J.R. Exp. Brain Res. **2002**, 143, 137.
[24]   Guan, X. L.; He, X.; Ong, W. Y.; Yeo, W. K.; Shui, G. H., Wenk, M. R. FASEB J. **2006**, 20, 1152.
[25]   He, X.; Guan, X. L.; Ong, W. Y.; Farooqui, A. A.; Wenk M. R.J. Neurosci. Res. **2007**, 85, 423.
[26]   Ong, W. Y.; Goh, E. W. S.; Lu, X. R.; Farooqui, A. A.; Patel, S. C.; Halliwell B. Brain Pathol. **2003**, 13, 250.
[27]   He, X.; Jenner, M.; Ong, W. Y.; Farooqui, A. A.; Patel, S. C. J. Neuropathol. Exp. Neurol. **2006**, 65:652.
[28]   Kohchi, C.; Ukena, K.; Tsutsui, K. Brain Res **1998**, 801, 233.
[29]   Chia, W.J.; Jenner, A.M.; Farooqui, A.A.; Ong W.Y. Exp. Brain Res. **2008**, 186, 143.
[30]   Bollinger, C.R.; Teichgrabe, V.; Gulbins E. Biochim. Biophys. Acta **2005**, 1746, 284.
[31]   Dumitru, C.A.; Zhang, Y.; Li X.; Gulbin E. Antioxid. Redox Signal. **2007**, 9, 1535.
[32]   Latorre, E.; Collado, M.P.; Fernández I.; Aragonés M.D.; Catalán R.E. Eur J Biochem. **2003**, 270, 36.
[33]   Megha, S.P.; London, E. Ceramide selectively displaces cholesterol from ordered lipid domains (rafts): implications for lipid raft structure and function. J. Biol. Chem. **2004**, 279, 9997.
[34]   Megha Sawatzki, P.; Kolter, T.; Bittman, R.,; London, E. Biochim. Biophys. Acta **2007**, 1768, 2205.
[35]   Zhang, J.; Ahea, Y.; el-Etr, M.; Baulieu, E.E. Biochem. J. **1997**, 322,175.
[36]   Kolsch, H.; Ludwig, M.; Lutjohann, D.; Rao M.L. J. Neural Transm. **2001**, 108, 475.

[37]   Lizard, G.; Miguet, C.; Bessède, G.; Monier, S.; Gueldry, S.; Neel, D.; Gambert P. Free Radic. Biol. Med. **2000**, 28:743.

[38]   Joffre, C.; Leclere, L.; Buteau, B.; Martine, L.; Cabaret, S.; Malvitte, L.; Acar, N.; Lizard, G.; Bron, A.; Creuzot-Garcher, C.; Bretillon, L. Curr Eye Res. **2007**, 32:271-280.

[39]   Rojo, L.; Sjöberg, M. K.; Hernández, P.; Zambrano, C.; Maccioni, R. B. J. Biomed. Biotechnol. **2006**, 200, 3976

[40]   Koudinov, A. R.; Koudinova, N. V. Neurobiol. Lipids **2003**, 1, 45.

[41]   Millanvoye-Van Brussel, E.; Topal, G.; Brunet A.; Do Phaw T.; Deckert V.; Rendu F.; David-Dufilho M. Biochem. J. **2004**, 380:533-539.

CHAPTER 5

Association of Mitochondrial Signaling in Alzheimer's Disease and Hypoxia

Cristina Carvalho[1,2], Sónia C. Correia[1,2], Renato X. Santos[1,2], Susana Cardoso[1,2], Paula I. Moreira[1,3], Xiongwei Zhu[4], Mark A. Smith[4] and George Perry[5*]

[1]*Center for Neuroscience and Cell Biology,* [2]*Department of Zoology – Faculty of Sciences and Technology,* [3]*Institute of Physiology – Faculty of Medicine, University of Coimbra, Coimbra, Portugal;* [4]*Department of Pathology, Case Western Reserve University, Cleveland, Ohio, USA;* [5]*UTSA Institute for Neuroscience and Department of Biology, College of Sciences, University of Texas, San Antonio, Texas, USA*

Abstract: Neurodegenerative diseases, particularly those associated with aging such as Alzheimer's disease, represent a significant public health concern. The development of effective treatments is, however, hindered by the complex, multigenic nature of these diseases and by their poorly understood molecular pathophysiology. Mitochondria seem to play a primary role in neurodegeneration, due to the high energy demand of the brain. These organelles are the main producers of energy through the tricarboxylic acid cycle and host a high number of biochemical pathways including those involved in storage and maintenance of intracellular calcium levels, cellular homeostasis and survival pathways. However, mitochondria are a double edge sword. In the presence of certain oxidative stimuli, for instance, when oxygen demand exceeds supply (hypoxia), mitochondria can activate several death pathways. Indeed, hypoxia has been implicated in several neurodegenerative diseases including Alzheimer's disease. Current knowledge supports the idea that during hypoxic events mitochondrial complex III produces high levels of reactive oxygen species (ROS), which play a key role in the regulation of the transcription factor hypoxia inducible factor 1 that triggers several death effectors. In this chapter we will discuss the involvement of mitochondria in AD putting focus on the mitochondrial pathways activated by hypoxia, which could eventually lead neurodegenerative events.

Keywords: Alzheimer's disease; mitochondrial dysfunction; reactive oxygen species; cerebral amyloid angiopathy; β-secretase; Aβ neurotoxicity

INTRODUCTION

Diseases resulting from degenerative changes in the nervous system markedly impact the lives of millions and pose growing public health challenges. Alzheimer's disease (AD), the most common form of dementia, is affecting an increasing number of individuals each year with the number of patients reaching upwards 26.6 million people worldwide in 2006, a number that could quadruple by 2050 [1,2]. Therefore, the prevention and treatment of AD represents one of the critical goals of medical research.

AD is the most common cause of dementia among older people. This degenerative brain disease typically begins with a subtle decline in memory and progresses to global deterioration in cognitive and adaptive functioning. The neuropathological features associated with the disease include the presence of extracellular senile plaques, intracellular neurofibrillary tangles (NFT), and the loss of basal forebrain cholinergic neurons that innervate the hippocampus and the cortex. NFT are formed from paired helical filaments composed of neurofilaments and hyperphosphorylated tau protein. Senile plaques are formed mostly from the deposition of the amyloid-β (Aβ) peptide, a 39–43 amino acid peptide generated through the proteolytic cleavage of the amyloid-β precursor protein (AβPP) by the β- and γ-secretases [3,4]. Aβ deposits may also be found in brain parenchyma and in the walls of small brain arteries, leading to cerebral amyloid angiopathy (CAA). CAA is associated with local loss of neurons, synaptic abnormalities, microglial activation and microhaemorrhage. Such alterations will impact neuronal and synaptic function and, even at its earliest stage, Aβ deposits around brain vessels could certainly interfere with the dynamic adaptation of cerebral blood flow (CBF) to changing brain function.

Accumulating evidence shows that vascular changes play an important role in AD pathogenesis [5,6]. It has been

*Address correspondence to: Institute for Neuroscience and Department of Biology, College of Sciences, University of Texas, San Antonio, Texas 78249, USA. Tel: (210) 458-4450; Fax: 210-458-4445. Email: george.perry@utsa.edu

shown that atherosclerosis, stroke and cardiac disease may cause cerebrovascular dysfunction and trigger AD pathology [7]. Magnetic resonance imaging (MRI), transcranial doppler measurements, and single photon excitation computed tomography (SPECT) in humans showed that the resting CBF is significantly reduced and is an early event in AD. Arterial spin-labeling MRI has demonstrated cerebral hypoperfusion in AD patients [8]. A previous study performed with functional MRI (fMRI) that use blood oxygenation level dependent (BOLD) contrast to measure increases in CBF during a task that assesses episodic memory have established that there is a delay in the CBF response in patients with mild cognitive impairment (MCI), this delay being more pronounced in AD patients [9]. As MCI may represent a prodromal state for AD, these results suggest that CBF reductions are present in the early stages of AD pathophysiology.

A reduction in blood flux leads to hypoxia (a decrease in oxygen levels) in brain tissue [10]. Hypoxia alters the synaptic plasticity and promotes mitochondrial dysfunction, oxidative stress, and apoptosis in several regions of the brain including the cerebral cortex, hippocampus and striatum [11-15]. It has also been shown that disruption of calcium homeostasis, following hypoxia, may contribute to the neurotoxicity of Aβ and subsequent development of AD [16]. Sun and colleagues [17] reported that hypoxia leads to increased β-secretase activity and production of Aβ. Similarly, Guglielmotto and collaborators [18] demonstrated that hypoxia up-regulates β-secretase potentiating the production of Aβ. The authors reported that this effect is mediated by mitochondrial reactive oxygen species (ROS) [18]. In this chapter, we will give an overview about the role of mitochondria in physiologic and neuropathologic conditions, particularly in AD. The hypoxia-mediated mitochondrial pathways, particularly that involving the hypoxia inducible factor 1α (HIF-1α), will also be discussed.

MITOCHONDRIA AND THE BRAIN

The brain has a high energy demand, and although it represents only 2% of the body weight, it receives 15% of cardiac output and accounts for 20% of total body oxygen consumption [4,19]. Physiological demand for oxygen can also vary depending on brain tissue requirements at a given moment [20]. Brain average oxygen consumption is, with exception of myocardial tissue, higher than oxygen consumption in other body tissues and averages approximately 3,5ml/100g/min [21]. As such, complex cellular oxygen sensing systems have evolved for tight regulation of oxygen homeostasis and avoid or, at least, minimize brain damage [22].

Glucose, the major fuel in the brain, is transported across the cell membranes by facilitated diffusion mediated by glucose transporter proteins. More than any other organ, the brain is entirely dependent on a continuous supply of glucose from the circulation since glucose is almost the sole substrate for energy metabolism [23]. This extraordinary energy requirement is largely driven by energy needed to maintain ion gradients across the neuronal plasma membrane that is critical for the the generation of action potentials. This intense energy requirement is continuous and even brief periods of oxygen or glucose deprivation can result in neuronal dysfunction or death [24]. Despite its high energy demand to maintain "housekeeping" functions, the brain cannot store energy very well. Cerebral energy only sustains brain function for a few minutes before irreversible injury, which result from metabolic failure [25].

Mitochondria are ubiquitous and dynamic organelles responsible for many crucial cellular processes in eukaryotic organisms. In fact, mitochondria are considered the "gatekeepers of life and death". These organelles play a key role in cellular energy production [26-29]. The metabolism of glucose through the tricarboxylic acid (TCA) cycle generates the electron donors NADH and succinate that donate electrons to complexes I and II of the respiratory chain, respectively. Electrons from these complexes are transferred to coenzyme Q, Complex III, cytochrome c, complex IV and finally to molecular oxygen that is reduced to water. The electron transport system is organized in this way in order to regulate ATP production (Fig. **1**). Indeed, part of the energy of those electrons is used to pump protons to the mitochondrial matrix. These protons (voltage gradient) are then used by ATP synthase to generate ATP from ADP (Fig. **1**). The ATP generated during this cycle is utilized to provide the energy necessary to carry out active cellular processes.

Although these intracellular organelles are mainly devoted to energy production they are also an important source of reactive oxygen species (ROS) [30-33]. The oxidative phosphorylation system (OXPHOS) is not 100% efficient and approx. 20% of protons undergo regulated proton leak leading to the generation of ROS and, possibly, thermogenic processes [34-37]. Thus a basal production of ROS occurs when OXPHOS is uncoupled from ATP synthesis [20].

Indeed, ROS and reactive nitrogen species (RNS) are products of normal cellular metabolism [38, 39]. These reactive species may play a dual role, deleterius or beneficial depending on their levels (Fig. **2**) [40].

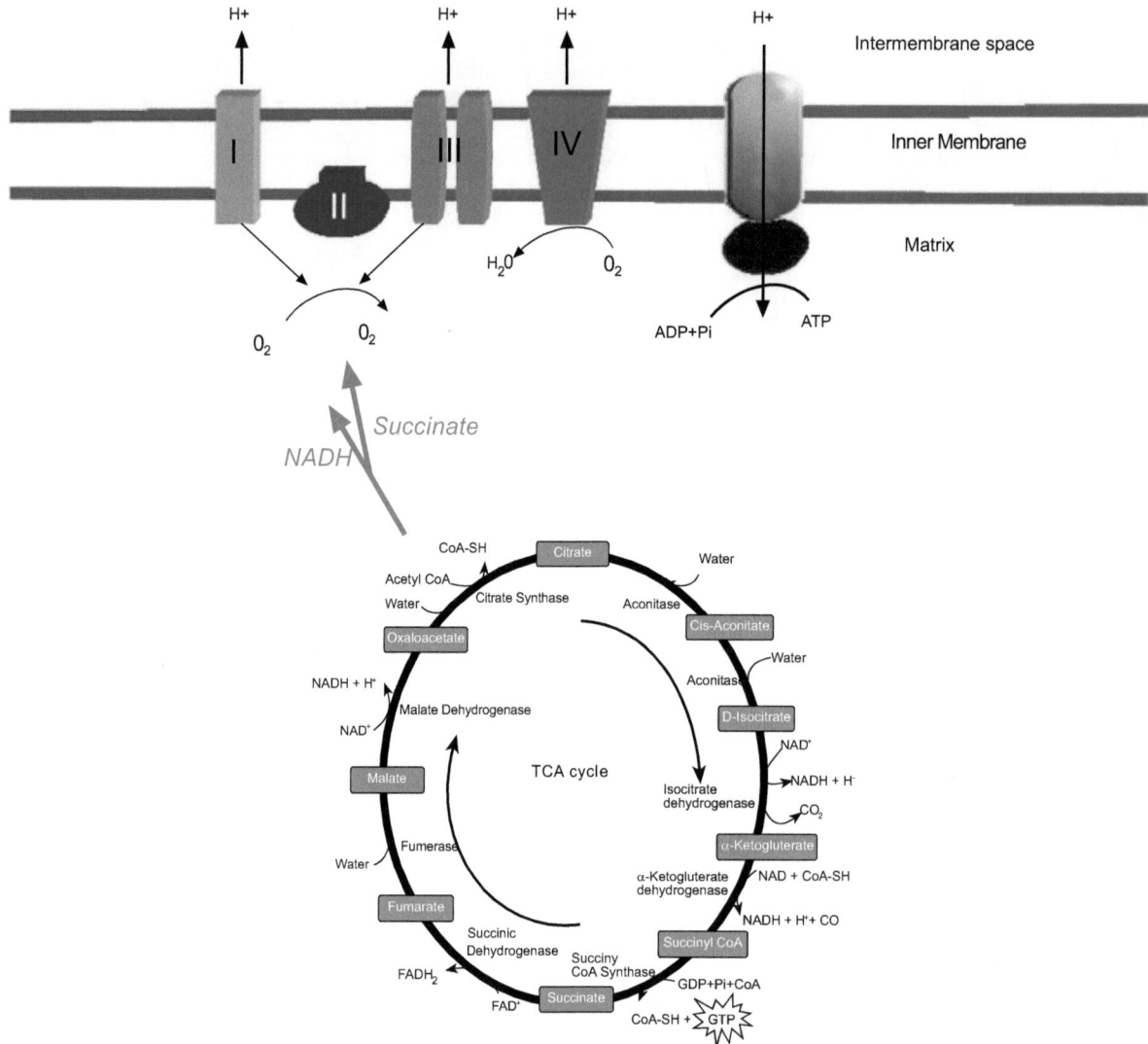

Figure 1: Mitochondria and energy production. The primary physiological function of mitochondria is to generate adenosine triphosphate (ATP), through oxidative phosphorylation via the electron transport chain. Glucose enters in tricarboxylic acid (TCA) cycle where NADH and succinate are produced in order to provide electrons to the complex I and II, respectively, of the respiratory chain. Electrons from these complexes are transferred through the respiratory chain, with concomitant basal production of reactive oxygen species namely superoxide $(O_2^{\cdot-})$ and, finally transferred to molecular oxygen that is reduced to water. Part of the energy of these electrons is used to pump protons to the mitochondrial matrix creating a voltage gradient used by ATP synthase to generate ATP from ADP.

Low or moderate levels of reactive species are involved in physiological processes like the expression of several genes involved in antioxidant defence and survival and regulation pathways (Fig. **2**) [41]. It was also shown that most cell types elicit a small oxidative burst that generates low levels of ROS when they are stimulated by cytokines, growth factors and hormones [42]. This led to the assumption that the initiation and/or proper functioning of several signal transduction pathways rely on the action of reactive species as signalling molecules which may act on different levels in the signal transduction cascade. Reactive species can thus play a very important physiological role as secondary messengers [43-47].

However, high levels of reactive species promote oxidative imbalance and activate anomalous signaling mechanisms related to various disease states [48,49]. The term ''oxidative stress'' describes the adverse interactions of molecular oxygen (O_2), or its reactive derivatives, with biomolecules causing a disequilibrium between the generation of cellular damaging molecules and the cellular capacity for detoxification [39]. During aging and pathological conditions, the production of reactive species exceeds the scavenging capacity of endogenous systems, resulting in the damage of cellular components such as proteins, lipids, and nucleic acids. Besides being one major source of ROS, mitochondria are also one of the preferential targets of reactive species. The mitochondrial DNA (mtDNA) is particularly susceptible to oxidative damage due to its lack of protective histones, limited repair capabilities, and proximity to the electron transport chain [50-53]. Oxidative damage of mitochondrial biomolecules causes the impairment of these organelles, which potentiates the release of certain apoptogenic factors that may activate the intrinsic death pathway (Fig. **2**) [54].

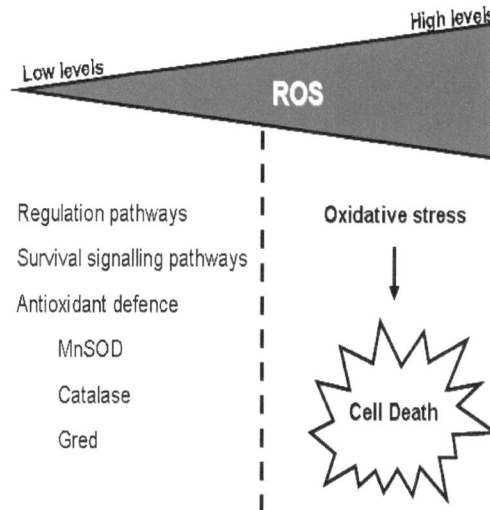

Figure 2: Dual role of mitochondrial reactive oxygen species. The continuous electron leak from the respiratory chain leads to the generation of damaging reactive oxygen species (ROS) that play a dual role. Low levels of ROS induce the expression of several genes involved in antioxidant defence including manganese superoxide dismutase (MnSOD), catalase, glutathione reductase (Gred), and intracellular signalling and regulation. However, high levels of ROS promote oxidative stress and activate anomalous signalling mechanisms that will ultimately lead to cell death.

MITOCHONDRIAL DYSFUNCTION IS A KEY EVENT IN NEURODEGENERATION

Considerable data support the hypothesis that mitochondrial abnormalities link gene defects and/or environmental insults to neurodegeneration [55]. It is now well known that subtle alterations in energy metabolism can lead to insidious pathological changes in neuronal cells [56-58]. Indeed, the literature shows that reduced glucose utilization and energy metabolism and oxidative stress are key players involved in the onset and progression of AD [4,59-63], oxidative stress occurring prior to cytopathology [64,65]. Studies performed in postmortem AD brain and fibroblasts show a reduction in the activity of pyruvate dehydrogenase, isocitrate dehydrogenase and α-ketoglutarate dehydrogenase, three TCA complexes [66,67]. A reduction in the activity of the mitochondrial complexes I, III and IV have also been found in platelets and lymphocytes from AD patients and postmortem brain tissue [68-71]. Several *in vitro* studies corroborate the idea that mitochondria are key players in AD. It has been previously shown that Aβ requires functional mitochondria to induce toxicity [72]. Furthermore, Hansson *et al.* [73] identified an active γ-secretase complex in rat brain mitochondria. Being composed by nicastrin (NCT), anterior pharynx-defective-1 (APH-1), and presenilin enhancer protein 2 (PEN2), this γ-secretase complex cleaves, among other substrates, amyloid β protein precursor (AβPP) generating Aβ and AβPP-intracellular domain. Furthermore, the presence of AβPP was detected in mitochondrial membranes of PC12 cells bearing the Swedish double mutation in AβPP gene [74]. It was also shown that Aβ potentiates the opening of the mitochondrial permeability transition pore (PTP) induced by Ca^{2+} [75,76]. The PTP is a non-selective, high-conductance channel that spans the inner and outer mitochondrial membranes [77-79] and is modulated by several physiological factors [80,81]. The sudden increase in the permeability of the inner mitochondrial membrane plays a key role in apoptotic cell death by facilitating the release of apoptogenic factors such as cytochrome c that will activate the apoptotic cell death pathway (Fig. **3**). Du and collaborators [82] reported that interaction of cyclophilin D, an integral part of the PTP, with with mitochondrial Aβ potentiates mitochondrial, neuronal and synaptic stress. It was also observed that cyclophilin D deficiency substantially improves learning and memory and synaptic function in an AD mouse model and alleviates Aβ-mediated reduction of long-term potentiation [82].

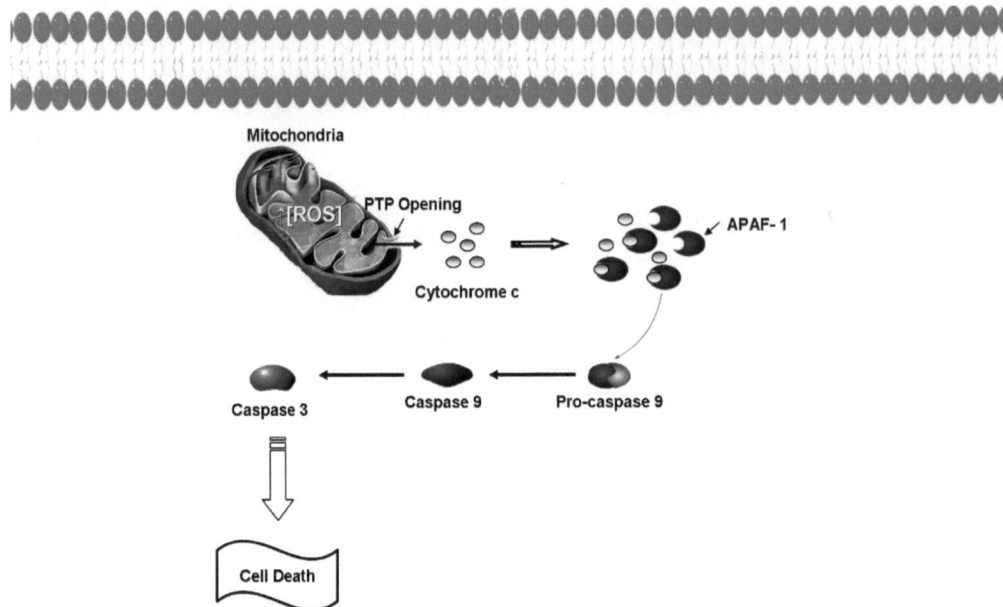

Figure 3: Mitochondrial-mediated cell death. The impairment of mitochondria is intimately associated with an increase in reactive oxygen species (ROS) levels, a decrease in ATP levels and calcium (Ca^{2+}) dyshomeostasis. One important phenomenon associated with mitochondrial dysfunction and oxidative stress is the induction of the permeability transition pore (PTP). The sudden increase in the permeability of the inner mitochondrial membrane plays a key role in apoptotic cell death by facilitating the release of apoptogenic factors such as cytochrome c. Once released to the cytosol, cytochrome c interacts with apoptotic protease activating factor-1 (APAF-1) which cleaves pro-caspase 9 into an active form, caspase-9, that in turn activates caspase-3, resulting in the activation of apoptotic cell death pathway.

The neurodegenerative processes occurring in AD are intimately associated with the apoptotic pathway. Previous studies performed in AD brains found an imbalance between pro-apoptotic (Bax, Bak and Bad) and anti-apoptotic (Bcl-2 and Bcl-x_L) proteins [83,84] and the initiator caspases 8 and 9 and the effector caspases 3 and 6 [85-88]. Other studies demonstrated a marked decrease in the expression of some anti-apoptotic gene such as NCKAP1 [89]. It was also show the existence of active caspases and caspase-cleaved substrates in neurons, around senile plaques and NFT [84,90,91], and also in postsynaptic densities [92]. Both caspase-cleaved AβPP and activated caspase 3 have been shown to be present and associated to granulovacuolar degeneration, a diagnostic AD neuropathological sign in brains of affected patients [93]. Furthermore, a marked co-localization of pathological hyperphosphorylated tau, cleaved caspase-3 and caspase-6 have been recently reported in TUNEL-positive neurons in the brainstem of AD patients [94]. These studies show that in AD a close association between mitochondrial dysfunction, oxidative stress and apoptotic cell death occurs.

IS THE HYPOXIA-INDUCED MITOCHONDRIAL IMPAIRMENT A CAUSE OF AD?

Cerebrovascular insufficiency such as reduced blood supply to the brain or disrupted microvascular integrity in cortical regions may play a key role in the chain of events ending in cognitive failure. Endothelial cells from small vessels are characterized by a relatively high number of mitochondria that can produce the energy necessary for the functioning of specific blood brain barrier (BBB) transport proteins [95].

The aged and degenerative brain is characterized by a decreased CBF, lower metabolic rates of glucose and oxygen and a compromised structural integrity of the cerebral vasculature particularly that of the microvessels (Fig. **4**). It has been shown a profound involvement of endothelial dysfunction in AD-related cerebral hypoperfusion and AD pathophysiology [96]. Furthermore, Aβ deposition in brain vessels occurs in many AD patients and results in CAA and decreased blood fluxes in the brain **(Fig.4)** [4,97]. Data from the literature showed that application of exogenous Aβ to normal blood vessels *ex vivo* causes endothelium-dependent vasoconstriction with decrease in blood fluxes and, consequently, oxygen levels [98,99].

Cells utilize oxygen as the final electron acceptor in the aerobic metabolism of glucose to generate ATP which fuels most active cellular processes. The occurrence of hypoxia leads rapidly to metabolic crisis and represents a severe threat to ongoing physiological function and, ultimately, viability [20]. Hypoxia has been implicated in several brain pathologies including stroke, head trauma, neoplasia and neurodegenerative disease [22, 100,101].

The pathways underlying hypoxic neurotoxicity and cell death are complex and multifaceted and involve several cellular responses, including oxidative stress, altered ionic homeostasis, mitochondrial dysfunction and activation of apoptotic cascades [12,1-2,103]. It has been shown that neuronal apoptosis associated with hypoxic/ischemic injury, aging, and neurodegenerative diseases is due to calcium overload resulting from the mobilization of extracellular calcium through *N*-methyl D-aspartate (NMDA) receptors [104-107]. Hypoxia also triggers free radical generation and depletion of antioxidant status, thus leading to oxidative damage [108,109]. Changes in synaptic efficacy occur very early during hypoxia and may indeed, be the first response of the neuron to ischemic insult [110,111].

Studies have suggested that hypoxia can induce apoptosis dependent on transcriptional activation of apoptotic factors [112]. HIF-1 is a heterodimeric protein composed of a constitutively expressed HIF-1β subunit and an inducible HIF-1α subunit. Under normoxic conditions, HIF-1α is hydroxylated by prolyl hydroxylase enzymes (PHDs) and rapidly degraded by the ubiquitin-proteasome system. On the other hand, during hypoxic conditions, enzymatic inhibition of PHDs abrogates HIF-1α proteasomal degradation and results in HIF-1α stabilization and translocation to the nucleus (Fig. 4). In the nucleus, HIF-1α recruits HIF-1β and modulates the expression of a wide range of genes involved in angiogenesis, metabolism, apoptosis, and cell survival [113,114].

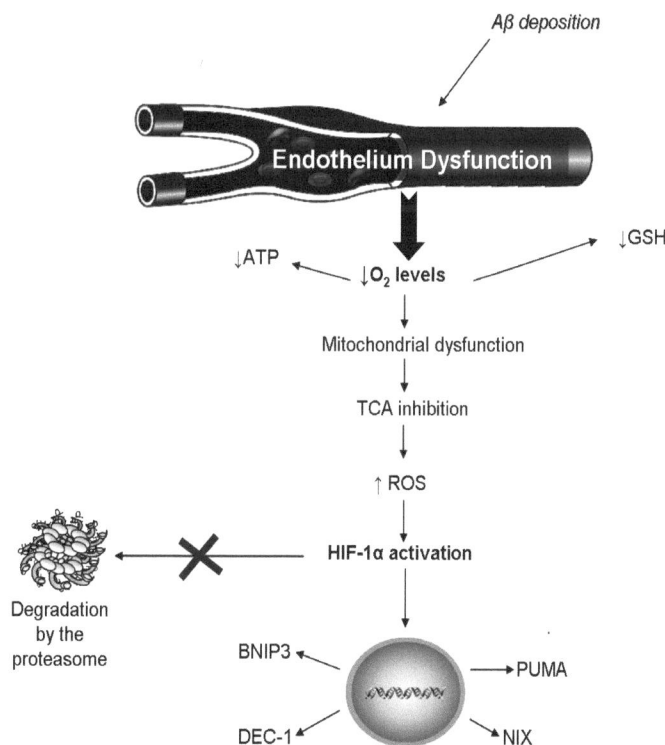

Figure 4: Hypoxia-mediated cell death in neurodegeneration. Hypoxia has been implicated in several pathologies of the central nervous system. In Alzheimer's disease the deposition of the amyloid β (Aβ) protein in brain vessels potentiate the occurrence of hypoxic phenomena. A drop in tissue oxygen levels to the point where oxygen demand exceeds supply rapidly leads to a metabolic crisis putting in danger the ongoing physiological functions. This metabolic crisis comprises a severe energy (ATP) drop that results from an impairment of mitochondria function, including the inhibition of the tricarboxylic acid (TCA) cycle. These

alterations are intimately associated with an increase in the production of reactive oxygen species (ROS) as well as reactive nitrogen species, namely nitric oxide (NO) and a concomitant decrease in antioxidants namely reduced glutathione (GSH), the first line of defense against oxidative stress. Hypoxia-inducible factor-1 (HIF-1 α) is a transcription factor that is oxygen sensitive. In physiological conditions HIF-1α is continuously degraded by proteasome. However, in the presence of low levels of oxygen (O_2) and increased levels of ROS, HIF-1α is activated and translocated to the nucleus where it will bind to hypoxia response elements (HREs) increasing the expression of pro-apoptotic proteins, namely the defective chorion-1 (DEC-1), Bcl2/adenovirus E1B 19kD-interacting protein-3 (BNIP3), its orthologue Nip3-like protein X (NIX), PUMA and cyclin G2 leading to cell death. See text for more complete information

Previous studies demonstrated that under hypoxic conditions mitochondrial ROS, produced at complex III, are necessary and sufficient to stabilize HIF-1α avoiding its degradation by the proteasome (Fig. **4**) [115-117]. Other studies, however, reported a general decrease in ROS levels under hypoxic conditions [118] and showed that a functional respiratory chain may not be necessary for HIF-1α regulation [22,119,120]. More recently, Serra-Pérez and collaborators [121] reported that postischemic metabolic alterations in TCA metabolites impair HIF-1 α degradation in the presence of oxygen by decreasing its hydroxylation, and highlight the involvement of metabolic pathways in HIF-1α regulation besides the well known effects of oxygen.

Whether and to what extent the HIF system may participate in the disease process remains to be elucidated. Indeed, the literature supports a dual role of the HIF system, depending on whether it is the cause or the consequence [22]. Previous studies reported that initially HIF activates a survival pathway that involves the expression of angiogenic and vasodilators genes such as vascular endothelial growth factor (VEGF), inducible nitric oxide synthase and erythropoietin [122-126]. However, sustained and prolonged activation of the HIF pathway may lead to a transition from neuroprotective to cell death responses. The long-lasting activation includes responses with adverse effects on cell function by inducing cell-cycle-arrest-specific and pro-apoptotic proteins such as defective chorion-1 (DEC-1), Bcl2/adenovirus E1B 19kD-interacting protein-3 (BNIP3), its orthologue Nip3-like protein X (NIX), PUMA and cyclin G2 expression (Fig. **4**). In addition, direct stabilization through the pro-apoptotic protein p53 has been suggested by studies demonstrating physical and functional interactions between HIF-1α and p53 (Fig. **4**) [127]. The protein p53 is a master regulator of cell death by inducing apoptosis through the control of apoptosis-related gene expression [128]. In response to certain death stimuli, a fraction of stabilized p53 rapidly translocates to mitochondria launching a rapid pro-apoptotic response in a transcription-independent manner that jump-starts and amplifies the slower transcription dependent response [129-131].

Several studies have been conducted to establish the role of hypoxia in neurodegeneration and, specifically, in AD. It has been previously shown that cerebral hypoxia results in increased activity of caspase-9 and caspase-3 in the cerebral cortex of newborn piglets [132,133]. The mechanism of activation of caspase-9 during hypoxia that leads to initiation of programmed cell death in mammalian brain tissue is not known, but data indicate that the decrease in ATP levels and cytochrome c release had primary roles in this process (Fig. **4**) [134]. Also the increase in nitric oxide levels induced by hypoxia has been shown to activate caspase-9 through a transcription-dependent mechanism. Indeed, there is a nitric oxide-mediated increase in pro-apoptotic proteins such as Bax and Bad during hypoxia that may lead to APAF-1 activation resulting in the conversion of procaspase-9 into active caspase-9 and subsequent activation of caspase-3. It has been also reported that hypoxic stress induces a down-regulation of anti-apoptotic proteins of the Bcl-2 family promoting the apoptotic cell death [135]. Additionally, expression of HIF-1-regulated "pro-death" BH3-only family members, such as BNIP3, has also been shown to increase following cerebral ischemia [136-139]. Recently, Chen and collaborators [140] reported that the silencing of HIF-1α inhibits the expression of VEGF and apoptotic-related proteins such as p53 and caspase-3 protecting neurons against ischemia-reperfusion.

The inhibition of TCA cycle and depletion in glutathione (GSH) levels as a result of its greater use for quenching the accelerated free radical generation accompanied by a concomitant increase in glutathione disulfide was also observed in hippocampal cells under hypoxia (Fig. **4**) [141]. Recently, Sarada *et al.* [142] reported that neuroblastoma cells exposed to hypoxia present increased free radical production and apoptosis and decreased GSH content and glutathione reductase, glutathione peroxidase and superoxide dismutase activities.

Wang and collaborators [143] demonstrated that the expression of APH-1A, a component of the γ-secretase complex, and the γ-secretase mediated Aβ and Notch intracellular domain generation are regulated by HIF-1. Another study showed a functional hypoxia-responsive element in the β-site AβPP cleavage enzyme 1 (BACE1) gene promoter [17]. The authors report that hypoxia up-regulated γ-secretase cleavage of AβPP and Aβ production by increasing BACE1 gene transcription and expression both *in vitro* and in vivo. Hypoxia treatment markedly increased Aβ deposition and neuritic plaque formation and potentiated the memory deficit in Swedish mutant AβPP transgenic mice [17]. These results clearly demonstrate that hypoxia can facilitate AD pathogenesis, and they provide a molecular mechanism linking vascular factors to AD. Zhang and collaborators [144] also showed that acute hypoxia increases the expression and the enzymatic activity of BACE1 by up-regulating the level of BACE1 mRNA, resulting in increases in the AβPP C-terminal fragment-β and Aβ Guglielmotto and collaborators [18] observed that hypoxia significantly increased BACE1 gene transcription through an early up-regulation dependent on the release of mitochondrial ROS and a late up-regulation due to the overexpression and activation of HIF-1α, resulting in increased BACE1 activity and Aβ production. Furthermore, the authors reported that the oxidative stress-mediated up-regulation of BACE1 is mediated by c-jun N terminal kinase pathway [18]. This study strengthens the hypothesis that oxidative stress is a basic common mechanism of Aβ accumulation. A recent study demonstrated that salidroside is able to attenuate abnormal processing of AβPP induced by hypoxia in SH-SY5Y cells, providing a new insight into prevention and treatment of AD [145]. Furthermore, Fang and collaborators [146] reported that hypoxia promotes the phosphorylation of tau protein via ERK pathway suggesting that hypoxia may potentiate NFT formation.

CONCLUDING REMARKS

Mitochondrial ROS have a dual role since at high levels they potentiate cell death pathways and at low/moderate levels they activate survival pathways. At low levels of oxygen cells are unable to generate sufficient energy for survival so, a mechanism for sensing a decrease in the oxygen level before it reaches a critical point is crucial for cells survival. Under low levels of oxygen levels, HIF-1 is activated leading to adaptations to the hypoxic environment. It has been shown that under hypoxic conditions, mitochondrial ROS are required for HIF-1 activation. Although the exact mechanism remains unclear, it seems that HIF system activates several pathways that cover a wide array of responses to hypoxia, ranging from mechanisms that increase cell survival to those inducing cell cycle arrest or even apoptosis. Recent studies show that hypoxia potentiates the development of AD since it favours Aβ production and tau phosphorylation. Although AD is intimately associated with mitochondrial dysfunction, oxidative stress and hypoxic episodes, the role of HIF is this pathology remains largely unknown. More studies should be done to clarify the molecular mechanisms that regulate HIF-1 in order to evaluate the value of this transcription factor as a therapeutic target in neurodegenerative diseases associated to hypoxia such as AD.

REFERENCES

[1]   Brookmeyer, R.; Johnson, E.; Ziegler-Graham, K.; Arrighi, M.H. *Alzheimers Dement*, **2007**, *3*, 186.
[2]   Zawia, N.H.; Lahiri, D.K.; Cardozo-Pelaez, F. *Free Radic. Biol. Med.*, **2009**, *46*, 1241.
[3]   Bojarski, L.; Herms, J.; Kuznicki, J. *Neurochem. Int.*, **2008**, *52*, 621.
[4]   Moreira, P.I.; Duarte, A.I.; Santos, M.S.; Rego, A.C.; Oliveira, C.R. *J Alzheimer's Dis.,***2009**, *16*, 741.
[5]   de la Torre, J.C. *Lancet Neurol.*, **2004**, *3*, 184.
[6]   Bell, R.D.; Zlokovic, B.V. *Acta Neuropathol.*, **2009**, *118*, 103.
[7]   Rocchi, A.; Orsucci, D.; Tognoni, G.; Ceravolo, R.; Siciliano, G. *Curr. Alzheimer Res.*, 2009, *6*, 224.
[8]   Johnson, N.A.; Jahng, G.H.; Weiner, M.W.; Miller, B.L.; Chui, H.C.; Jagust, W.J.; Gorno-Tempini, M.L.; Schuff, N. *Radiology*, **2005**, *234*, 851.
[9]   Rombouts, S.A.; Goekoop, R.; Stam, C.J.; Barkhof, F.; Scheltens, P. *Neuroimage*, **2005,** *26*, 1078.
[10]   Roy, S.; Rauk, A. *Med. Hypotheses,***2005**, *65*, 123.
[11]   Askew, E.W. *Toxicology,* **2002**, *180*, 107.
[12]   Maiti, P.; Singh, S.B.; Sharma, A.K.; Muthuraju, S.; Banerjee, P.K.; Ilavazhagan, G. *Neurochem. Int.*, **2006**, *49*, 709.
[13]   Maiti, P.; Singh, S.B.; Muthuraju, S.; Veleri, S.; Ilavazhagan, G. *Brain Res.*, **2007**, *1175C*, 1.
[14]   Maiti, P.; Muthuraju, S.; Ilavazhagan, G.; Singh, S.B. (2008a), 'Hypobaric hypoxia induces dendritic plasticity in cortical and hippocampal pyramidal neurons in rat brain', *Behav. Brain Res.,* **2008a**, *189*, 233.
[15]   Maiti, P.; Singh, S.B.; Mallick, B.N.; Muthuraju, S.; Ilavazhagan, G. *J. Chem. Neuroanat.*, **2008b**, *36*, 227.
[16]   Kawahara, M.; Kuroda, Y. *Brain Res. Bull.*, **2000**, *53*, 389.
[17]   Sun, X.; He, G.; Qing, H.; Zhou, W.; Dobie, F.; Cai, F.; Staufenbiel, M.; Huang, L.E.; Song, W. *Proc. Natl. Acad. Sci. U.S A.,* **2006**, *103*, 18727.

[18]   Guglielmotto, M.; Aragno, M.; Autelli, R.; Giliberto, L.; Novo, E.; Colombatto, S.; Danni, O.; Parola, M.; Smith, M.A.; Perry, G.; Tamagno, E.; Tabaton, M. *J. Neurochem.*, **2009**, *108*, 1045.

[19]   Nunomura, A.; Hofer, T.; Moreira, P.I.; Castellani, R.J.; Smith, M.A.; Perry, G. *Acta Neuropathol.*, **2009**, *118*, 151.

[20]   Taylor, C.T. *Biochem. J.*, **2008**, *409*, 19.

[21]   Ronnett, G.V.; Ramamurthy, S.; Kleman, A.M.; Landree, L.E.; Aja, S. *J. Neurochem.*, **2009**, *109*, 17.

[22]   Acker, T.; Acker, H. *J. Exp. Biol.*, **2005**, *207*, 3171.

[23]   Duelli, R.; Kuschinsky, W. *News Physiol. Sci.*, **2001**, *16*, 71.

[24]   Simpkins, J.W.; Dykens, J.A. *Brain Res. Rev.*, **2007**, *57*, 421.

[25]   Issam, A. In *AANS publications committee neurosurgical topics;* Awad, M.D., Ed.; American Association of Neurological Surgeons, **1993**, 327.

[26]   Dykens, J.A. In *Neurodegenerative Diseases: Mitochondria and Free Radicals in Pathogenesis,* : Beal, M.F., Bodis-Wollner, I., Howell, N., Eds.; John Wiley & Sons, **1997**, 29.

[27]   Green, D.R.; Kroemer, G. *Science,* **2004**, *305*, 626.

[28]   Beal, M.F. *Annu. Neurol.*, **2005**, *58*, 495.

[29]   Niizuma, K.; Endo, H.; Pak, H. *J. Neurochem.*, **2009**, *109*, 133.

[30]   Shigenaga, M.K.; Hagen, T.M.; Ames, B.N. *Proc. Natl. Acad. Sci. U.S A*, **1994**, *91*, 10771.

[31]   Kroemer, G.; Petit, P.; Zamzami, N.; Vayssiere, J.L.; Mignotte, B. *FASEB J.,* **1995**, *9*, 1277.

[32]   Benzi, G.; Moretti, A. *Neurobiol. Aging*, **1995**, *16*, 661.

[33]   Ozawa, T. *Physiol.Rev.*, **1997**, *77*, 425.

[34]   Richter, C.; Kass, G.E. *Chemico-biol. Interact.,* **1991**, *77*, 1.

[35]   Beal, M.F. *Curr. Opin. Neurobiol.,* **1996**, *6*, 661.

[36]   Cay, J.; Jones, D.P. *J. Bioenerg. Biomembr.*, **1999**, *31*, 327.

[37]   Ricquier, D.; Bouillard, F. *J. Physiol.,* **2000**, *529*, 3.

[38]   Valko, M.; Leibfritz, D.; Moncol, J.; Cronin, M.T.; Mazur, M.; Telser, J. *Int. J. Biochem. Cell Biol.*, **2007**, *39*, 44.

[39]   Goetz, M.; Luch, A. *Cancer Lett.*, **2008**, *266*, 73.

[40]   Valko, M.; Rhodes, C.J.; Moncol, J.; Izakovic, M.; Mazur, M. *Chemico-biol. Interact.*, **2006**, *160*, 1.

[41]   Dröge, W. *Physiol. Rev.*, **2002**, *82*, 47.

[42]   Thannickal, V.J.; Fanburg, B.L. *Amer. J. of Physiol. Lung Cell. Mol. Physiol.* **2000**, *279*, L1005.

[43]   Perry G.; Epel D. *Dev. Biol.* **1985a**, *107,* 47.

[44]   Perry G.; Epel D. *Dev. Biol.* **1985b**, *107,* 58.

[45]   Lowenstein, C.J.; Dinerman, J.L.; Snyder, S.H. *Ann. of Intern. Med.,* **1994**, *120*, 227.

[46]   Storz, P. *Front. Biosci.*, **2005**, *10*, 1881.

[47]   McBride, H.M.; Neuspiel, M.; Wasiak, S. *Curr. Biol.,* **2006**, *16*, R551.

[48]   Brown, G.C.; Borutaite, V. *IUBMB Life*, **2001**, *52*, 189.

[49]   Sheu, S.; Nauduri, D.; Anders, M. *Biochim et Biophysica Acta,* **2006**, *1762*, 256.

[50]   Wallace, D.C. *Science,* **1992**, *256*, 628.

[51]   Linnane, A.W.; Marzuki, S.; Ozawa, T.; Tanaka, M. *Lancet,* **1989**, *1*, 642.

[52]   Miquel, J. *Arch. Gerontol.Geriatr.*, **1991**, *12*, 99.

[53]   Moreira, P.I.; Nunomura, A.; Nakamura, M.; Takeda, A.; Shenk, J.C.; Aliev, G.; Smith, M.A.; Perry, G. *Free Rad. Biol. Med.*, **2008a**, *44*, 1493.

[54]   Hoye, A.; Davoren, J.; Wipf, P. *Acc. Chem. Res.*, **2008**, *41*, 87.

[55]   Gibson, G.E.; Karuppagounder, S.S.; Shi, Q. *Ann. New York Acad. Sci.*, **2008**, *1147*, 221.

[56]   Swerdlow, R.H.; Parks, J.K.; Miller, S.W.; Tuttle, J.B.; Trimmer, P.A.; Sheehan, J.P.; Bennett Jr., J.P.; Davis, R.E.; Parker Jr., W.D. *Ann. Neurol.*, **1996**, *40*, 663.

[57]   Swerdlow, R.H.; Parks, J.K.; Cassarino, D.S.; Maguire, D.J.; Maguire, R.S.; Bennett Jr, J.P.; Davis, R.E.; Parker Jr, W.D. *Neurology*, **1997**, *49*, 918.

[58]   Swerdlow, R.H.; Parks, J.K.; Cassarino, D.S.; Trimmer, P.A.; Miller, S.W.; Maguire, D.J.; Sheehan, J.P.; Maguire, R.S.; Pattee, G.; Juel, V.; Phillips, L.H.; Tuttle, J.B.; Bennett Jr., J.P.; Davis, R.E.; Parker Jr., W.D. *Exp. Neurol.*, **1998**, *153*, 135.

[59]   Moreira, P.I.; Cardoso, S.M.; Santos, M.S.; Oliveira, C.R. *J. Alzheimer's Dis.*, **2006**, *9*, 101.

[60]   Moreira, P.I.; Santos, M.S.; Oliveira, C.R. *Antioxid. Redox Signal.*, **2007a**, *9*, 1621.

[61]   Moreira, P.I.; Santos, M.S.; Seiça, R.; Oliveira, C.R. *J. Neurolog. Sci.*, **2007b**, *257*, 206.

[62]   Zhu, X.; Perry, G.; Moreira, P.I.; Aliev, G.; Cash, A.D.; Hirai, K.; Smith, M.A. *J. Alzheimer's Dis.,* **2006**, *9*, 147.

[63]   Moreira, P.I.; Santos, M.S.; Oliveira, C.R.; Shenk, J.C.; Nunomura, A.; Smith, M.A.; Zhu, X.; Perry, G. *CNS & Neurolog. Disord. Drug Targets,* **2008b**, *7*, 3.

[64] Nunomura, A.; Perry, G.; Aliev, G.; Hirai, K.; Takeda, A.; Balraj, E.K.; Jones, P.K.; Ghanbari, H.; Wataya, T.; Shimohama, S.; Chiba, S.; Atwood, C.S.; Petersen, R.B.; Smith, M.A. *J Neuropathol. Exp. Neurol.*, **2001**, *60*, p. 759.

[65] Hirai, K.; Aliev, G.; Nunomura, A.; Fujioka, H.; Russell, R.L.; Atwood, C.S.; Johnson, A.B.; Kress, Y.; Vinters, H.V.; Tabaton, M.; Shimohama, S.; Cash, A.D.; Siedlak, S.L.; Harris, P.L.; Jones, P.K.; Petersen, R.B.; Perry, G.; Smith, M.A. *J Neurosci,* **2001**, *21*, 3017.

[66] Huang, H.M.; Ou, H.C.; Xu, H.; Chen, H.L.; Fowler, C.; Gibson, G.E. *J Neurosci. Res.*, 2003, *74*, 309.

[67] Bubber, P.; Haroutunian, V.; Fisch, G.; Blass, J.P.; Gibson, G.E. *Ann. Neurol.*, **2005**, *57*, 695.

[68] Kish, S.J.; Bergeron, C.; Rajput, A.; Dozic, S.; Mastrogiacomo, F.; Chang, L.J.; Wilson, J.M.; DiStefano, L.M.; Nobrega, J.N. *J. Neurochem.* **1992**, *59*, 776.

[69] Parker Jr., W.D.; Parks, J.; Filley, C.M.; Kleinschmidt-DeMasters, B.K. *Neurology*, **1994**, *44*, 1090.

[70] Bosetti, F.; Brizzi, F.; Barogi, S.; Mancuso, M.; Siciliano, G.; Tendi, E.A.; Murri, L.; Rapoport, S.I.; Solaini, G. *Neurobiol. Aging*, **2002**, *23*, 371.

[71] Valla, J.; Schneider, L.; Niedzielko, T.; Coon, K.D.; Caselli, R.; Sabbagh, M.N.; Ahern, G.L.; Baxter, L.; Alexander, G.; Walker, D.G.; Reiman, E.M. *Mitochondrion,* **2006**, *6,* 323.

[72] Cardoso, S.M.; Santos, S.; Swerdlow, R.H.; Oliveira, C.R. *FASEB J.* **2001**, *15*, 1439.

[73] Hansson, C.A.; Frykman, S.; Farmery, M.R.; Tjernberg, L.O.; Nilsberth, C.; Pursglove, S.E.; Ito, A.; Winblad, B.; Cowburn, R.F.; Thyberg, J.; Ankarcrona, M. *J. Biolog. Chem.,* **2004**, *279*, 51654.

[74] Keil, U.; Bonert, A.; Marques, C.A.; Scherping, I.; Weyermann, J.; Strosznajder, J.B.; Muller-Spahn, F.; Haass, C.; Czech, C.; Pradier, L.; Muller, W.E.; Eckert, A. *J. Biolog. Chem,* **2004**, *279*, 50310.

[75] Moreira, P.I.; Santos, M.S.; Moreno, A.; Oliveira, C. *Biosci. Rep.,* **2001**, *21*, 789.

[76] Moreira, P.I.; Santos, M.S.; Moreno, A.; Rego, A.C.; Oliveira, C. *J. Neurosci. Res,* **2002**, *69*, 257.

[77] Bernardi, P.; Broekemeier, K.M.; Pfeiffer, D.R. *J. Bioenerg. Biomembr.*, **1994**, *26*, 509.

[78] Zoratti, M.; Szabo, I. *Biochim. et Biophysica Acta*, **1995**, *1241*, 139.

[79] Bernardi, P.; Petronilli, V. *J. Bioenerg. Biomembr.*, **1996**, *28*, 131.

[80] Rosser, B.G.; Gores, G.J. *Gastroenterol.*, **1995**, *108*, 252.

[81] Bernardi, P.; Scorrano, L.; Colonna, R.; Petronilli, V.; Di Lisa, F. *Eur. J. Biochem.*, **1999**, *264*, 687.

[82] Du, H.; Guo, L.; Fang, F.; Chen, D.; Sosunov, A.A.; McKhann, G.M.; Yan, Y.; Wang, C.; Zhang, H.; Molkentin, J.D.; Gunn-Moore, F.J.; Vonsattel, J.P.; Arancio, O.; Chen, J.X.; Yan, S.D. *Nature Med.*, **2008** *14*, 1097.

[83] Su, J.H.; Deng, G.; Cotman, C.W. *J. Neuropathol. Exp. Neurol.,* **1997**, *56*, 86.

[84] Kitamura, Y.; Shimohama, S.; Kamoshima, W.; Ota, T.; Matsuoka, Y.; Nomura, Y.; Smith, M.A.; Perry, G.; Whitehouse, P.J.; Taniguchi, T. *Brain Res.*, **1998**, *780*, 260.

[85] Stadelmann, C.; Deckwerth, T.L.; Srinivasan, A.; Bancher, C.; Brück, W.; Jellinger, K.; Lassmann, H. *Amer. J. Pathol.,* **1999**, *155*, 1459.

[86] Behl, C. *J Neural Transm.,* **2000**, *107*, 1325.

[87] Albrecht, S.; Bourdeau, M.; Bennett, D.; Mufson, E.J.; Bhattacharjee, M.; LeBlanc, A.C. *Amer. J. Pathol.,* **2007**, *170*, 1200.

[88] Calissano, P.; Matrone, C.; Amadoro, G. *Commun. Integrative Biol.,* **2009**, *2*, 163.

[89] Suzuki, T.; Nishiyama, K.; Yamamoto, A.; Inazawa, J.; Iwaki, T.; Yamada, T.; Kanazawa, I.; Sakaki, Y. *Genomics,* **2000**, *63*, 246.

[90] Gastard, M.C.; Troncoso, J.C.; Koliatsos, V.E. *Ann. Neurol.,* **2003**, *54*, 393.

[91] Cribbs, D.H.; Poon, W.W.; Rissman, R.A.; Blurton-Jones, M. *Amer. J. Pathol.,* **2004**, *165*, 353.

[92] Louneva, N.; Cohen, J.W.; Han, L.Y.; Talbot, K.; Wilson, R.S.; Bennett, D.A.; Trojanowski, J.Q.; Arnold, S.E. *Amer. J. Pathol.,* **2008**, *173*, 1488.

[93] Su, J.H.; Kesslak, J.P.; Head, E.; Cotman, C.W. *Acta Neuropathol.,* **2002**, *104*, 1.

[94] Wai, M.S.; Liang, Y.; Shi, C.; Cho, E.Y.; Kung, H.F.; Yew, D.T. *Biogerontology,* **2009**, *10*, 457.

[95] Farkas, E.; Luiten, P.M. *Prog. Neurobiol.*, **2001**, *64*, 575.

[96] Lange-Asschenfeldt, C.; Kojda, G. *Exp. Gerontol.*, **2008**, *43*, 499.

[97] Xiong, H.; Callaghan, D.; Jones, A.; Bai, J.; Rasquinha, I.; Smith, C.; Pei, K.; Walker,D.; Lue, L.F.; Stanimirovic, D.; Zhang, W. *J. Neurosci.,* **2009**, *29*, 5463.

[98] Thomas, T.; Thomas, G.; McLendon, C.; Sutton, T.; Mullan, M. *Nature,* **1996**, *380*, 168.

[99] Townsend, K.P.; Obregon, D.; Quadros, A.; Patel, N.; Volmar, C.; Paris, D.; Mullan, M. *Ann. New York Acad. Sci.,* **2002**, *977*, 65.

[100] Bharke, M.; Hale, S.B. *Sports Med.*, **1993**, *16*, 97.

[101] Hainsworth, R.; Drinkhil, M.J.; Rivera-Chira, M. *Clin. Auton. Res.*, **2007**, *17*, 13.

[102] Jayalakshmi, K.; Sairam, M.; Singh, S.B.; Sharma, S.K.; Ilavazhagan, G.; Banerjee, P.K. *Brain Research*, **2005**, *10469*, 97.

[103] Hota, S.K.; Barhwal, K.; Singh, S.B.; Ilavazhagan, G. *Neurochem. Int.*, **2007**, *51*, 384.

[104] Erecinska, M.; Silver, I.A. *Adv. Neurology*, **1996**, *71*, 119.

[105] Albers, D.S.; Beal, M.F. *J. Neural Transm. Suppl.*, **2000**. *59*, 133.

[106] Weinberg, J.M.; Venkatachalam, M.A.; Roeser, N.F.; Nissim, I. *Proc. Natl. Acad. Sci. U.S.A.*, **2000**, *97*, 2826.

[107] Wang, J.; Green, P.S.; Simpkins, J.W. *J. Neurochem.*, **2001**, *77*, 804.

[108] Adams, J.A. *Br. J. Anaesth.*, **1975**, *47*, 1221.

[109] Chandel, N.S.; Maltepe, E.; Goldwasser, E.; Mathieu, C.E.; Simon, M.C.; Schumacker, P.T. *Proc. Natl. Acad. Sci. U.S.A.*, **1998**, *95*, 11715.

[110] Aoyagi, A.; Saito, H.; Abe, K.; Nishiyama, N. *Brain Res.*, **1998**, *799*, 130.

[111] Fleidervish, I.; Gebhardt, C.; Astman, N.; Gutnick, M.J.; Heinemann U. *J. Neurosci.*, **2001**, *21*, 4600.

[112] Harris, A.L. *Nat. Rev. Cancer*, **2002**, *2*, 38.

[113] Correia, S.C.; Moreira, P.I. *J Neurochem.*, **2010**, *112*, 12.

[114] Correia, S.C.; Moreira, P.I. *J Alzheimer's Dis.*, **2010**, *20*, 475.

[115] Bell, E.L.; Chandel, N.S. *Essays Biochem.*, **2007**, *43*, 17.

[116] Chandel, N.S.; McClintock, D.S.; Feliciano, S.E.; Wood, T.M.; Melendez, J.A.; Rodriguez, A.M.; Schumacker, P.T. *J. Biol. Chem.*, **2000**, *275*, 25130.

[117] Hirota, K.; Semenza, G.L. *J. Biol. Chem.*, **2001**, *276*, 21166.

[118] Görlach, A.; Berchner-Pfannschmidt, U.; Wotzlaw, C.; Cool, R.H.; Fandrey, J.; Acker, H.; Jungermann, K.; Kietzmann, T. *Thromb. Diath. Haemorrh.*, **2003**, *89*, 926.

[119] Srinivas, V.; Leshchinsky, I.; Sang, N.; King, M.P.; Minchenko, A.; Caro, J. *J. Biol. Chem.*, **2001**, *276*, 21995.

[120] Vaux,.C.; Metzen, E.; Yeates, K.M.; Ratcliffe, P.J. *Blood*, **2001**, *98*, 296.

[121] Serra-Pérez, A.; Planas, A.M.; Núñez-O'Mara, A.; Berra, E.; García-Villoria, J.; Ribes, A.; Santalucia, T. *J Biol. Chem.*, **2010**, *285*, 18217.

[122] Bernaudin, M.; Marti, H.H.; Roussel, S.; Divoux, D.; Nouvelot, A.; MacKenzie, E.T.; Petit, E. *J. Cereb. Blood Flow Metab.*, **1999**, *19*, 643.

[123] Brines, M.L.; Ghezzi, P.; Keenan, S.; Agnello, D.; de Lanerolle, N.C.; Cerami, C.; Itri, L.M.; Cerami, A. *Proc. Natl. Acad. Sci. U.S.A.*, **2000**, *97*, 10526.

[124] Jin, K.L.; Mao, X.O.; Greenberg, D.A. *Proc. Natl. Acad. Sci. U.S.A.*, **2000**, *97*, 10242.

[125] Jin, K.; Zhu, Y.; Sun, Y.; Mao, X.O.; Xie, L.; Greenberg, D.A. *Proc. Natl. Acad. Sci. U.S.A*, **2002**, *99*, 11946.

[126] Aminova, L.R.; Chavez, J.C.; Lee, J.; Ryu, H.; Kung, A.; LaManna, J.C.; Ratan, R.R. *J. Biol. Chem,.* **2005**, *280*, 3996.

[127] Acker, T.; Plate, K.H. *J. Mol. Med.*, **2002**, *80*, 562.

[128] Schmitt, C.A.; Fridman, J.S.; Yang, M.; Baranov, E.; Hoffman, R.M.; Lowe, S.W. *Cancer Cell*, **2002**, *1*, 289.

[129] Marchenko, N.D.; Zaika, A.; Moll, U.M. *J. Biol. Chem.*, **2000**, *275*, 16202.

[130] Mihara, M.; Erster, S.; Zaika, A.; Petrenko, O.; Chittenden, T.; Pancoska, P.; Moll, U.M. *Molecular Cell*, **2003**, *11*, 577.

[131] Erster, S.; Mihara, M.; Kim, R.H.; Petrenko, O.; Moll, U.M. *Mol. Cell. Biol.*, **2004**, *24*, 6728.

[132] Khurana, P.; Ashraf, Q.M.; Mishra, O.P.; Delivoria-Papadopoulos, M. *Neurochem. Res.*, **2002**, *27*, 931.

[133] Mishra, O.P.; Delivoria-Papadopoulos, M. *Neurosci. Lett.*, **2006**, *401*, 81.

[134] Delivoria-Papadopoulos, M.; Ashraf, Q.M.; Mishra, O.P. *Neurosci. Lett.*, **2008**, *438*, 38.

[135] Barhwal, K.; Singh, S.B.; Hota, S.K.; Jayalakshmi, K.; Ilavazhagan, G. *Eur. J. Pharmacol.*, **2007**, *570*, 97.

[136] Chen, M.; He, H.; Zhan, S.; Krajewski, S.; Reed, J.C.; Gottlieb, R.A. *J. Biol. Chem.*, **2001**, *276*, 30724.

[137] Yin, X.M.; Luo, Y.; Cao, G.; Bai, L.; Pei, W.; Kuharsky, D.K.; Chen, J. *J. Biol. Chem.*, *277*, 42074.

[138] Shibata, M.; Hattori, H.; Sasaki, T.; Gotoh, J.; Hamada, J.; Fukuuchi, Y. *J. Cereb. Blood Flow Metab.*, **2002**, *22*, 810.

[139] Schmidt-Kastner, R.; Aguirre-Chen, C.; Kietzmann, T.; Saul, I.; Busto, R.; Ginsberg, M.D. *Brain Res.*, **2004**, *1001*, 133.

[140] Chen, C.; Hu, Q.; Yan, J.; Yang, X.; Shi, X.; Lei, J.; Chen, L.; Huang, H.; Han, J.; Zhang, J.H.; Zhou, C. *Neurobiol Dis.* **2009**, *33*, 509.

[141] Barhwal, K.; Hota, S.K.; Prasad, D.; Singh, S.B.; Ilavazhagan, G. (2008), 'Hypoxia-induced Deactivation of NGF-mediated ERK1/2 Signaling in Hippocampal Cells: Neuroprotection by Acetyl-L-Carnitine', *J. Neurosci. Res.*, **2008**, *86*, 2705.

[142] Sarada, S.K.; Himadri, P.; Ruma, D.; Sharma, S.K.; Pauline, T.; Mrinalini. *Brain Res.*, 2008, *1209*, 29.

[143] Wang, R.; Zhang, Y.W.; Zhang, X.; Liu, R.; Zhang, X.; Hong, S.; Xia, K.; Xia, J.; Zhang, Z.; Xu, H. *FASEB J.,* **2006**, *20*, 1275.

[144] Zhang, X.; Zhou, K.; Wang, R.; Cui, J.; Lipton, S.A.; Liao, F.F.; Xu, H.; Zhang, Y.W. *J. Biol. Chem.*, **2007**, *282*, 10873.

[145] Li, Q.Y.; Wang, H.M.; Wang, Z.Q.; Ma, J.F.; Ding, J.Q.; Chen, S.D. *Neurosci. Lett.*, **2010**, (epub ahead of print).

[146] Fang, H.; Zhang, L.F.; Meng, F.T.; Du, X.; Zhou, J.N. *Neurosci. Lett.*, **2010**, *474*, 173.

CHAPTER 6

Dopamine-Mediated Oxidative Stress Associated with Neurodegeneration in Parkinson Disease

Tahira Farooqui[*]

The Ohio State University, Department of Entomology/Center for Molecular Neurobiology, Columbus, OH 43210, USA

Abstract: Dopamine-mediated neurotoxicity has been speculated as a potential contribution to the pathogenesis of Parkinson disease (PD), including a diffuse protein aggregation pathology but relatively selective death of dopaminergic neurons in the substantia nigra. The dopamine-mediated oxidative stress, produced by dopamine oxidation resulting in reactive oxygen species (ROS) and reactive dopamine quinones, is hypothesized to be the key event in the specific cell death of dopaminergic neurons in the pathogenesis of sporadic PD and neurotoxin-induced parkinsonism. The cytotoxity in dopaminergic neurons occurs primarily due to the generation of highly reactive cyclized *O*-quinones, which damage mitochondia by opening the mitochondrial permeability transition pore, leading to cell death. These toxic quinones conjugate with several key PD pathogenic molecules, such as tyrosine hydroxylase, α-synuclein and parkin, forming a complex of protein-bound-quinone (quinoprotein), consequently inhibiting enzyme/protein function. Furthermore, cyclized dopamine quinones also inhibit proteasome activity, resulting in protein aggregation that may facilitate Lewy body formation in PD. Dopamine quinone formation is also closely linked to other representative hypotheses for PD. However, reductants such as glutathione (GSH), ascorbic acid (AA), and superoxide dismutase (SOD) may protect dopaminergic neurons from dopamine-induced toxicity or by various other biochemical insults associated with PD. The chaperone heat-shock protein 70 (HSP70) reduces protein misfolding and aggregation in cells, implicating its protective role against a variety of insults including oxidative stress. There are several pathogenic mechanisms possibly involved with death of dopaminergic neurons, but this overview focuses on dopamine-mediated oxidative stress that may contribute to selective neurodegeneration of dopaminergic neurons in PD.

Keywords: Parkinson disease; dopamine; dopaminergic neurons, dopamine quinones; quinoprotein; oxidative stress; neuroinflammation; glutathione conjugation

INTRODUCTION

Parkinson disease (PD) is a progressive neurodegenerative disorder with a prevalence of ~1–2% at 65 years and ~5% at 85 years of age, implicating age as one of the major risk factor in the sporadic PD. Only fewer findings (~5%) at a young and/or juvenile age of onset have been reported with familial form of the disease, which is caused by the mutation in certain genes. The major symptom of PD was originally described in 1817 by James Parkinson as Shaking Palsy [1]. Clinically, PD is manifested by resting tremor, motor rigidity, bradykinesia (slowness of movement) and loss of postural reflexes [2]. The pathological hallmarks of PD are the progressive loss of dopaminergic neurons, most crucially in the substantia nigra pars compacta (SNpc), and the presence of fibrillar inclusions called 'Lewy bodies' in the remaining neuronal cytoplasm in the SNpc [3-5]. Furthermore, Lewy bodies are predominantly composed of fibrillar α-synuclein (monomers of α-synuclein aggregate converted into fibrils), implicating a direct correlation of protein aggregation with neurodegeneration in PD [6]. The loss of dopaminergic neurons in the SNpc of PD patients results in massive depletion of striatal dopamine, followed by motor impairments [7-9]. A number of pathological and pharmacological studies on PD and many other *in vivo* and *in vitro* studies using dopaminergic neurotoxins to induce parkinsonism in animal model systems have hypothesized that mitochondrial dysfunction, inflammation, oxidative stress, dysfunction of the ubiquitin proteasome system (UPS), and protein phosphorylation may be the key molecular mechanisms responsible for the pathogenesis and progress of sporadic and familial PD [10-17]. However, among these pathogenic mechanisms thought to contribute to the demise of these cells, none of them can explain the death of specific (dopaminergic) neurons in PD. This overview focuses on the molecular aspects of selective neurodegeneration of dopaminergic neurons in PD.

*Address correspondence to: Department of Entomology/Center for Molecular Neurobiology, The Ohio State University, Columbus, OH 43210-1220, Telephone: (614) 783-4369, Email: farooqui.2@osu.edu

DOPAMINE QUINONES SYNTHESIS AND PD

The enzymic oxidation or autoxidation (via reduction of ferric iron or other metals) of the excess amount of dopamine, its precursor 3,4-dihydroxy-L-phenylalanine (L-DOPA), and its metabolite 3,4-dihydroxyphenylacetic acid (DOPAC) containing 2 hydroxyl residues exert cytototoxity in dopaminergic neurons, primarily due to the generation of highly reactive cyclized quinones, such as aminochrome, DOPAchrome, and furanoquinone (Fig. **1**).

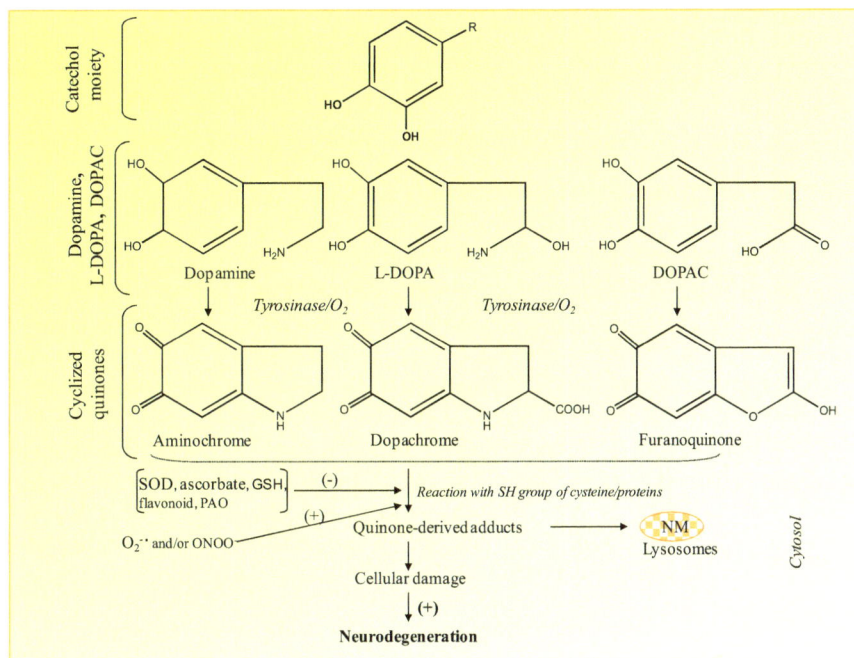

Figure 1: Cyclized quinones of dopamine, L-DOPA, and DOPAC form quinone-derived adducts via nucleophilic addition reactions with SH of cysteine/protein, causing cellular damage. NM, neuromelanin, DA-derived quinones and DA adducts in the cytosol and organelles are phagocytosed in bilamellar autophagic vacuoles lysosomes where they are permanently stored as NM. GSH, glutathione; SOD, Superoxide dismutase, PAO, phenolic antioxidants.

The cyclized *o*-quinones derived from dopamine, DOPA, and DOPAC, or reactive species derived from these quinones have been shown to cause proteasomal impairment, providing a potential basis for the selectivity of dopaminergic neuron damage in PD [18]. The autoxidation of dopamine at neutral pH (Eqs. 1–3) producing quinones is facilitated by the Fenton reaction (4), oxidizing ferrous iron (Fe^{2+}) by H_2O_2 to ferric iron (Fe^{3+}), a hydroxyl radical (OH·) and hydroxyl anion (OH⁻), as shown below:

1. $QH_2 + OH^- \leftrightarrow QH^- + H_2O$

2. $QH^- + O_2 \rightarrow \,^{\cdot}QH + O_2^{\cdot-}$

3. $QH + O_2 \leftrightarrow Q + O_2^{\cdot-} + H^+$

4. $Fe^{2+} + H_2O_2 \rightarrow Fe^{3+} + OH^{\cdot} + OH^-$

In these reactions, dopamine quinones are represented as dihydroquinone, QH_2; semiquinone,.QH; and quinone, Q. QH_2 is oxidized to QH and then to Q, respectively [19]. Dopamine undergoes a very complex series of metabolic events in dopaminergic cells: (1) tyrosinase mediated generation of dopamine *o*-quinone, (2) cyclization to leukoaminochrome at physiological pH, and (3) subsequent oxidation to the cyclized *o*-quinone, aminochrome (Fig. 1). Similar pathways exist for DOPA, norepinephrine, epinephrine, and DOPAC, generating dopachrome, noradrenochrome, adrenochrome, and furanochrome. The cyclized quinones of dopamine, L-DOPA, and DOPAC are synthesized by oxidation of catechol ring. If this process takes place within the neuronal cytosol, it allows quinones to react with cytosolic components (such as cysteine residues), whereas in neuronal lysosomes quinone-derived adducts contribute to the formation of neuromelanin (NM), the insoluble pigment regarded as the most important iron storage system in catecholaminergic neurons. NM synthesis occurs due to excess cytosolic catecholamines. The dying population of dopaminergic neurons in PD displays intra-lysosomal neuromelanin [20].

High dose of methamphetamine (METH; a drug of abuse and neurotoxin) shows PD-like pathology [21]. METH increases extracellular dopamine from both vesicular and nerve terminal stores, targeting dopaminergic neurons [21]. The persistent dopaminergic deficits may be caused by formation of dopamine reactive oxygen and nitrogen species (ROS and RNS) that show their extreme reactivity with proteins, lipids, and nucleotides, and therefore considered as mediators of neurotoxicity.

METH has been recently shown to increase striatal glutamate and mediate long-term dopamine toxicity [22]. Dopamine dependent oxyradical formation may be linked to METH-mediated toxicity as well as other specific forms of neurodegeneration [23]. The oxidation of dopamine, and its metabolites to quinone derivatives adduced by cysteine inhibits ubiquitin proteasomal activity, suggesting a link with METH-mediated toxicity [24]. Collective evidence suggests that dopamine cyclized-*o*-quinones provide a potential basis for selective neurotoxic dopamine action in PD.

DOPAMINE-QUINONE INTERACTION ALTERS PROTEIN FUNCTIONS

Dopamine quinones readily participate in nucleophilic addition reactions with sulfhydryl (SH) groups on free cysteine, CySH, or cysteine found in protein [21, 25-27]. The formation of covalent bonding between dopamine quinone and nucleophilic -SH groups on proteins may cause changes in protein structural cofiguration, altering enzyme activity. Thus, conjugation of quinones with cysteine residues in proteins results in the formation of quinoproteins that have impaired function, reducing cell viability [28]. A summarized list containing proteins that are covalently modified by dopamine quinones in living dopaminergic neurons [29-36] is shown in Table **1**. The oxidation of 5-S-cysteinyldopamine (CYS-DA) has also been demonstrated to produce neurotoxic products (dihydrobenzothiazines and benzothiazines). These substances show lethal effects when injected into mice brains [37]. Further more, postmortem studies have shown significant increase in cysteinyl adducts of dopamine, L-DOPA, and DOPAC in SNpc dopaminergic neurons in PD brains [38, 39], implicating a direct correlation between the depigmentation/degeneration of dopaminergic neurons and enhanced rates of autoxidation.

Table 1. Dopamine-quinones covalently modify cellular enzymes/proteins.

Enzymes/Proteins	DAQs-Mediated Effect	Reference
TH	DAQs covalently modify TH enzyme and inhibit its activity	[29, 30]
DAT	DAQs covalently bind to SH group on DAT cysteinyl residues	[31, 32]
TPH	Inhibition of TPH results in serotonergic neuronal toxicity	[33]
α-Synuclein	-DAQs interact with α-Syn and inhibits conversion of protofibril to fibril, resulting in cell death -DAQs promote the oligomer formation of mutant α-synuclein	[10, 34]
Parkin	-DAQs covalently modify parkin, which inactivates E3 ubiquitin ligase enzyme	[35]
Dopamine receptors	-DAQs regulate METH-induced dopaminergic neurotoxicity	[36]
MtCK*	-DAQs cause rapid loss of specific protein, increase in protein aggregation and/or crosslinking	[38]
Mitofilin*	-DAQs cause rapid loss of specific protein, increase in protein aggregation and/or crosslinking	[38]
Mortalin	-DAQs cause rapid loss of specific protein, increase in protein aggregation and/or crosslinking	[38]
75 kDa subunit NADH d'nase	-DAQs cause rapid loss of specific protein, increase in protein aggregation and/or crosslinking	[38]
SOD2	-DAQs cause rapid loss of specific protein, increase in protein aggregation and/or crosslinking	[38]

DAQ, dopamine quinone; METH, Methamphetamine; TPH, Tryptophan hydroxylase; TH, Tyrosine Hydroxylase; MtCK, Mitochondrial creatine kinase; SOD2, Superoxide dismutase 2. NADH d'nase, *NADH dehydrogenase. *abundance reduction was confirmed by western blot analysis.*

There is substantial evidence for the deficiency of complex I activity of the mitochondrial electron transport chain and increased number of activated microglia in PD brain, supporting the hypothesis that mitochondrial dysfunction and oxidative stress may be linked to dopaminergic neurodegeneration in PD [17, 40, 41]. Several studies have shown significant changes in relative abundance and/or decreased functions of mitochondrial-associated proteins compared to the controls in animal model system of PD [42-[42-45].

Using the proteomic approach following exposure to dopamine quinones, loss in protein levels of specific mitochondrial proteins, such as mitochondrial creatine kinase (MtCK), mitofilin, mortalin, the 75 kDa subunit of NADH dehydrogenase (NADH d'nase), and superoxide dismutase 2 (SOD 2) has been reported (Table 1) in rat brain mitochondria [46]. These proteins are engaged in mitochondrial functions: (1) MtCK is associated with ADP-ATP exchange and the permeability of transition pore, (2) mitofilin maintains mitochondrial cristae, (3) mortalin plays a chaperone role in protein import and folding, (4) NADH d'nase transfers electrons from NADH to the respiratory chain, and (5) SOD2 that acts as mitochondrial enzymic antioxidant [46]. The significant loss in protein levels was confirmed by western blot analyses [46]. Following exposure to dopamine quinone, immunoeactive bands of proteins (e.g. MtCK and mitofilin) show higher molecular weight than controls, suggesting the possible protein aggregation and/or cross-linking, as a potential cosequence of dopamine quinone oxidation. The alterations in key mitochondrial proteins support the hypothesis that dopamine oxidation results in the selective loss of a subset of proteins that may contribute to mitochondrial dysfunction, leading to increased ROS, causing severe oxidative stress within dopaminergic neurons, and ultimately resulting to PD. The exact nature of such protein loss remains to be elucidated.

GLUTATHIONE CONJUGATION: AN ANTIOXIDANT PATHWAY

Glutathione, a tripeptide molecule of three amino acids (glycine, glutamate and cysteine), is a naturally occurring compound in animal tissues. Reduced glutathione (GSH) is the highest concentration ($\sim$0.5 to 10 mM) non-protein thiol in mammalian cells. GSH is involved in many biological processes such as proteins and DNA synthesis, amino acid transport, and providing protection for sulfhydryl groups (SH) of proteins against oxidation. GSH acts as an antioxidant and free radical scavenger.

GSH can be oxidized to glutathione disulfide (GSSG) by glutathione peroxidase and then recycled back to GSH by glutathione reductase as shown in the following reactions:

1. $2GSH + H_2O_2 \rightarrow GSSG + 2H_2O$

2. $GSSG + NADPH + H^+ \rightarrow 2\ GSH + NADP^+$

The ratio of reduced/oxidized glutathione (GSH/ GSSG) is considered as a marker of oxidative stress. GSH deficiency or decrease in GSH/GSSG ratio manifests increased susceptibility to oxidative stress, resulting in neurodegenerative diseases such as PD and Alzheimer disease. Several studies have demonstrated a deficiency in reduced glutathione (GSH) in SNpc of patients with PD [47, 48]. The activities of enzymes related to glutathione synthesis, degradation, and function analyzed in various brain regions from patients dying with pathologically proven PD have demonstrated selective elevation in the glutathione degradative pathway.

Glutathione peroxidase-positive microglia are more abundant in PD and dementia with Lewy bodies tissues, and they make contacts with neurons [50]. Glutathione transferases, enzyme catalyzing conjugation of dopamine and DOPA-*O*-quinone with glutathione, prevent the formation of neurotoxic aminochrome and dopachrome and eliminate reactive chemical species generated from oxidation of catecholamines [51]. Glutathione transferases help in the detoxification of electrophilic decomposition products and prevent oxygen mediated toxicity by redox cycling of catecholamine derivatives.

NMR analysis of dopamine and DOPA *o*-quinone-glutathione conjugates has revealed that the addition of glutathione occurs at C-5, forming unreactive glutathione conjugates, 5-*S*-glutathionyl-dopamine and 5-*S*-glutathionyldopa. Both conjugates are found to be resistant to oxidation by biological oxidizing agents, such as O_2, H_2O_2, and superoxide anion ($O^{\bullet-}_2$), implicating that glutathione transferases play a neuroprotective role by preventing catecholamine-dependent formation of ROS [51, 52].

Using a new tandem HPLC plus ESI-MS procedure, it has been recently reported that GSH conjugation with amino-chrome-derived dopamine quinones forms short-lived intermediate GSH conjugates [53], via condensation reactions without enzymatic catalysis (Fig. **2**). The aminochrome-derived dopamine quinones can also undergo internal rear-rangement forming 5,6-di- hydroxyindole and aggregating to melanin, a macromolecular compound produced by the polymerization of the quinone product (Fig. **2**). The conjugation of quinones with GSH depends on the GSH concen-tration, reaction time, and presence of ambient reductants (oxygen or nitrogen) in the solution [53]. Following con-densation these conjugates tend to polymerize (aggregate) possibly due to structural changes that can be blocked by GSH, therefore it has been hypothesized that conjugation of proteins by cyclized dopamine quinones may induce Lewy body formation in dopaminergic neurons via similar polymerization reaction [53]. Conjugation of dopamine quinones with cysteine residues in proteins has been shown to form dopamine-cysteine adducts, allowing modified proteins to aggregate in the presence of Cu^{2+} [54]. Therefore it is possible that GSH may inhibit the aggregation of GSH-aminochrome conjugates. If so then the decreased GSH level in SNPC of PD patients may participate in facili-tating protein aggregation and even Lewy body formation of PD [53, 54]. 6-Hydroxydopamine (6-OHDA) is a se-lective catecholaminergic neurotoxin [55-58].

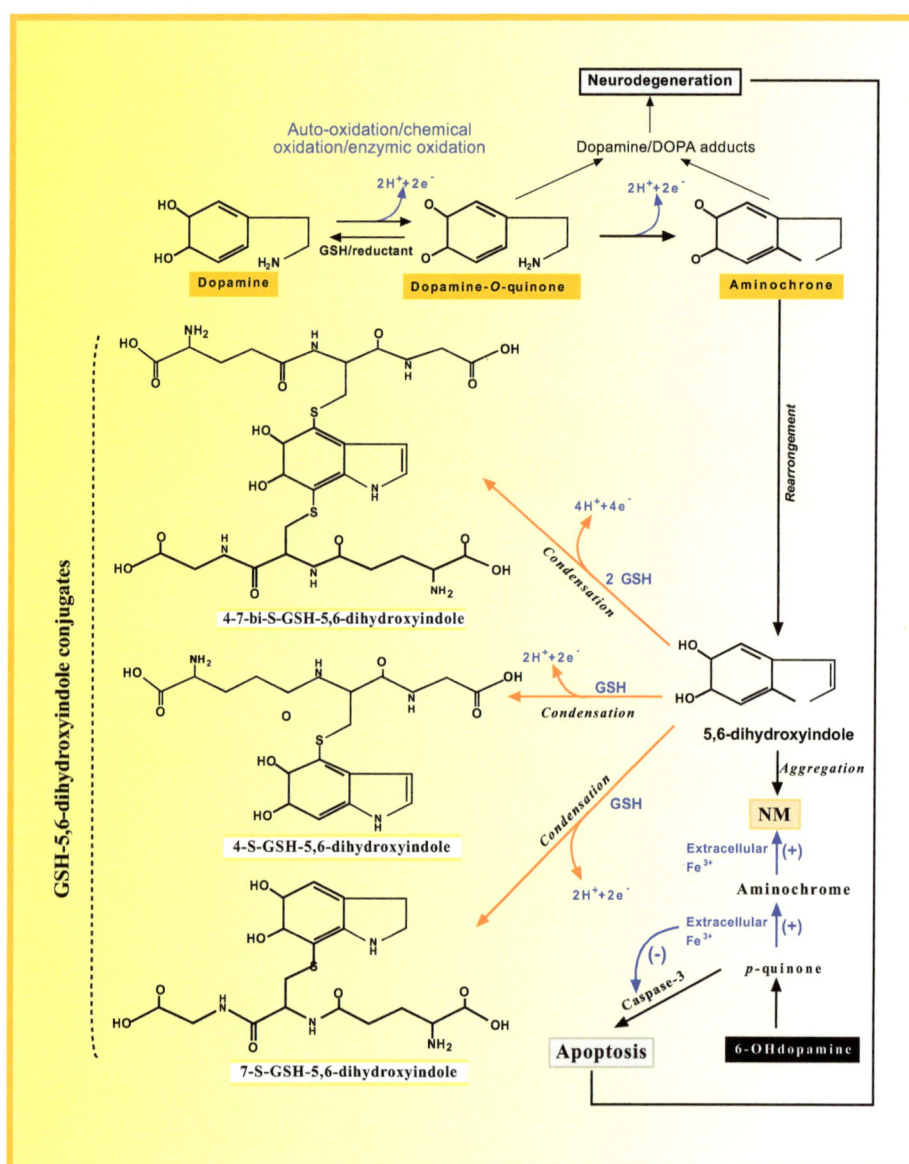

Figure 2: Hypothetical scheme of dopamine oxidation resulting in neurodegeneration and GSH-mediated conjugation of qui-nones leading to neuroprotection. Dopamine can be oxidized by auto-oxidation, metal ions or enzymic reactions. Dopamine oxi-dation results in reactive dopamine quinones (such as dopamine-*O*-quinone, aminochrome and 5,6-dihydroxyindole) and de-

crease reduced glutathione level, which facilitates dopamine oxidation. According to this scheme, GSH reacts with 5,6-dihydroxyindole, derived from aminochrome, and forms various GSH-aminochrome conjugates. Aggregation of 5,6-dihydroxyindole forms melanin, whereas conjugation of 5,6-dihydroxyindole with GSH forms various GSH-5,6-dihydroxyindole conjugates. GSH-quinone conjugates are prone to aggregate in solution with time, but this can be inhibited by excess GSH/exogenous reductants. Extracellular ferrous iron also promotes formation of neuromelanin (NM) and prevents apoptosis by inhibiting caspase-3 enzyme. Information adapted from Zhou and Lim, 2009 [53] and Izumi *et al., 2005* [55].

Its injection into the nigrostriatal pathway results in loss of nigrostriatal dopaminergic neurons, therefore 6-OHDA has been used widely in animal models of PD. 6-OHDA is produced by dopamine hydroxylation in the presence of Fe^{2+} iron and H_2O_2 [59]. 6-OHDA has been detected as a naturally occurring amine in human brains [60] as well as in the urine of PD patients [61]. 6-OHDA is oxidized rapidly by molecular oxygen, forming $O^{\cdot-}_2$, H_2O_2, and the corresponding *p*-quinones. GSH has been reported to attenuate 6-OHDA-mediated cytotoxicity [62]. Normal levels of iron have been reported in Lewy bodies or in SNPC neurons of control brain specimens, however significantly higher iron levels observed in SNPC of PD brain [59]. Due to iron accumulation in SNPC of PD brains, it is suggested that accumulated iron increases levels of oxidative stress by producing hydroxyl radicals via the Fenton reaction [59, 63].

Generation of ROS during dopamine metabolism can cause apoptosis, which may play a role in the pathogenesis of PD [64]. Apoptotic biochemical cascades (including alterations in mitochondria and the endoplasmic reticulum and activation of cysteine proteases, caspases), have also been suggested to be involved in the dysfunction/death of neurons in neurodegenerative diseases, including PD [65]. Mitochondrial apoptosis is regulated by the Bcl-2 protein family (Bcl-2, Bax, and Bim), and suppressed by GSH and its precursor, N-acetylcysteine [66, 67]. Thus depletion of endogenous antioxidant may induce mitochondrial oxidative stress, leading to neuronal apoptosis and ultimately to neural death. However, in SH-SY5Y cells, a human neuroblastoma cell line, 6-OHDA-mediated toxicity is determined by the amount of *p*-quinone produced from auto-oxidation of 6-OHDA rather than by ROS because GSH conjugation with *p*-quinones provides significant protection, whereas catalase enzyme that detoxifies H_2O_2 and superoxide anions does not prevent 6-OHDA-mediated neuronal death [68].

According to Izumi *et al* (2005), addition of ferrous iron to the culture medium inhibits caspase-3 activation and apoptotic nuclear morphologic changes, and blocks 6-OHDA-induced cytotoxicity in SH-SY5Y cells and primary cultured mesencephalic dopaminergic neurons, implicating that extracellular ferrous iron in contrast to intracellular iron may play a neuroprotective role in conversion of *p*-quinone into neuromelanin and reduce apoptosis (Fig. **2**). Under normal conditions, melanin is considered as an effective radical scavenger and protective of SNPC. However, iron-melanin interaction (iron-melanin complex) may be crucial in altering lipid peroxidation (amount, rate of formation, and distribution of OH·) leading to neurodegeneration of melaninized dopamine neurons in SNPC, implicating for PD [69]. Elevated iron concentration in the SNPC of PD patients has been implicated to induce loss of dopaminergic neurons in PD because iron, a transitional metal, promotes free radical formation [69, 70, 71]. Iron chelators have been found to be neuroprotective against 6-OHDA lesions [72, 73]. Thus, iron accumulation in the SNpc of patients may contribute to oxidative stress during PD.

GSH exhibits several functions in the brain, chiefly acting as an antioxidant and a redox regulator. There is a significant progress in research showing a connection between imbalance in GSH homeostasis and PD [54, 64]. Dopamine-induced cell death can also been prevented with other thiol containing compounds, such as N-acetylcysteine and dithiothreitol in PC12 cells, implicating possible treatment of PD [64]. Thus, intracellular GSH as well as other thiol containing compounds may constitute an important natural defense against oxidative stress.

Other Neuroprotective Approaches

Another possibility of neuroprotection is involvement of heat-shock protein 70 (HSP70). The level of chaperone, HSP70, depends on the concentration of carboxy terminal of α-synuclein (TAT-α-synuclein), suggesting its involvement in the inhibition of α-synuclein-mediated toxicity [74]. Free radicals can also be produced by 1-methyl-4-phenyl-1,2,3,6-tetrahydropyridine (MPTP). HSP70 is shown to reduce MPTP-induced apoptosis in the SNpc [75]. The unilateral protection of the dopaminergic system by HSP70 increases amphetamine induced turning toward the uninjected side [75]. These findings suggest that HSP70 may play a protective role in idiopathic PD.

There is substantial evidence that mitochondrial dysfunction and oxidative damage may play a role in the pathogenesis of PD. In animal model, a selective inhibitor of complex I of the electron transport gene can mimic both the biochemical and histopathological findings of PD [76]. Several other agents such as creatine, coenzyme Q_{10}, Ginkgo biloba, nicotinamide, and acetyl-L-carnitine also exert anti-oxidative effects in animal models of neurodegenerative diseases [77]. However, the combination of CoQ$_{10}$ and creatine produces additive neuroprotective effects on improving motor performance and extending survival in mice, suggesting that combination therapy may be more useful in the treatment of PD [78]. Other neuroprotection approaches including blockade of dopamine transporter by uptake inhibitors (cocaine, mazindol), blockade of NMDA receptors by dizocilpine maleate, increased neuronal survival by brain-derived neurotrophic factors (BDNF), and inhibition of monoamine oxidase B (MAO-B) enzyme by selegiline have not been discussed in this chapter [79-83].

CONCLUSIONS

PD is a complex multifactor neurodegenerative disorder in the CNS. It is characterized by loss of dopaminergic neurons in the SNpc along with reduction in the striatal dopamine content. The morphological hallmark of PD is the presence of dystrophic neurites (Lewy neurites) and eosinophilic inclusions (Lewy bodies) [3-5]. It is now well accepted that selective loss of dopaminergic neurons in PD may occur due to autoxidation of dopamine as well as enhanced metabolism of dopamine leading to increased generation of detrimental ROS (Fig. 3). The etiological factors involved with the pathogenesis of PD are still elusive, but collective evidence supports that combination of genetic and environmental factors is likely responsible for the selective loss of dopaminergic neurons [11, 84, 85].

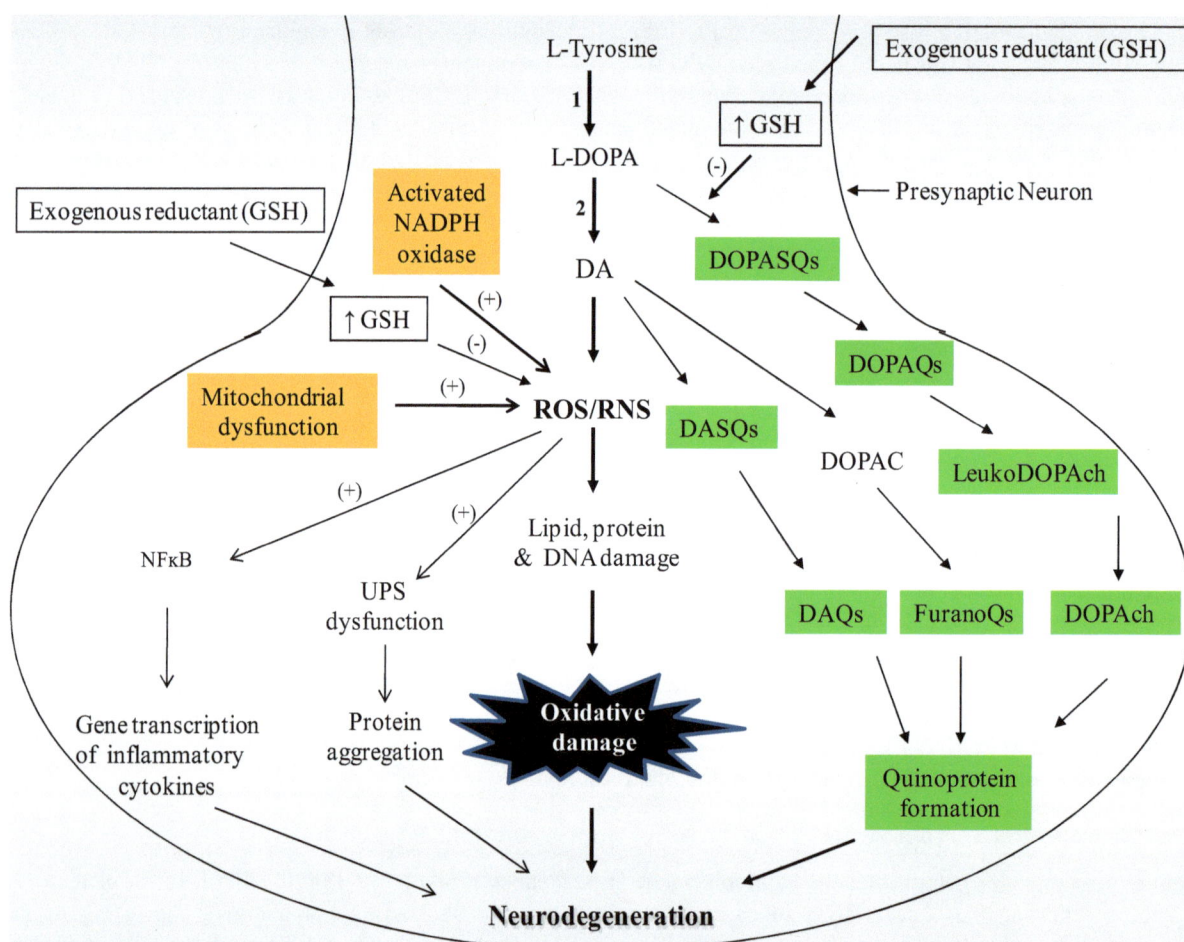

Figure 3: Proposed scheme showing possible pathways involved with the neurodegeneration of dopaminergic neurons in PD and their neuroprotection by exogenous reductants such as GSH. Auto-oxidation of dopamine, L-DOPA and DOPAC results in formation of quinoproteins, contributing to neurodegeneration dopaminergic neurons. ROS can be produced from dopamine metabolism, mitochondrial dysfunction, and activation of NADPH oxidase. Increased levels of ROS induce oxidative stress, inflammation, and impair ubiquitin proteasome system function, causing neurodegeneration. (1), tyrosine hydroxylase; (2), aromatic amino acid decarboxylase; DA, Dopamine; L-DOPA, 3,4-dihydroxy-*L*-phenylalanine; ROS, reactive oxygen species; RNS, reactive nitrogen species; GSH, reduced glutathione; DASQs, dopamine semiquinones; DAQs, dopamine quinones; FuranoQs, Fura-

no quinones; DOPASQs, DOPA semiquinones; LeukoDOPAch, LeukoDOPAchrome; DOPAch, DOPAchrome. NF-кB, nuclear factor; (+), stimulation; (-) inhibition.

The cyclic quinones produced from tyrosinase mediated oxidation of dopamine, DOPA, and DOPAC have been shown to inhibit proteasomal activity, implicating these intermediates play role in the pathogenesis of PD [18]. Furthermore, the autoxidation of dopamine and subsequent polymerization generating quinoproteins also contribute to cell death (Fig. **3**).

Quinone concentration in dopaminergic neurons does not only depend on their rate of synthesis but can be controlled by the levels of cellular antioxidants, such as GSH, closely associated with the pathogenesis of PD (Fig. **3**). The relative contribution of each mechanism (Fig. **3**) in the selective death of dopaminergic neurons remains unknown. However, collective evidence implicates the involvement of all above mechanisms in the pathogenesis of PD. Several approaches have been designed to attenuate the effects of oxidative stress and thus protecting striatal dopaminergic neurons in PD [90].

There is a growing connection between imbalance in GSH homeostasis and a number of neurodegenerative diseases, therefore in this chapter author has mainly focused on glutathione-mediated neuroprotection. GSH alters the redox state of the cell, therefore it may be used as a potential way to prolong the survival of PD patients.

In summary, catecholamines have been demonstrated to be toxic both *in vivo* and *in vitro* via oxidative stress-generated mechanisms [19, 21, 27]. In addition to dopamine oxidation products (such as quinones, superoxide, H_2O_2, and hydroxyl radicals in the presence of transitional metals), other mechanisms such as mitochondrial dysfunction, UPS dysfunction, activation of NADPH oxidase, and neuroinflammation have also been held responsible for the degeneration of dopaminergic neurons in PD. Further studies are required to evaluate the relative contribution of each pathway in the neurodegenerative process in PD. Understanding of this pathogenic network possibly involved with death of dopaminergic neurons, and percent contribution of each pathway in selective death of these neurons in PD brain may help with future therapeutic aspects of the disease.

REFERENCES

[1]   Parkinson J: An Essay on the Shaking Palsy. London, Sherwood, Neely and Jones, 1817.
[2]   Jankovic, J. *J. Neurol. Neurosurg. Psychiatry.* **2008**, 79, 368.
[3]   Forno, L.S. *J. Neuropathol. Exp. Neurol.* **1996**, 55, 259.
[4]   Dunnett, S.B.; Björklund, A. *Nature Suppl.* **1999**, 399, A32.
[5]   Dawson, T.M. *Cell.* **2000**. 101, 115.
[6]   Spillantini, M.G.; Schmidt, M.L.; Lee, V.M.; Trojanowski, J.Q.; Jakes, R.; Goedert, M. *Nature.* **1997**, 388, 839.
[7]   Lang, A. E.; Lozano, A. M. *N. Engl. J. Med.* **998**, 339, 1044.
[8]   Damier, P.; Hirsch, E.C.; Agid, Y.; Graybiel, A.M. *Brain.* **1999**, 122, 1437.
[9]   Pakkenberg, B.; Moller, A.; Gundersen, H. J.; Mouritzen, D. A.; Pakkenberg, H. *J. Neurol. Neurosurg. Psychiatry.* **1991**, 54, 30–33
[10] Asanuma, M.; Miyazaki, I.;Diaz-Corrales, F.J.; Ogawa, N. *Acta Med. Okayama.* **2004**, 58, 221.
[11] Thomas, B.; Beal, M.F. *Hum. Mol. Genet.* **2007**, 16, R183.
[12] Bogaerts, V.; Theuns, J.; van Broeckhoven, C. *Genes Brain Behav.* **2008**, 7, 129.
[13] Halliwell, B. *J. Neurochem.* **2006**, 97, 1634.
[14] Liani, E.; Eyal, A.; Avraham, E.; Shemer, R.; Szargel, R.; Berg, D.; Bornemann, A.; Riess, O.; Ross, C.A.; Rott, R.; Engelender, S. *PNAS.* **2004**, 101, 5500.
[15] Wood-Kaczmar, A.; Gandhi, S.; Wood, N.W. *Trends Mol. Med.* **2006**, 12, 521.
[16] Yang, Y.X.; Wood, N.W.; Latchman, D.S. *Neuroreport.* **2009**. 20, 150.
[17] Farooqui, T. In: Biogenic Amines: Pharmacological, Neurochemical and Molecular Aspects in the CNS. Farooqui, T. and Farooqui, A.A. (Eds), Nova Science Publishers, **2010**, in press.
[18] Zafar, K.S.; Siegel, D., Ross, D. *Mol. Pharmacol.* **2006**, *70*, 1079.
[19] Graham, DG. *Mol. Pharmacol.* **1978**. 14, 633.
[20] Sulzer, D.; Zecca, L. *Neurotox. Res.* **2000**, 1, 181.
[21] LaVoie, M.J.; Hastings TG. *J. Neuroscience.* **1999**. *19.* 1484.
[22] Mark, KA,; Soghomonian, JJ.; Yamamoto, BK. *J. Neurosci.* **2004**. 24, 11449.

[23]   Zhou, Z.D.; Lim, T.M. *Free radic. Res.* **2009**, 43, 417.

[24]   Riddle, EL.; Fleckenstein, AE.; Hanson, GR. *AAPS J.* **2006. 8.** E413.

[25]   Fornstedt, B.; Rosengren, B.; Rosengren, E.; Carlsson, A. *Neuropharmacology.* **1986**, 25, 451.

[26]   Hasting, T.G.; Zigmond, M.J. *J. Neurochem.* **1994**, 63, 1126.

[27]   Hasting, T.G.; Lewis, D.A.; Zigmond, M.J. *PNAS.* **1996**, 93, 1956.

[28]   Miyazaki, I.; Asanuma, M.; Hozumi, H.; Miyoshi, K.; Sogawa, N. *FEBS Lett.* **2007**, 581, 5003.

[29]   Xu, Y.; Stokes, A.H.; Roskoski, R. Jr; Vrana, K.E. *J. Neurosci. Res.* **1998,** 54, 691.

[30]   Kuhn, D.M.; Arthur, R.E.; Thomas, D.M.; Elfrink, L.A. *J. Neurochem.* **1999**, 73, 1309.

[31]   Berman, S.B.; Zigmond, M.J.; Hastings, T.G. J. Neurochem. **1996,** 67, 593.

[32]   Whitehead, R.E.; Ferrer, J.V.; Javitch, J.A.; Justice, J.B. *J. Neurochem.* **2001**, 76, 1242.

[33]   Kuhn, D.M.; Arthur, R. Jr. *J. Neurosci.* **1998**, 18, 7111.

[34]   Takeda, A.; Hasegawa, T.; Matsuzaki-Kobayashi, M.; Sugeno, N.; Kikuchi, A.; Itoyama, Y.; Furukawa, K. *J. Biomed. Biotechnol.* **2006**, 2006,19365.

[35]   LaVoie, M.J.; Ostaszewski, B.L.; Weihofen, A.; Schlossmacher, M.G.; Selkoe, D.J. *Nature Med. 2005, 11, 1214.*

[36]   Miyazaki, I.; Asanuma, M.; Diaz-Corrales, F.J.; Fukuda, M.; Kitaichi, K.; Miyoshi K.; Ogawa, N. *FASEB. J.* **2006.** 20, 571.

[37]   Shen, X.M.; Dryhurst, G. *Chem. Res. Toxicol.* **1996.** 9, 751.

[38]   Fornstedt, B.; Brun, A.; Rosengren, E.; Carlsson, A. *J. Neural. Transm. Park. Dis. Dement. Sect.* **1989.** 1, 279.

[39]   Spencer, JP.; Jenner, P.; Daniel, SE.; Lees AJ.;, Marsden, DC.; Halliwell, B. *J. Neurochem.* **1998.** 71, 2112.

[40]   Beal, MF.2003. Ann. N.Y. Acad. Sci. **2003.** 991, 120.

[41]   Schapira, AH. *Neurol. Clin.* **2009.** 27, 583.

[42]   Jin, J.; Meredith, GF.; Chen, L.; Zhou, Y.; Xu, J.; Shie, FS.; Lockhart, P.; Zhang, J. *Brain Res. Mol. Brain Res.* **2005.** 134, 119.

[43]   Poon, HF.; Frasier, M.; Shreve, N.; Calabrese, V.; Wolozin, B;, Butterfield, DA. *Neurobiol Dis.* **2005.** 18, 492.

[44]   Palacino, JJ.; Sagi, D.; Goldberg, MS.; Krauss, S.; Motz, C.; Wacker, M.; Klose, J.; Shen, J. *J. Biol. Chem.* **2004.** 279. 18614.

[45]   Periquet M, Corti O, Jacquier S, Brice A. J. Neurochem. **2005.** 95, 1259.

[46]   Van Laar, VS.; Dukes, AA.; Cascio M.; Hastings, TG. *Neurobiol. Dis.* **2008.** 29, 477.

[47]   Jenner, P. *Acta. Neurol. Scand. Suppl.* **1993.** 146, 6.

[48]   Martin, HL.; Teismann, P. FASEB. J. **2009.** In press.

[49]   Sian, J.; Dexter, DT.; Lees, AJ.; Daniel, S.; Jenner, P.; Marsden, CD. *Ann. Neurol.* **1994.** 36, 356.

[50]   Power, JH.; Blumbergs, PC. Acta. Neuropathol. **2009.** 117, 63.

[51]   Dagnino-Subiabre, A.; Cassels, BK.; Baez, S.; Johansson, AS.; Mannervik, B.; Segura-Aguilar, J. *Biochem. Biophys.Res. Commun.* **2000.** 274, 32.

[52]   Baez, S.; Segura-aez, J.; Widersten, M.; Johansson, A-S.; Mannervik, B. *Biochem. J.* **1997.** 324, 25.

[53]   Zhou, Z.D.; Lim, T.M. *Free radic. Res.* **2009**, 43, 417.

[54]   Akagawa, M.; Ishii, Y.; Ishii, T.; Shibata, T.; Yotsu-Yamashita, M.; Suyama, K.; Uchida, K. *Biochemistry.* **2006.** 45, 15120.

[55]   Ungerstedt, U. *Eur. J. Pharmacol.* **1968.** 5, 107.

[56]   Uretsky, N.J.; Iversen, L.L. *Nature.* **1969.** 221, 557.

[57]   Iversen, L.L.; Uretsky, N.J. *Brain Res.* **1970.** 24, 364.

[58]   Kondoh, T.; Bannai, M.; Nishino, H.; Torii, K. *Exp. Neurol.* **2005.** 192, 194.

[59]   Kienzl, E.; Puchinger, L.; Jellinger, K.; Linert, W.; Stachelberger,H.; Jameson, RF. *J. Neurol. Sci.* **1995.** 134, 69.

[60]   Curtius, HC.; Wolfensberger, M.; Steinmann, B.; Redweik,U.; Siegfried, J. *J. Chromatogr.* **1974.** 99, 529. 1974)

[61]   Andrew R, Watson DG, Best SA, Midgley JM, Wenlong H, Petty RK.1993. The determination of hydroxydopamines and other trace amines inthe urine of parkinsonian patients and normal controls. *Neurochem Res.* **1993.** 18, 1175.

[62]   Shimizu, E.; Hashimoto, K.; Komatsu, N.; Iyo, M. *Neuropharmacology.* **2002.** 43, 434.

[63]   Fenton, H.J.H. *J. Chem. Soc., Trans.* **1984.** 65, 899.

[64]   Offen, D.; Ziv, I.; Sternin, H.; Melamed, E.; Hochman, A. *Exp Neurol.* **1996.** 141, 32

[65]   Mattson, M.P. *Antiox. & Redox Sig.* **2006.** 8, 1997.

[66]   Zimmermann AK, Loucks FA, Le SS, Butts BD, Florez-McClure ML, Bouchard RJ, Heidenreich KA, Linseman DA. *J.Neurochem.* **2005.** 94, 22.

[67]   Loucks, F.A.; Schroeder, E.K.; Zommer, A.E.; Hilger, S.; Kelsey, N.A.; Bouchard, R.J.; Blackstone, C.; Brewster, J.L.; Linseman, D.A. *Brain Res.* **2009.** 1250, 63..

[68]   Izumi, Y.; Sawada, H.; Sakka, N.; Yamamoto, N.; Kume, T.; Katsuki, H.; Shimohama, S.; Akaike, A. *J. Neurosci. Res.* **2005**. 79, 849.

[69]   Ben-Shachar, D.; Riederer, P.; Youdim, M.B. *J. Neurochem.* **1991**. 57, 1609.

[70]   Sofic, E.; Paulus, W.; Jellinger, K.; Riederer, P.; Youdim, M.B. *J. Neurochem.* **1991**. 56, 978.

[71]   Ben-Shachar, D.; Youdim, M.B. *J. Neurochem.* **1991**. 57, 2133.

[72]   Ben-Shachar, D.; Eshel,G.; Finberg, J.P.; Youdim,M.B. *J. Neurochem.* **1991**. 56, 1441.

[73]   Shachar, D.B.; Kahana, N.; Kampel, V.; Warshawsky, A.; Youdim, M.B. *Neuropharmacology.* **2004**. 46, 254.

[74]   Albani, D.; Peverelli, E.; Rametta, E.; Batelli, S.; Veschini, L.;Negro, A.; *FASEB J.* **2004**, 18, 1713.

[75]   Dong, Z.; Wolfer, D.P.; Lipp, H-P.; Büeler, H. *Mol. Therapy.* **2006**. 11, 80.

[76]   Beal, M.F. *Ann. Neurol.* **2003**. 53 Suppl 3:S39-47; discussion S47-8.

[77]   Ebadi, M.; Brown-Borg, H.; EI Refaey, H.; Singh, B.B.; Garrett, S.; Shavali, S.; Sharma, S.K. *Brain Res. Mol. Brain Res.* **2005**. 134, 65.

[78]   Yang, L.; Callingasan, N.Y.; Wille, E.J.; Cormier, K.; Smith, K.; Ferrante, R.J.; Beal, M.F. *J. Neurochem.* **2009**. 109, 1427.

[79]   Ebadi, M.; Srinavasan, S.K.; Baxi, M.D. *Prog. Neurobiol.* **1996**. 48, 1.

[80]   Ebadi, M.; Brown-Borg, H.; Ren, J.; Sharma, S.; Shavali, S.; El ReFaey, H.; Carlson, E.C. *Curr. Drug Targets.* **2006**. 7, 1513.

[81]   Volpe, B.T.; Wildmann, J.; Altar, C.A. *Neuroscience.* **1998**. 83, 741.

[82]   Armentero, M.T.; Fancellu, R.; Nappi, G.; Bramanti, P.; Blandini, F. *Neurobiol. Dis.* **2006**. 22, 1.

[83]   Blandini, F.; Armentero, M.T.; Fancellu, R.; Blaugrund, E.; Nappi, G. *Exp. Neurol.* **2004**. 187, 455.

[84]   Ren, Y.; Liu, W.; Jiang, H.; Jiang, Q.; Feng, J. *J. Biol. Chem.* **2005**. 280, 34105.

[85]   Büeler H. *Exp. Neurol.* 2009. 218, 235.

[86]   Mena, M.A.; Rodriguez-Navarro, J.A.; Ros, R.; de Yebenes, J.G. *Curr. Med. Chem.* **2008**. 15, 2305.

[87]   Halliwell, B. *Antioxid. Redox Signal.* **2006**. 8, 2007.

[88]   Farooqui, A.A.; Ong, W-Y.; Horrocks. In: Neurochemical Aspects of Excitotoxicity. **2008**. Springer, New York

[89]   Farooqui, A.A. In: Hot topics in Neural Membrane Lipidology. **2009**. Springer, New York.

[90]   Thomas B. *Antioxid. Redox Signal.* **2009**. 11, 2077.

CHAPTER 7

Contribution of Complement in Neurodegenerative and Neuroinflammatory Diseases

Annapurna Nayak[1,2], Uday Kishore[1] and DM Bonifati[3*]

[1]Centre for Infection, Immunity and Disease Mechanisms, Biosciences, School of Health Sciences and Social Care, Brunel University, Uxbridge West London UB8 3PH United Kingdom; [2]Jawaharlal Nehru Institute of Advanced Studies (JNIAS), Secunderabad, Andhra Pradesh, India and [3*] Unit of Neurology, Department of Neurological disorders, Santa Chiara Hospital, Largo Medaglie d'Oro 1, Trento, Italy

Abstract: The complement system is a powerful and vital component of the innate immune system that plays a dual role in neurodegenerative diseases. When present at an optimum level in the normal brain, the complement system plays a neuroprotective role as it is involved in a number of processes like clearance of apoptotic cells, opsonisation of pathogens,etc. However factors such as oxidative stress and age can modify this protective ability can lead to chronic inflammation resulting in neurodegeneration. In a diseased brain, aggregated polypeptides can potentially present their different charge patterns to C1q, which is a vital charge pattern recognition molecule of the complement system. Consequently activation of complement leads to microglial activation which in turn leads to defective clearance of the aggregated polypeptides by macrophages leading to chronic inflammation, especially in age-related neurodegenerative disease (e.g., Alzheimer's disease). The current article aims at discussing the role of the complement system (especially C1q) and its consequences in initiation/progression in neurodegenerative diseases such as amyloid-associated dementias (Alzheimer's disease, Down's syndrome), non-amyloid associated dementia (Familial dementia, Huntington's disease, Parkinson's disease) and pathogen-induced dementia (prion diseases). Evidences point towards the existence of an over-activated complement system in a diseased brain can directly or indirectly lead to neuroinflammation which subsequently leads to neurodegeneration, the effects of which are manifested through the various clinical signs and symptoms. As C1q is the initiation molecule of the classical pathway, C1q-inhibitors that down regulate the complement cascade without negatively affecting the protective functions of complement can pave way for potential future immunotherapeutic approaches.

Keywords: Neuroinflammation; Neurodegeneration; Complement; Neurodegenerative diseases

INTRODUCTION

The complement system is a powerful and vital component of the innate immune system which is also capable of priming and augmenting adaptive immunity. It is involved in a range of functions that involves host defence against the action of pathogenic microorganisms, removal of immune complexes and apoptotic cells and facilitating adaptive immune responses [1, 2].

The complement system is also known to mediate the production of anaphylatoxins (C3a, C4a & C5a) which in turn trigger degranulation, cell lysis and phagocytosis by educing chemotaxis and cell activation [1]. The complement system is activated by three pathways with different target recognition components. However the common aim of all the three pathways is to activate the central pivotal component of the complement system, i.e. the C3 component (Fig. **1**). The three pathways are:

A. The Classical Pathway

This cascade involves a sequentially acting multistep cascade in which the complement components C1q, C1s, C1r, C4, C2 & C3 play very important roles. C1r and C1s, the two serine protease proenzymes, along with C1q constitute C1, the first component of the classical complement pathway [3] The activation of the C1q complex (C1q + C1s–C1r–C1r–C1s) subsequently cleaves C4 and C2 to yield the central molecule C3 convertase that cleaves C3. Then the C2–C9 components are activated and the terminal membrane attack complex (MAC) [3] may bind to cell membranes and cause cell lysis.

*****Address correspondence to:** Unit of Neurology, Department of Neurological disorders, Santa Chiara Hospital, Largo Medaglie d'Oro 1, Trento, Italy. Email: dmbonifati@alice.it

Akhlaq A. Farooqui & Tahira Farooqui (Eds.)

B. The Alternative Pathway

The alternative pathway is started by low-level activation of C3 by hydrolysed C3 and activated factor B. Factor B and the activated C3b bind together and factor B is cleaved by factor D to form C3 convertase. The difference between classical and alternative pathway is that the alternative pathway does not depend on the presence of immune complexes.

C. The Mannose-Binding Lectin Pathway

This pathway is activated through the binding of mannose-binding lectin (MBL) to some pathogen-associated molecular patterns (PAMPs) made up by repetitive carbohydrate patterns on pathogen surfaces. Then lectin activates complement through the MBL-associated serine protease (MASP-2), that in turn leads to the activation of complement components C4, C2 and C3. The activation is similar to the classical pathway [114, 116]. MASP-2 is similar to C1s in its ability to generate C3 convertase cleaving C4 and C2 [130]. At the end the insertion of the MAC into the pathogens cell membrane leads to their lysis.

Altered levels of the activation of the complement system are considered important causative factors in inflammatory, neurodegenerative and cerebrovascular diseases [2]. However recent findings have suggested the existence of a fourth pathway that can generate C5a in the absence of C3 thus leading to the terminal cascade [4].

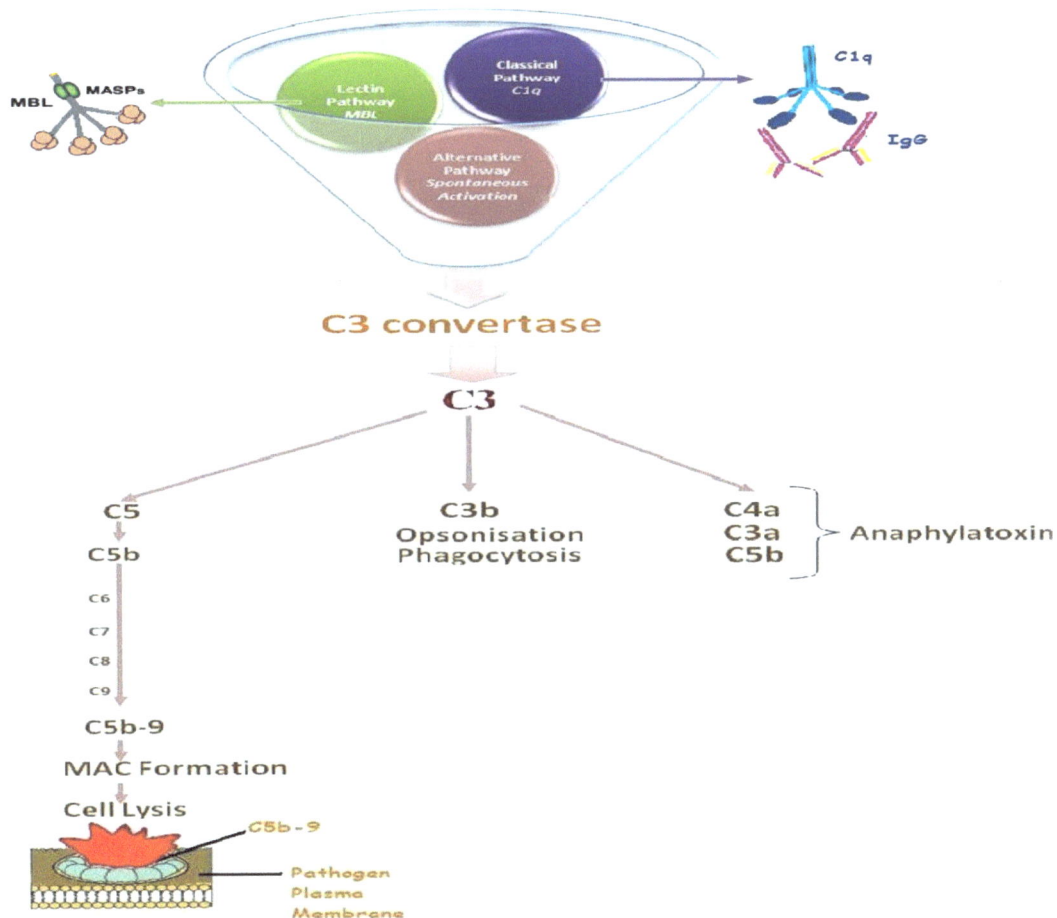

Figure 1: Diagram illustrating the complement system. The complement system involves three pathways namely classical pathway, alternative pathway and mannose binding lectin pathway. The funnel depicts the complement system and the common aim of the three pathways is to yield C3 convertase. The classical and lectin pathway, upon binding to their respective activation subcomponents, cleaves C4 and C2 to C4a/b and C2a/b respectively. C4b and C2b form a complex i.e. C3 convertase. This convertase facilitates the cleavage of C3 which in turn cleaves C5 to yield C5b. C5b hence forms a complex with C6, C7, C8 and C9 to form C5b-9, also known as the Membrane Attack Complex (MAC) that leads to lysis of the target cells.

CENTRAL NERVOUS SYSTEM (CNS) AND THE COMPLEMENT CASCADE

The blood-brain barrier (BBB) as the name suggests, is a protective physical barrier between the blood and the brain, which consists of tight intercellular junctions made up of endothelial cells and astrocytes associated with them. Because of this particular barrier, immune reactions such as inflammation are weaker and slower in the CNS.

Nonetheless contrary to popular belief that the CNS is an immune-privileged organ, it contains many immune system components including complement proteins that are synthesized by glial cells and neurons that also express receptors that are capable of recognising and processing apoptotic cells [131]. There are three categories of glial cells: Astrocytes (astroglia); Oligodendrocytes (Oligodendroglia); and Microglial cells (Microglia). Microglial cells constitute up to 20% of all glial cells and are distributed equally throughout the CNS and play a central role in CNS inflammation. When exposed to appropriate stimuli, microglia can switch to a reactive state and also can mediate an array of pro-inflammatory responses when challenged by the bioactive peptides produced by complement activation. For instance, the complement component C3, C3a and C3b may induce chemotaxis of phagocytic cells such as microglia.

Following injury, the number of cells with phagocytic activity (macrophages) increases in the CNS and this increase appears to be due to both invasion of monocytes from the bloodstream and to activation of local microglial cells. This activation of microglial cells and/or invasion of monocytes lead to tissue damage and inflammation. However the inflammatory reaction in the CNS is different than the inflammatory reactions in the other tissues of the body as there is often no invasion of neutrophil granulocytes and the activation of microglia and invading monocytes to macrophages may take several days. Microglial activation is known to precede and cause neuronal degeneration in various CNS diseases such as Alzheimer's disease and dementia [11, 12, 132]. CNS is also prone to form and accumulate protein aggregates inside neurons. However, it is likely that complement activation and inflammation in the CNS have also a neuroprotective role in clearing up apoptotic cells or debris during development and neuroplasticity.

In several neurodegenerative diseases, the neuronal cell loss, that characterise the disease, is associated with CNS inflammation and with the presence of inflammatory molecules and microglial activation. Amyloid diseases are a part of an emerging heterogeneous group of clinical conditions collectively known as disorders of protein folding. Classical amyloid lesions in the CNS are usually found in the form of parenchymal preamyloid lesions that are immuno-reactive with specific anti-amyloid antibodies, negative to Congo red or thioflavin S staining and amorphous nonfibrillar in structure under electron microscopy (EM). Some of the neurodegenerative diseases will be discussed in subsequent sections by describing the role of complement in the progression of diseases.

ALZHEIMER DISEASE (AD)

AD is the most frequent form of dementia in the elderly and is a growing public health problem. It affects about 6% of the population over the age of 65 years and its prevalence increases with age till 75-80 years of age. [5] It is a chronic severe neurodegenerative disease characterised by cognitive (memory loss, language and visuo-spatial difficulties, attention and executive dysfunction) and behavioural symptoms (depression, delusions, agitation) [6, 102]. It worsens over time and gradually affected individuals are severely disabled and not able to perform daily activities independently.

The neuropathological features are neuronal loss, extracellular deposition of amyloid in the brain parenchyma (senile plaques) and cerebral vessel walls and intracellular aggregation of paired helical filaments of hyperphosphorylated tau protein (neurofibrillary tangles). These features are present mainly in entorhinal cortex, hippocampus, and midtemporal gyrus along with the presence of macroscopically visible cerebral atrophy.

Senile plaques comprise of extracellular deposition of aggregated cleavage products of neuronal amyloid precursor protein (APP). The processing of APP through β and γ-secretase enzymes and the formation of toxic β-amyloid peptides (Aβ) by cleavage of CTFβ (a 99 residue membrane bound protein) is presumably the key step in the neuropathology of AD [72]. The pathogenic role of APP, and hence Aβ, is revealed by some forms of familial AD dominantly inherited that are linked to APP mutations. Aβ is an important component of the lesions found in the brains of individuals with AD. Different types of extracellular AP aggregates exist out of which the Aβ1-42 is more predominant as it is less soluble and aggregates more readily than the soluble Aβ1-40, a form which is shortened by two residues at the C-terminus [7]. Mutations of presenilins (PS1 and PS2) that influence APP metabolism and increase the production of Aβ1-42 also cause familial early onset AD (FAD) [103]. A strong genetic risk factor, shown in the sporadic late form of AD, is the presence of the ε4 allele of the apolipoprotein E (APOE) gene, which encodes a protein involved in cholesterol metabolism. APOE4 may contribute to AD by modulating the metabolism and aggregation of Aβ and by regulating brain lipid metabolism [8].

The pathological process that ultimately leads to neuronal loss is not clear and it is not known if the pathological findings are the consequences or the primary cause of the disease. Many biochemical processes have been described in the AD brains mainly oxidative stress and scaffolding dysfunction with β-amyloid peptide deposition. A role for excitotoxins has been considered. Inflammation and complement activation are likely to have an important role in neuronal loss and have been widely described in AD brains starting with the first description of Alois Alzheimer in 1911. Since then a consistent association of AD neuropathology with inflammation have been described [9, 10, 93]. Activated microglial has also been demonstrated in AD patients and frontotemporal dementia using positron emission tomography [11, 12].

An important role of inflammation in AD is also suggested by a number of epidemiological studies that appear to establish a direct correlation between the chronic use of non steroidal anti-inflammatory drugs (NSAIDs) and a reduced risk of developing AD [13,14]. Long-term use of anti-inflammatory medications has been associated with reduced microglial activation [15] and/or decreased generation of Aβ protein [16, 17]. This is further supported by the fact that NSAIDs have been demonstrated to have a protective action in animal models of AD [18].

However recently the effectiveness of NSAIDs in reducing risk of developing AD have been questioned when research failed to show a reduction in AD risk and also the AD anti-inflammatory prevention trial was discontinued early due to adverse events. On the contrary there was evidence of worsening cognition associated with NSAIDs when compared with a placebo [127]. Similar results were found in a large observational study [19]. The authors of this latter study argued that age differences in the cohorts under study may account in part for the discrepancy findings i.e. if NSAIDs exposure delays AD onset, younger cohorts may show a reduced risk while older cohorts may show no or increased risk.

Even if adaptive immunity and the formation of specific antibodies is probably not involved in AD pathogenesis, immunisation of murine models of AD with Aβ showed a prevention of β-amyloid plaques formation in young animals and a reduced neurite dystrophy and gliosis in the older animals with consequent improvement of memory and behavioural disturbances in the treated animals [133]. These initial animal experiments prompted clinical trials in humans. Unfortunately a Phase II trial in which a vaccine containing Aβ1 - 42 peptide was administered intramuscularly was discontinued early for the occurrence of an aseptic meningoencephalitis in around 5% of the treated patients [20]. Nonetheless the occurrence of this side effect confirms the potential capability of Aβ in triggering inflammation *in vivo*.

Many factors such as oxidative stress, vascular injury, fat intake and folic deficiency have been implicated in triggering Aβ deposition, microglia activation and neuroinflammation [21]. All these processes may contribute to neuronal dysfunction and result eventually in cell death. Activated microglia expressing the major histocompatibility class II antigens and complement receptor are present around amyloid plaques and dystrophic neurites. Thus aggregated Aβ peptides are potentially capable of inducing microglia to secrete proinflammatory cytokines, reactive oxygen species, complement factors, neurotoxicsecretory products and chemokines [22]. Proinflammatory cytokines such as TNF-α, IL-1β and IL6 converge to produce an abnormal processing and hyper-phosphorylation of the tau protein, another landmark of AD pathogenesis through the down regulation of the cdk5/p35 pathway [23]. There is increasing evidence supporting the intrinsic toxicity of hyperphosphorylated tau on neuronal degeneration.

Complement Activation and its Role in Progression of AD

Several clues point towards a significant role for complement activation in neuroinflammation and neurodegeneration. Aggregated Aβ peptides can activate alternative and classical pathway in an antibody-independent manner in vitro. Subsequently microglial cells expressing the complement receptors CR3 and CR4 may then be recruited [24]. It is possible that C1q, and hence classical pathway, is particularly important in this activation. C1q is capable of engaging a broad range of ligands and can bind to Aβ via its globular domain (gC1q) and more specifically via its B chain [25].

Post-mortem studies have demonstrated an upregulation of many complement proteins and mRNAs, such as C1q, C1r, C1s, C2, C3, C4, C5, C6, C7, C8, and C9 in AD brains. C4d and C3d mark tangles and plaques and MAC has been found in dystrophic neuritis [26]. In a recent study complement components especially C9 were detected on plaques and on neurofibrillary tangles in AD specimens. Both early and late-stage complement activation occurs on neocortical plaques in subjects across the cognitive spectrum [27]. In the entorhinal cortex, hippocampus, and midtemporal gyrus, where plaques and tangles are numerous, C1q mRNA levels were increased 11- to 80-fold over control levels [28] and the number of C1q-positive plaques are significantly higher than in control cases. Localization of C1q-positive plaques correlated with the expression of C1q gene demonstrating an *in loco* production of C1q protein [29].

Upregulation of C1q and its co-localization with fibrillar Aβ have also been observed in an animal model of AD [30]. To evaluate the role of C1q in AD a mouse model lacking C1q (Q-/-) was crossed with an AD mouse model (APP). At older ages, the APP and APPQ-/- mice accumulated similar amount of amyloid and fibrillar β-amyloid in frontal cortex and hippocampus; but activated glia around the plaques was significantly lower in the APPQ-/- mice. In another murine model containing transgenes for both APP and mutant presenilin 1 (APP/PS1), a similar decrease of pathology was found [31]. Thus it can be observed that absence of C1q in an AD brain could realistically save the brain from disease progression by hampering the development of amyloid plaques.

Neurofibrillary tangles present in the AD brain also associated with the potential to activate complement system. Also soluble, non-fibrillar Aβ1-42 may induce a dose-dependent activation of C4 through a C1q independent mechanism. This may be a first protective attempt to clean up amyloid fibrils before aggregation. Once Aβ aggregate the continuous inflammation and complement activation may be deleterious. Activated microglia and complement activation may increase oxidative stress as demonstrated by the presence of reactive oxygen species in neurons incubated with purified C1q [32]. Proteomic study of plasma of AD patients showed factorH (fH) and β-2-macroglobulin (A2M) as AD-specific plasma biomarkers and their presence was associated with disease progression in AD [33].

On the other hand few studies involving animal models suggest that microglia and complement activation may have a protective role against Aβ-induced neurotoxicity and may promote the clearance of amyloid and degenerating neurons. Amyloid precursor protein (APP) transgenic mouse models of AD that lack the ability to activate the classical complement pathway display less neuropathology than do the APPQ+/+ mice. Both C3 and C4 deposition increase with age in APPQ+/+ transgenic mice but while C4 is predominantly localized on the plaques and/or associated with oligodendrocytes only in APPQ+/+ mice, C3 immunoreactivity is seen in both animal models and, is higher in APPQ-/- than in APPQ+/+ mice, providing evidence for alternative pathway activation. This increase in C3 levels is associated with decreased neuropathology [34].

An increased production of transforming growth factor (TGF)- β1 resulted in a vigorous microglial activation that was accompanied by at least a 50% reduction in Aβ accumulation in human hAPP transgenic mice and higher levels of C3 [35]. To evaluate the role of complement in the pathogenesis of AD-like disease in these mice, Wyss-Coray(2002) inhibited C3 activation by expressing soluble complement receptor-related protein y (sCrry), a complement inhibitor. The resulting transgenic mice showed a 2-3 fold increase in Aβ accumulation and neuronal degeneration in 1-year-old mice compared to the original AD mice [35] and showed that fibrillar amyloid plaque burden correlated with increased levels of insoluble Aβ1-42 levels and reduced levels of soluble Aβ1-42 along with loss of neuronal-specific nuclear protein-positive neurons in the hippocampus. Mice genetically deficient in the complement component C5 have been shown to be more susceptible to hippocampal excitotoxic lesions [36]. All these results indicate that complement activation products can increase microglia activation and oxidative stress and damage neurons but also protect helping in cleaning up aggregated proteins, a process that seems strictly connected with aging.

A delicate balance between the two roles seems at work in healthy and AD brains. Inhibitor of complement such as inhibitor C4b-binding protein (C4BP) have been detected in Aβ plaques and on apoptotic cells in AD brain and C4BP levels in cerebrospinal fluid (CSF) of dementia patients and controls were low compared to levels in plasma and correlated with CSF levels of other inflammation-related factors [37]. Also, a lack of CD59, an inhibitor that reduce formation of MAC complex, may contribute to the pathogenesis of AD. Its expression was significantly decreased in the frontal cortex and hippocampus of AD brains and Aβ -peptide itself was found to downregulate CD59 expression at the mRNA level in the AD brain [37].

Thus in this cascade, the complement system especially the recognition subcomponent of the classical pathway C1q is particularly important as it acts as a vital mediator of Aβ-induced inflammatory reaction through complement activation. This activation thus leads to microglial activation which in turn leads to senile plaque formation and Aβ phagocytosis (Fig. **2**). As complement plays such a huge role in the neurodegeneration and neuroinflammation on AD, it is feasible to conclude that by aiming at curbing the activation of the complement cascade the progression of AD can be controlled and hence therapeutic advances that address the issue of complement activation can be considered as a feasible approach to treating AD.

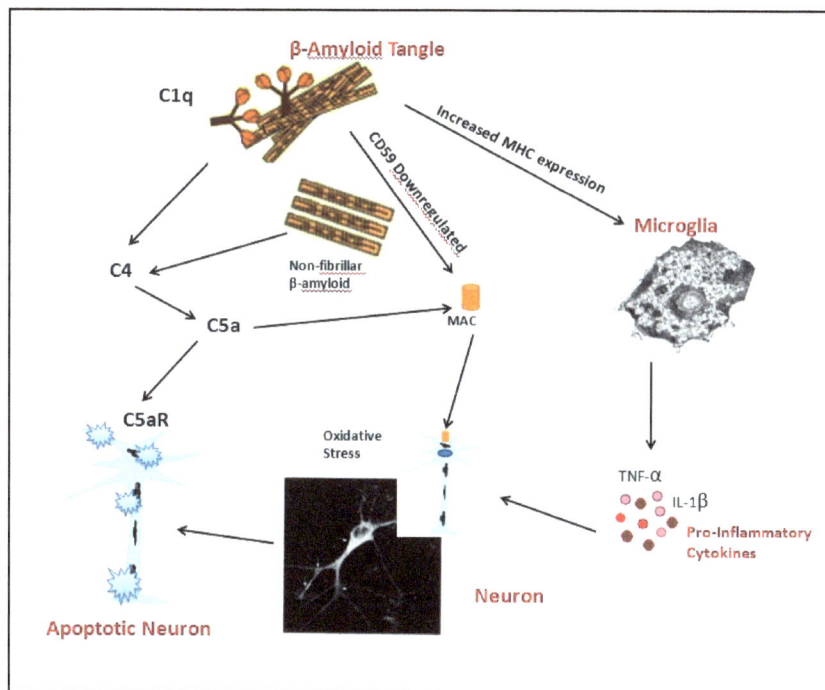

Figure 2: Illustration depicting the relationship between amyloid beta and complement proteins and their role in activating microglia and neurodegeneration. The deposition of β-amyloid leads to the activation of C1q which in turn leads to a cascade of events that activates microglia. The activated microglia releases pro-inflammatory cytokines like TNF-α and IL-1β. Inflammation caused by these cytokines potentially leads to neurodegeneration. The activation of complement by C1q also leads to MAC formation via classical pathway, which degenerate neurons.

PARKINSON DISEASE (PD)

After Alzheimer's disease, PD is the second most common neurodegenerative disorder that mainly affects the motor system and the classical symptons are tremor at rest, postural imbalance, slowness of movement and rigidity [38]. It is characterised by a slow degeneration of dopaminergic neurons in the substantia nigra pars compacta and in the striatum (astrogliosis) and by the presence of proteinaceous inclusions (Lewy bodies or Lewy neurites) that constitute of filamentous α-synuclein [110]. To a minor extent other non-dopaminergic systems such as norepinephrinergic neurons in the locus coeruleus and serotoninergic neurons in the raphe nuclei are also affected by the pathological process [39]. The primary cause of neuronal loss in PD is vague but several molecular and cellular mechanisms may be at work such as mitochondrial dysfunction, oxidative stress, protein handling excitotoxicity and apoptosis process. As in AD, a small number of PD patients have a familial form of the disease due to mutations in some genes such as parkin and α-synuclein (the main component of Lewy bodies) [40]. Many of the proteins in the familial form of the disease are involved in the degradation of misfolded or damaged proteins by the ubiquitin-proteasome pathway. Despite all these mechanisms involved, neuroinflammation has also been implicated in contributing to the cascade of events leading to neuronal degeneration.

Complement and its Role in Neuroinflammation in PD

Inflammation in the striatum and the substantia nigra, may aggravate the course of the disease [41]. In contrast with AD, recently Brochard et al showed higher densities of CD8+ and CD4+ cells in the brains of patients with PD [42 while] Saijodemonstrated that NURR1, an essential protein for dopaminergic neurons survival and mutations of which cause a familial for of PD, has a previously unknown function in microglia and astrocytes and can protect dopaminergic neurons from inflammation-induced death [43]. Post-mortem studies in humans have shown the presence of activated microglial within the substantia nigra [44] and locus coeruleus [45] of patients with PD and increased major histocompatibility complex expression on microglia have been shown [46].

Many cytokines and proinflammatory molecules have also been reported in the striatum of PD patients in post-portem studies and in serum and cerebrospinal fluid[47]. These data have been confirmed in animal models of the disease [48, 49]. Salutary effects of PPARγ-agonists were also seen in animal models of Parkinson's disease [50, 51].

As in AD, PET scan analysis with PK-11195 (a ligand indicative of microglial activation) showed an increased binding in the pons, basal ganglia and frontal and cortical regions in patients with PD [52, 53]. Epidemiological data shows that the risk of PD was decreased in patients taking NSAIDs and anti-inflammatory agents can inhibit dopaminergic neuronal loss in animal models of PD [38]. Yamadademonstrated all components of the MAC intracellularly either on Lewy bodies and on oligodendroglia in the substantia nigra of patients with sporadic [54, 55] and familial PD [56]. Extraneuronal Lewy bodies and dendritic spheroid bodies were also stained for C3d, C4d, C7 and C9, but not for C1q. In the same study complement-activated oligodendroglia were revealed but the staining for the alternative complement activation proteins was negative.

Substantia nigra specimens from PD patients showed that Lewy bodies as well as melanized neurons stained for iC3b and C9 and the staining was significantly increased respect aged normal and AD specimens. iC3b and C9 staining was not correlated with the remaining melanized neurons, nor with the duration of PD [57]. In the same study there was marked variation in the percentages of immunoreactive melanized neurons for different specimens and there was no correlation between the percentage of iC3b positive melanized neurons and the duration of the disease. iC3b on melanized neurons may also have a neuroprotective role by decreasing the production of inflammatory cytokines [58] and protecting neurons against excitotoxins [59]. Detection of C9, on the other hand; suggests that deposition of the MAC on dopamine neurons may have lytic effects and contribute to the neuronal loss. To confirm the role of complement in PD McGeer and McGeer [38] found an increased of mRNA complement levels in affected brain regions.

The mechanism by which complement is activated on PD is unknown. In contrast with AD, a cellular immune mediated response is possible. Recently T lymphocytes have been described in the substantia nigra of PD [42] and the presence of surface IgG have been reported by Orr et al [60] on 30% of dopamine neurons in the PD substantia nigra. Alternatively, complement activation could be secondary to cell injury, oxidative stress or the presence of aggregated α-synuclein or other proteins.the alternatively spliced alpha-Syn 112 form, but not full-length alpha-Syn 140, activated complement [61].

Together, these findings indicate that a small amount of inflammation with activation of resident microglia and complement system is present in the brain regions involved in PD. These findings may contribute to loss of dopaminergic neurons. Interestingly complement activation on melanized neurons appears to decrease with normal aging, suggesting a possible neuroprotective role of complement in the normal substantia nigra [57]. Lewy bodies in PD may have a similar role as amyloid and tangles in AD activating complement and microglia. The consequent release of neurotoxic products such as MAC and oxygen free radicals may damage dopaminergic neurons [15].

DEMENTIA WITH LEWY BODIES

Dementia with Lewy bodies (DLB) share many clinical and pathological features with PD and dementia is frequently found in PD. It is supposed that either the diseases depend on an underlying common process (Lewy body disease) related to dysregulation of the synaptic protein, α-synuclein [110]. The disease consist of a primary dementia characterised by visuoperceptual and executive dysfunction accompanied by prominent visual hallucinations, fluctuating attention and parkinsonism.

The neuropathology of DLB is similar to that of PD but a more severe diffuse load of Lewy bodies (eosinophilic, neuronal inclusions, which can be stained by actin, neurofilament, ubiquitin and α-synuclein), is present in brain stem, diencephalon, anterior cingulate, amygdala and cerebral cortex. Many cases have substantial AD pathology (amyloid plaque and neocortical tangle) and this could explain the reason why many cases with Lewy body pathology present with an insidious amnestic syndrome are more similar to AD. Cognitive impairment seems to correlate with the presence of an AD-like pathology along with cortical and limbic α-synuclein load which is the main constituent of Lewy body both in PD and DLB [62]. The contribution of the AD pathology to the dementia has been debated but mixed cases are more demented than pure cases.

Thus it is not surprising that varying degree of microglia activation in DLB have been reported. While an increase in the microglia activation was reported to be found even in the absence of any AD pathology [10] others observed no significant increase in microglia activation when compared with controls in post-mortem studies [63, 64]. However it is worth to take note that methodological differences and case selection may explain these contrasting results. There are limited studies that evaluate the role of complement in DLB. C3d and C4d staining on Lewy bodies was reported in the brain stem from subjects with DLB [65].

Complement component C3d was only occasionally been seen in diffusely ubiquinated neurons but late complement components have not been detected in these neurons [64] and the Lewy bodies were also negative for C5–9. Double staining for complement and alpha-synuclein was negative, suggesting the absence of complement in LBs in demented PD patients without AD pathology (pure LB dementia) [64]. In the same study LB bearing neurons were not associated with activated microglia cells in contrast to ubiquinated plaques and MHC class II and CD68 staining were comparable.

A recent study involved the staining of the substantia nigra specimens from patients with DLB for iC3b and C9. It was observed that the Lewy bodies in these specimens stained for both the early (iC3b) and late (C9) complement proteins [66]. This latter finding suggests that complement activation may contribute to loss of dopaminergic neurons in some individuals with DLB. Limitations to study such patients hamper research as pure DLB is rare and concomitant AD pathology (senile plaques or neurofibrillary tangles) is often present and the absence of an animal model of this disease adds to the limitations. Thus it is possible to speculate that an intermediate degree of neuroinflammation and complement activation is present in DLB depending on the α-synuclein (Lewy bodies) and AD pathology load but however further research is required to confirm this fact.

FAMILIAL DEMENTIA

Familial dementia is a form of dementia in which a certain mutation is inherited for generations and this mutation leads to early onset of dementia either in the dominant or non-dominant form. Overexpression of amyloid precursor protein (APP) as well as mutations in the APP and presenilin genes has been known to cause rare forms of AD. They cause AD by elevating levels of neurotoxic Aβ. Over expression of APP also causes defects in axonal transport [67]. However several other mutations have been known to cause early dementia along with neuronal loss and other similar features as AD. Some of the forms will be discussed subsequently.

FAMILIAL BRITISH DEMENTIA & FAMILIAL DANISH DEMENTIA (FBD AND FDD)

Worster-Drought reported a hereditary case of gradual dementia, spastic tetraparesis and ataxia in 1933 in which the majority of the generations suffered from the above symptoms leading to death following severe dementia and insanity [68]. However following research by several scientists on the successors of the same family, Vidal et al reported the discovery of a unique 4K protein subunit named ABri from isolated amyloid fibrils from a patient suffering from FBD [69]. ABri is a proteolytic product of a larger precursor molecule BriPP which is coded by the gene BRI2 (also known as ITMB2B) present on the long arm of the chromosome 13. FBD was formerly known as familial cerebral amyloid angiopathy – British type. Subsequently, another similar case of familial dementia was observed by Stromgrem et al in 1970 in which severe presenile dementia was observed in 5 generations followed by early death. After 3 decades of its actual discovery, Vidal et al [70] identified a defect in the BRI2 gene that lead to dementia and also isolated and characterized the amyloid protein implicated in FDD, but this time it was a different proteolytic fragment ADan of the same precursor molecule BriPP. FDD is also known as heredopathia ophthalmooto-encephalica. As the gene responsible for both the disorders is situated on the long arm of the chromosome 13, the disorders are collectively known as the "Chromosome 13 dementias". The only difference between ABri and ADan is that it arises from two different genetic defects i.e. ABri is a result of a Stop-to-Arg mutation and ADan is due to a ten nucleotide duplication-insertion immediately before the stop codon.

Both FBD and FDD share similarities with AD in terms of the neuropathological hallmarks and the disease progression except for the age at which the disease first manifests itself. The similarities include the presence of amyloid-associated proteins such as serum amyloid P component, apolipoprotein E, apolipoprotein J, vitronectin, glycosaminoglycans and extracellular matrix proteins along with the presence of neurofibrillary tangles which are found in AD brains is also found in FBD and FDD affected individuals and the tangles are immunohistochemically, ultrasturally and biochemically identical to that seen in AD [71].

Complement and the Dementia Peptides

As observed in most neurodegenerative diseases such as AD and DS (Down's syndrome) [72], incidence of amyloid leads to initiation of local inflammatory responses, especially complement activation that thus contributing towards the progression of the disease. Dementia in any neurodegenerative disease is highly associated with the presence of activated microglia, reactive astrocytes thus leading to increased levels of inflammatory cytokines and complement products especially around the amyloid plaques and diseased neurites. The amyloid deposition in both FBD and FDD has been associated with activated microglia and reactive astrocytes. Although activated microglia is an important evidence of complement activation, the presence of C1q, C4d, C3d and MAC further supports the existence of complement activation and its role in the disease progression.

It has also been observed that complement activation in FDD and FBD continues mostly through the classical pathway (70-75%) when compared to the activation through the alternative pathway (25-30%), thus suggesting higher binding efficiency of ABri and ADan to the recognition protein C1q [73]. However ABri and ADan have also been demonstrated to trigger the alternative pathway to some extent, hence explaining the low percentage of alternative pathway activation. This is due to the ability of ABri and ADan to aggregate rapidly almost similar to the aggregation ability of $A\beta 1$-42. However the accumulation of complement activation products is lower in FDD than in FBD [71]. This is because the lesions in FDD are pre-amyloid and non-fibrillar in nature and hence incapable to activate complement as extensively as FBD.

Therefore the existence of activated microglia, the deposition of complement products and formation of MAC all point towards the role of complement activation in the disease progression of FBD and FDD in the same manner as in AD, the only difference being the gene that causes the diseases.

Huntington's Disease

Huntington's disease (HD) was first discovered by George Huntington in 1872 and is a non-curable autosomal-dominant progressive neurodegenerative disorder that affects control over movement, cognition and also psychological symptoms. The prevalence of the clinical syndrome is 5-10:100000 whereas it has been observed that nearly 20:100000 are carriers of the gene responsible for the disease. The disease is characterised by the onset of midlife chorea (around 33-44 years of age) although the disease can potentially present itself at any time from childhood to old age [75]. Other important symptoms include decline in mental abilities leading to personality alteration (i.e. depression, suicidal tendencies and in some cases, violent behaviour), abnormal involuntary movements that heavily affect gait & dexterity, development of dementia, thus ultimately leading to death.

HD is caused by an abnormal expansion of otherwise normal CAG trinucleotide repeats on the N terminus of the *IT 15* gene which was discovered in 1993 and located on Chromosome 4p16.3 that encodes the protein huntingtin (htt) [74]. The number of these CAG repeats is directly proportional to the intensity of the disease for instance, in a normal person the number of CAG repeats is around 8-39 whereas in HD the repeats can range from 36-120 in number [75]. The abnormal CAG repeats are responsible for the neuronal dysfunction that leads to the manifestation of clinical symptoms.

The neuropathological hallmark of HD is the degeneration of the nuclei of the basal ganglion situated in the lateral ventricle brain i.e. the caudate nuclei. The intensity of the degeneration The intensity of the degeneration can vary from mild to severe as a consequence of which there is an acute loss of neurons in the caudate along with less prominent neuronal loss in the putamen (Fig. **3**). As the disease progresses, there is a dramatic increase in neuronal loss from caudate along with presence of both reactive astrocytes and microglia in the grey matter in the caudate as opposed to the early stages in which no significant gliosis is observed [74,76]. In the neostriatum, huntingtin is found in the cell bodies and synaptic processes of surviving neurons and glial cells [76].

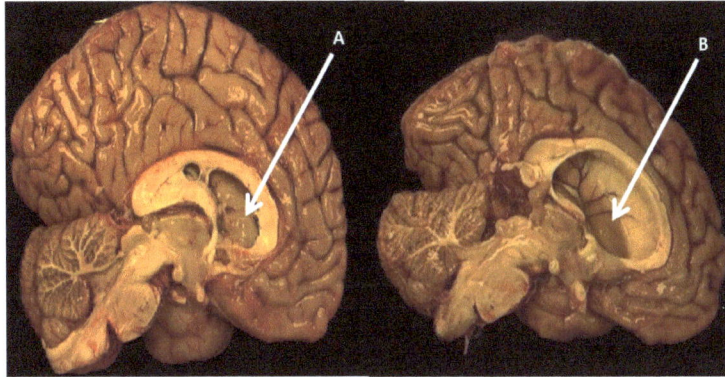

Figure 3: The above image illustrates the difference between the normal brain (A) and HD brain (B). In the normal brain the ventricle is significantly smaller when compared to B (the HD brain). The ventricle is denoted by the arrows. (Resource: Harvard Brain Tissue Resource Centre).

Although the genetic causes underlying the disease have now been discovered, the actual disease progression involved in HD remains a conundrum. However the complement cascade has been implicated to play a huge role in the development of HD.

Immune Activation and Complement in HD

In various neurodegenerative diseases such as AD [77] and familial dementia, the prevalence of an inflammatory response has been highly associated with the presence of reactive astrocytes and microglia. Besides the association with an inflammatory response, local biosynthesis of an array of inflammatory molecules including components of complement by activated glial cells has also been observed. Nevertheless one of the several abnormalities associated with HD is the up-regulation of immunoglobulins in the early stages that suggests that there is a possible link between HD and an overactive immune response especially in the CNS [78].Consistent with this finding, the presence of complement components in HD brains have been proved by a number of studies in the past two decades [76, 77, 79,80].

As mentioned before, the caudate and the striatum are the main areas of the brain affected in HD. Complement subcomponents such as C1q along with C4 and C3 were observed to be abundant in neurons in the HD caudate and striatum and the respective mRNAs were also present in the HD brain [76]. These findings suggest that there is recruitment of the complement cascade in the HD brain. Subsequently by considering the scenario in AD and PD, presence of complement components implies inflammation which is being engineered by glial cells. Reactive astrocytes in the HD caudate grey matter, Wilson's pencil and white matter in the internal capsule were positive for C1q, C3 and C4. Myelin sheaths in HD white matter were positive for C1q as well along with the complement activation specific iC3b that confirms the presence of C3b.[76]

Further evidence for the involvement of the complement system in the pathogenesis of HD in the CNS is provided by microglial activation. Microglial cells migrate to the site of damage or trauma as soon as it is inflicted and hence activation of microglia is an important biomarker for CNS damage [81]. It has been shown that C1q plays a pivotal role in activating these microglial cells. When microglia are activated, they release inflammatory cytokines such as IL-6 and TNF-α along with other substances such as chemokines and nitric oxide and C1q has been shown to induce the release of such inflammatory cytokines along with increasing nitric oxide release and oxidative release which are all markers of microglial activation. In the case of HD, upregulation of the major cytokines of the innate immune system is observed both centrally and peripherally [78]. These changes are observed even in presymptomatic HD mutation carriers long before they express clinical symptoms such as motor abnormalities [78, 82]. Microglial activation correlates with release of cytokines and severity of the disease and the pro-inflammatory cytokines such as IL-6 and IL-8 are observed in premanifest and early stages of HD whereas anti-inflammatory cytokines IL-4 and IL-10 are produced in the later stages. IL-4, IL-10 and IL-8 might be produced due to mutant huntingtin within the microglia but expression of IL-6, which is the first cytokine to be produced and is the earliest plasma abnormality detected in HD patients to date, is known to be enhanced by C1q [83].

Thus it can be seen how an over active immune system fuelled by both complement activation and local complement biosynthesis can be a vital clue to the overall pathogenesis of HD.

PRIONS DISEASE

Prions are unique infectious pathogens that cause fatal neurodegenerative diseases called transmissible spongiform encephalopathies (TSEs) also known as prion diseases and the term "prion" arises from "proteinaceous infectious" The main features of prion diseases besides being extremely fatal are that they have an unpredictable occurrence and have unique properties (eg: replication) [84]. Prion diseases are found in both humans and animals and the human prion diseases include Kuru disease [85,86] various forms of Creutzfeldt-Jakob disease (CJD) [87] Gerstmann-Straussler-Scheinker disease (GSS) [88], fatal familial insomnia [FFI] [89] and sporadic fatal insomnia (SFI).

The infections cause of scrapie is described as the scrapie isoform of the prion protein i.e. PrP^{Sc}. These are protease resistant oligomers that proliferate by promoting the misfolding and polymerization of the endogenous cellular PrP isoform PrP^c [90]. Infection occurs through oral ingestion of the pathogen following which the PrP^{Sc} penetrates the intestinal epithelium, spreads to the lymphoid tissues presumably via follicular dendritic cells (FDCs) and then migrates to the CNS leading the neuroinflammation and neurodegeneration [91]. This trafficking from the gut to mesenteric lymph nodes and then to the peripheral neurons is thought to be carried out by intestinal dendritic cells and FDCs are also implicated for PrP^{Sc} proliferation within the lymphoid tissues [92].

Role of Complement in Prion Diseases

Prion diseases like AD and DS are neurodegenerative diseases and the pathology is characterized by neuronal loss, glial activation and extracellular accumulations of the protease-resistant isoforms PrP^{Sc} of the cell-surface expressed -prion protein (PrP^c) leading to formation of amyloid plaques [93]. An interesting association is made between the presence of complement proteins and development of prion disease. In scrapie pathogenesis, C1q is implicated in the development of the disease. A report by Flores-Langarica and team [94] suggests that C1q contributes towards the PrP^{Sc} uptake by conventional dendritic cells which is followed by PrP^{Sc} accumulation within the follicular dendritic cells (FDCs) network. This is confirmed by mouse models [94, 95], lacking the recognition component C1q, in which there was reduction in PrP^{Sc} accumulation in FDCs. Depletion of C3 or genetic deficiency of C1q significantly delay the onset of the disease and leads to reduction in accumulation of PrP^{Sc} in the spleen in the early stages [93] In the initial stages of TSE infection, when PrP^{Sc} is administered orally or intraperitoneally, it activates the host complement cascade and thus becomes opsonised which in turn is recognised by migrating intestinal DCs and after being associated with the DCs they are transported within host tissues. FDCs express both the complement receptor CR2 (CD21) and PrP^c in high abundance wherein the complement-opsonized PrP^{Sc} encounters the normal prion protein PrP^c and induces it to acquire the PrP^{Sc} conformation [92].

PrP^{Sc} is present in the brain tissue in the form of scrapie amyloid fibrils (SAF) in mouse models of scrapie infection. Complement proteins are known to be associated with amyloid deposits. PrP^{Sc} is not available in the pure form as they are found colocalized with SAF and hence researching on PrP^{Sc} is challenging. Despite these challenges, Blanquet-Grossard in 2005 [96] demonstrated that human C1q interacts strongly with recombinant mouse PrP that was immobilised on a Biacore ® sensor chip and this interaction was facilitated by the gC1q domain.

Therefore it can be deduced that complement deficiency is a boon in the case of prion diseases as the deficiency curbs the progression of the disease. Therefore therapeutically the control of complement cascade has to be aimed in order to treat or atleast control the disease progression as it would stop formation of amyloid fibrils and hence neurodegeneration and neuroinflammation.

DOWNS SYNDROME (DS)

Down syndrome (DS) is a birth defect prevalent in nearly 0.45% of all human conceptions and is associated with a various detrimental phenotypes that include learning disabilities, fatal heart defects, early-onset Alzheimer's disease and childhood leukaemia. Down's syndrome is caused by trisomy of human chromosome 21 (Hsa21) [97].

The neuropathological lesions of DS are similar to that of AD, and this similarity has facilitated the possibility of studying AD through DS brain models. DS affected individuals over express a number of genes including the APP gene which is over-expressed at 150% than normal expression throughout their lifetime [98] Due to this overexpression, most patients end up developing the typical neuropathological lesions of AD such as NFTs and amyloid plaques by the age of 40. The patients also begin to express AD-related symptoms such as dementia, major change to personality, seizures and loss of independent skills as reported from a neurology clinic in 1982 [99].

Complement Activation and Progression of DS

As the complement cascade is highly implicated in the progression of AD and PD, it is feasible to compare AD brains to DS brains to study complement involvement. As mentioned earlier, studies have demonstrated the presence of inflammatory markers such as activated microglia [6], astrocytes [100] and upregulated complement proteins and their mRNAs especially the vital subcomponents like C1q, C1r, C1s, C2, C3, C4, C5,C6, C7, C8 and C9 in AD brains [28]. In comparison, a report by Stoltzner et al in 2000 suggests that complement immunoreactivity is highly observed in DS brains along with the association of C1q, C3 and C4d with the amyloid plaques. However C5b-9 is not seen to be colocalized with these plaques but nonetheless is found in subsets of neurons, NFTs and some dystrophic neurites [101]. It is also observed that the deposition of these complement proteins were dependent to the state of maturation and the quantity of the plaques.

AD and DS brains are feeding grounds for complement-mediated inflammation as there are several negative effects of the complement cascade in the CNS. First and foremost, C1q, which is the initiation subcomponent of the classical complement pathway, boosts Aβ aggregation in AD brains and hence also in DS brains [93]. This boost promotes plaque maturation and consequently triggers associated inflammatory responses and thus decrease the possibility of clearance of Aβ. The proliferation of C3 molecules could facilitate phagocytosis of adjacent healthy cells [102]. However if the C5b-7 complex which is the precursor to MAC formation, attaches to cell membranes it will cause cell lysis thus explaining the loss of neurons in these diseased brains. It will also lead to "bystander lysis" i.e. lysis of the neighbouring healthy neurons [101]. This might lead to speculations as to why there is extracellular Aβ deposition if it can be phagocytosed? The possible reason for this is that MAC formation fails to take place in extracellular Aβ deposits as they lack a cell membrane (101). Following the initiation of the complement cascade, microglial activation takes place leading to oxidative damage along with deposition of microglial cells which further facilitates neuronal degeneration [81].

Thus AD and DS brains have similar neuropathological features in which complement deposition leads to microglial activation which in turn leads to neuronal degeneration which is the major pathological feature in both brains. Hence therapeutic approaches should aim towards controlling the complement cascade to curb Aβ deposition and therefore slow down the development of AD in both AD and DS patients.

CONCLUSION

In most neurodegenerative diseases, it can thus be observed that the complement system plays an important role in the progression of disease. The summary of cascade of events that lead to neuroinflammation and neurodegeneration is illustrated in Fig. **4**. The complement system is a double-edged sword with dual contrasting properties i.e. both neuroprotective and neuroinflammatory and hence in neurodegeneration. When the complement levels are normal, it acts as a boon to the immune system through aiding in various processes including recognition of pathogens, opsonisation and clearance of apoptotic cells. However factors such as oxidative stress due to the presence of excess free radicals and aging can reverse this protective role and hence bring about the destructive aspect of complement i.e. lead to Neurodegeneration. This takes place especially in the presence of aggregated polypeptides that can present themselves to vital charge pattern recognition molecules of the complement system, especially C1q. This aggregation leads to augmentation of the microglial activity and hence leads to microglial activation, initiated by C1q. This defect in the efficiency of the complement system leads to defective clearance of the aggregated polypeptides by macrophages which in turn lead to chronic inflammation. Chronic inflammation plays a major role in age-related neurodegenerative diseases (for e.g.: late onset AD). Thus immunotherapeutic approaches that aim at curbing complement activation especially by introducing C1q-inhibitors that would down regulate the complement cascade without interfering with the beneficial effects of complement would alleviate control of progression of a number of CNS diseases that are plaguing mankind, especially the older population.

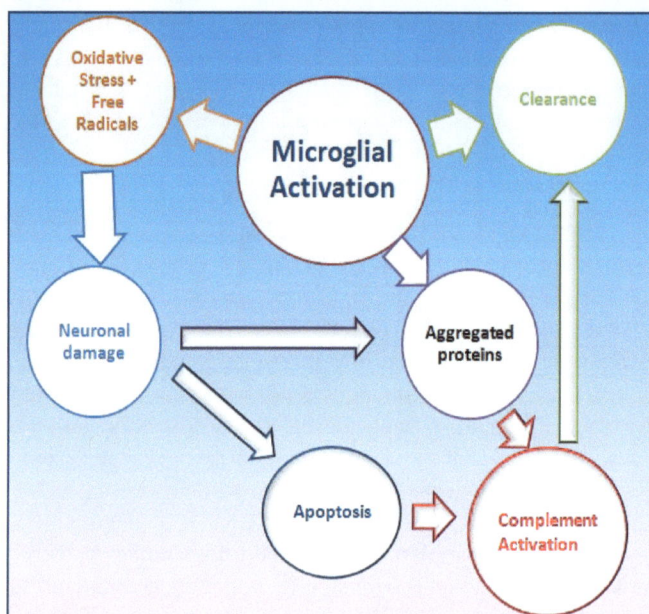

Figure 4: Role of complement in neuroinflammation and neurodegeneration. The relationship is interlinked and one cascade leads to another with the end product being neuroinflammation and hence neural degeneration. This is true for most of the neurodegenerative diseases and is the main cause for the pathologies and hence the signs and symptoms.

ACKNOWLEDGEMENTS

U.K. acknowledges BRIEF award by Brunel University. A.N. is funded by Future Genetics Ltd, Future House, London, UK.

REFERENCES

[1]   Janssen, Bert J. C. (2005) Structures of complement component C3 provide insights into the function and evolution of immunity. Nature, Vol. 437, pp. 505-511.

[2]   Rus, H., Cudrici, C. and David, S. & Niculescu, F. (2006) The complement system in central nervous system diseases. 39 (5), 395-402. 5, Autoimmunity, Vol. 39, pp. 395-402.

[3]   Arlaud, GJ,(2002) Structural biology of the C1 complex of complement unveils the mechanisms of its activation and proteolytic activity. Mol. Immunol., Vol. 39, pp. 383–394.

[4]   Huber-Lang, M,(2006) Generation of C5a in the absence of C3: a new complement activation pathway. 6, Vol. 12, pp. 682-87.

[5]   Ferri, CP, Prince, M and Brayne, C(2005) Alzheimer's Disease International. Global prevalence of dementia: a Delphi consensus study.. 9503, 2005, Lancet, Vol. 366, pp. 2112-7.

[6]   Afagh, A.,1996 Localization and cell association of C1q in Alzheimer's disease brain, Experimental Neurology, Vol. 138, pp. 22-32.

[7]   Roher, AE,(1993) beta-Amyloid-(1-42) is a major component of cerebrovascular amyloid deposits: implications for the pathology of Alzheimer disease. 22, Proc Natl Acad Sci U S A., Vol. 90, pp. 10836-40.

[8]   Bu, G. (2009) Apolipoprotein E and its receptors in Alzheimer's disease: pathways, pathogenesis and therapy. 5, Nat Rev Neurosci., Vol. 10, pp. 333-44.

[9] Lue, L.F.,(1999)Inflammation, Ab deposition, and neurofibrillary tangle formation as correlates of Alzheimer's disease neurodegeneration 10, J Neuropathol Exp Neurol, Vol. 55, pp. 1083-1088.

[10] Mackenzie, IR. (2000)Activated microglia in dementia with Lewy bodies. Neurology, Vol. 55, pp. 132-134.

[11] Cagnin, A,(2001) In-vivo measurement of activated microglia in dementia., Lancet, Vol. 358 (9280), pp. 461-467.

[12] Cagnin, A.,(2004)detection of microglial activation in frontotemporal dementia. 6, Ann Neurol, Vol. 56, pp. 894-897.

[13] Akiyama, H., Barger, S. and Barnum, S.(2000) Inflammation and Alzheimer's disease. 3, Neurobiol Aging, Vol. 21 (3), pp. 383-421.

[14] McGeer, P.L., Schulzer, M. and McGeer, E.G (1996) Arthritis and anti-inflammatory agents as possible protective factors for Alzheimer's disease: a review of 17 epidemiologic studies. 1996, Neurology, Vol. 47 (2), pp. 425-432.

[15] McGeer, P.L. and McGeer, E.G. (1998) Glial cell reactions in neurodegenerative diseases: pathophysiology and therapeutic interventions. Alzheimer Dis Assoc Disord, Vol. 12 (2), pp. S1-S6.

[16] Lim, G.P.,(2000) Ibuprofen suppresses plaque pathology and inflammation in a mouse model for Alzheimer's disease., J Neurosci, Vol. 20., pp. 5709-5714.

[17] Weggen, S,(2001) A subset of NSAIDs lower amyloidogenic Abeta42 independently of cyclooxygenase activity. Nature, Vol. 414(6860), pp. 212-6.

[18] Yan, Q,(2003) Anti-Inflammatory Drug Therapy Alters -Amyloid Processing and Deposition in an Animal Model of Alzheimer's Disease. J. Neurosci., Vol. 23, pp. 7504-7509.

[19] Breitner, JC, Haneuse, SJ and Walker, R(2009) Risk of dementia and AD with prior exposure to NSAIDs in an elderly community-based cohort. Neurology, Vol. 72(22) pp. 1899-905.

[20] Karkos, J. (2004) Immunotherapy of Alzheimer's disease. Results of experimental investigations and treatment of perspectives. Fortschr Neurol Psychiatr, Vol. 72 (4), pp. 204-219.

[21] Maccioni, RB,(2009)The role of neuroimmunomodulation in Alzheimer's disease., Ann N Y Acad Sci., Vol. 1153, pp. 240-6.

[22] Heneka, MT and O'Banion, MK. (2007) Inflammatory processes in Alzheimer's disease, J Neuroimmunol, Vol. 184 (1-2), pp. 69-91.

[23] 23 Quintanilla, RA,(2004) Interleukin-6 induces Alzheimer-type phosphorylation of tau protein by deregulating the cdk5/p35 pathway. Exp Cell Res., Vol. 295 (1), pp. 245-257.

[24] Rozemuller, JM,(1989) A4 protein in Alzheimer's disease: primary and secondary cellular events in extracellular amyloid deposition. J Neuropathol Exp Neurol., Vol. 48 (6), pp. 674-91.

[25] Kishore, U. and Reid, K.B. (2000),C1q: structure, function, and receptors. Immunopharmacology, Vol. 49 (1-2), pp. 159-170.

[26] McGeer, P.L. and McGeer, E.G. (2002) The possible role of complement activation in Alzheimer disease. Trends Mol Med, Vol. 8 (11), pp. 519-523.

[27] Loeffler, DA, Camp, DM and Bennett, DA. (2008)Plaque complement activation and cognitive loss in Alzheimer's disease. J Neuroinflammation, Vol. 5 (9).

[28] Yasojima, K,(1999) Up-regulated production and activation of the complement system in Alzheimer's disease brain. Am. J. Pathol., Vol. 154 (3), pp. 927-936.

[29] Tooyama, I.,(2001) Correlation of the expression level of C1q mRNA and the number of C1q-positive plaques in the Alzheimer Disease temporal cortex. analysis of C1q mrna and its protein using adjacent or nearby sections. Dement. Geriatr.Cogn.Disord, Vol. 12(4), pp. 237-242.

[30] Matsuoka, Y,(2001)Inflammatory responses to amyloidosis in a transgenic mouse model of Alzheimer's disease. Am J Pathol, Vol. 158(4), pp. 1345-1354.

[31] Fonseca, M.I.,(2004)Absence of C1q leads to less neuropathology in transgenic mouse models of Alzheimer's disease., J Neurosci, Vol. 24 (29), pp. 6457-6465.

[32] Luo, X.,(2003)C1q-calreticulin induced oxidative neurotoxicity: relevance for the neuropathogenesis of Alzheimer's disease., J Neuroimmunol, Vol. 135, pp. 62-71.

[33] Thambisetty, M,(2008) Proteome-based identification of plasma proteins associated with hippocampal metabolism in early Alzheimer's disease. J Neurol., Vol. 255(11), pp. 1712-20.

[34] Zhou, J,(2008)Complement C3 and C4 expression in C1q sufficient and deficient mouse models of Alzheimer's disease. J Neurochem., Vol. 106(5), pp. 2080-92.

[35] Wyss-Coray, T.,(2002)Prominent neurodegeneration and increased plaque formation in complement-inhibited Alzheimer's mice, Proc. Natl. Acad. Sci. U S A., Vol. 99(16), pp. 10837-10842.

[36] Pasinetti, G.M. (1996)Inflammatory mechanisms in neurodegeneration and Alzheimer's disease: the role of the complement system., Neurobiol. Aging., Vol. 1996 (17), pp. 707-716.

[37] Trouw, LA,(2008)C4b-binding protein in Alzheimer's disease: binding to Abeta1-42 and to dead cells., Mol Immunol, Vol. 45(13), pp. 3649-60.

[38] McGeer, P.L. and McGeer, E.G. (2004)Inflammation and neurodegeneration in Parkinson's disease., Parkinsonism. Relat. Disord., Vol. 10(1), pp. S3-S7.

[39] Hirsch, EC,(2003)Nondopaminergic neurons in Parkinson's disease., Adv Neurol., Vol. 91, pp. 29-37.

[40] Mizuno, Y,(2001)Familial Parkinson's disease. Alpha-synuclein and parkin., Adv Neurol., Vol. 86, pp. 13-21.

[41] Czlonkowska, A.,(2002)Immune processes in the pathogenesis of Parkinson's disease - a potential role for microglia and nitric oxide., Med Sci Monit, Vol. 8(8).

[42] Brochard, V,(2009) Infiltration of CD4+ lymphocytes into the brain contributes to neurodegeneration in a mouse model of Parkinson disease., J Clin Invest, Vol. 119(1), pp. 182-92.

[43] Saijo, K,(2009) A Nurr1/CoREST pathway in microglia and astrocytes protects dopaminergic neurons from inflammation-induced death., Cell, Vol. 137, pp. 47-59.

[44] McGeer, PL,(1988) Reactive microglia are positive for HLA-DR in the substantia nigra of Parkinson's and Alzheimer's disease brains. Neurology, Vol. 38(8), pp. 1285-91.

[45] Mirza, B,(2000)The absence of reactive astrocytosis is indicative of a unique inflammatory process in Parkinson's disease. Neuroscience, Vol. 95(2), pp. 425-432.

[46] Imamura, K,(2003) Distribution of major histocompatibility class II-positive microglia and cytokine profile of Parkinson's disease brains. Acta Neuropathol (Berl), Vol. 106, pp. 518-26.

[47] Hirsch, EC and Hunot, S. (2009) Neuroinflammation in Parkinson's disease: a target for neuroprotection? Lancet Neurol, Vol. 8, pp. 382-97.

[48] McGeer, P.L.,(2003) Presence of reactive microglia in monkey substantia nigra years after 1-methyl-4-phenil-1,2,3,6-tetrahydropyridine administration. Ann Neurol, Vol. 54, pp. 599-604.

[49] Depino, AM,(2003) Microglial activation with atypical proinflammatory cytokine expression in a rat model of Parkinson's disease., Eur J Neurosci, Vol. 18, pp. 2731-42.

[50] Breidert, T,(2002)Protective action of the peroxisome proliferator-activated receptor-gamma agonist pioglitazone in a mouse model of Parkinson's disease., J Neurochem, Vol. 82(3), pp. 615-24.

[51] Dehmer, T,(2004)Protection by pioglitazone in the MPTP model of Parkinson's disease correlates with I kappa B alpha induction and block of NF kappa B and iNOS activation., J Neurochem., Vol. 88(2) pp. 484-501.

[52] Gerhard, A, Pavese, N and Hotton, G. (2006)In vivo imaging of microglial activation with [11C](R)-PK11195 PET in idiopathic Parkinson's disease, Neurobiol Dis, Vol. 21(2), pp. 404-12.

[53] Ouchi, Y,(2005)Microglial activation and dopamine terminal loss in early Parkinson's disease. Ann Neurol., Vol. 57(2), pp. 168-75.

[54] Yamada, T., McGeer, P.L. and McGeer, E.G. (1991) Relationship of complement-activated oligodendrocytes to reactive microglia and neuronal pathology in neurodegenerative disease., Dementia, Vol. 2, pp. 71-77.

[55] Yamada, T., McGeer, P.L. and McGeer, E.G. (1992) Lewy bodies in Parkinson's disease are recognized by antibodies to complement proteins., Acta. Neuropathol. (Berl)., Vol. 84(1), pp. 100-104.

[56] Yamada, T,(1993) Histological and biochemical pathology in a family with autosomal dominant parkinsonism and dementia., Neurol. Psychiatry. Brain. Res., Vol. 2, pp. 26-35.

[57] Loeffler, DA, Camp, DM and Conant, SB. (2006) Complement activation in the Parkinson's disease substantia nigra: an immunocytochemical study. J Neuroinflammation, Vol. 29(6).

[58] Gasque, P,(2000)Complement components of the innate immune system in health and disease in the CNS. Immunopharmacology, Vol. 49, pp. 171-86.

[59] van Beek, J,(2001)Complement anaphylatoxin C3a is selectively protective against NMDA-induced neuronal cell death. Neuroreport, Vol. 12, pp. 289-93.

[60] Orr, CF,(2005)A possible role for humoral immunity in the pathogenesis of Parkinson's disease., Brain, Vol. 128, pp. 2665-74.

[61] Klegeris, A and McGeer, PL. (2007)Complement activation by islet amyloid polypeptide (IAPP) and alpha-synuclein 112. Biochem Biophys Res Commun, Vol. 357(4), pp. 1096-9.

[62] Jellinger, KA. (2009) Significance of brain lesions in Parkinson disease dementia and Lewy body dementia. Front Neurol Neurosci, Vol. 24, pp. 114-25.

[63] Shepherd, CE,(2000) Cortical inflammation in Alzheimer disease but not dementia with Lewy bodies. Arch neurol, Vol. 57, pp. 817-822.

[64] Rozemuller, A.J., et al.(2000)Activated microglial cells and complement factors are unrelated to cortical Lewy bodies., Acta. Neuropathol. (Berl)., Vol. 100(6), pp. 701-708.

[65]   Iseki, E and Marui W, Akiyama H, Ueda K, Kosaka K (2000) Degeneration process of Lewy bodies in the brains of patients with dementia with Lewy bodies using α-synuclein-immunohistochemistry. Neurosci Lett, Vol. 286, pp. 69-73.

[66]   Loeffler, DA, Camp, DM and Conant, SB. (2006)Complement activation in the Parkinson's disease substantia nigra: an immunocytochemical study. J Neuroinflammation., Vol. 3, p. 29.

[67]   Stokin, GB,(2008) Amyloid precursor protein-induced axonopathies are independent of amyloid-beta peptides. Human Molecular Genetics, Vol. 17(22), pp. 3474-86.

[68]   Worster-Drought, C, Hill, TTR and Mcmenemey, MH. (1933) Familial presenile dementia with spatic paralysis., The Journal of Neurology and Psychopathology, pp. 27-34.

[69]   Vidal, R.,(1999)A stop-codon mutation in the BRI gene associated with familial British dementia. Nature, Vol. 399, pp. 776-781.

[70]   Vidal, R.,(2000) A decamer duplication in the 3-prime region of the BRI gene originates an amyloid peptide that is associated with dementia in a Danish kindred. Proc. Nat. Acad. Sci., Vol. 97, pp. 4920-4925.

[71]   Rostagno, A,(2005)Chromosome 13 dementias., Cell Mol Life Sci., Vol. 62(16), pp. 1814-25.

[72]   Hardy, J and Allsop, D. (1991)Amyloid deposition as the central event in the aetiology of Alzheimer's disease. Trends Pharmacol Sci, Vol. 12(10), pp. 383-8.

[73]   Rostagno, A,(2002)Complement activation in chromosome 13 dementias. Similarities with Alzheimer's disease. J Biol Chem., Vol. 277(51), pp. 49782-90.

[74]   Purdon, SE, Mohr, E and Ilivitsky, V and Jones,BD. (1994), Huntington's disease: pathogenesis, diagnosis and treatment. J Psychiatry Neurosci, Vol. 19(5), pp. 359–367.

[75]   Saleh, N,(2009)Neuroendocrine Disturbances in Huntington's Disease. PLoS ONE., Vol. 4(3), pp. 1-7.

[76]   Singhrao, SK,(1999)Increased Complement Biosynthesis By Microglia and Complement Activation on Neurons in Huntington's Disease. Experimental Neurology, Vol. 159, pp. 362–376.

[77]   Webster, S.,(2000)Complement component C1q modulates the phagocytosis of Aβ by microglia. Experimental Neurology, Vol. 161(1), pp. 127-138.

[78]   Björkqvist, M,(2008) A novel pathogenic pathway of immune activation detectable before clinical onset in Huntington's disease. The Journal of Experimental Medicine, Vol. 205(8), pp. 1869-1877.

[79]   Vanguri, P and Shin, ML. (1986) Activation of complement by myelin: identification of C1-binding proteins of human myelin from central nervous tissue. Journal of Neurochemistry, Vol. 46(5), pp. 1535-41.

[80]   Pavese, N,(2006)Microglial activation correlates with severity in Huntington disease: a clinical and PET study. Neurology, Vol. 66(11), pp. 1638-43.

[81]   Streit, WJ, Mrak, RE and Griffin, WT. (2004)Microglia and neuroinflammation: a pathological perspective. Journal of Neuroinflammation, Vol. 1(14).

[82]   Tai, YF,(2007)Microglial activation in presymptomatic Huntington's disease gene carriers. Brain, Vol. 130(7), pp. 1759-66.

[83]   Fraser, AD,(2006)C1q and MBL, components of the innate immune system, influence monocyte cytokine expression., Journal of leukocyte biology, Vol. 80, pp. 107-128.

[84]   Ridley, Rosalind M and Baker,Harry F. (1996) The Paradox of Prion Disease. [book auth.] Rosalind M. Ridley Harry F. Baker. Prion Diseases. s.l. : Humana Press, pp. 1-2.

[85]   Brandner, S,(2008)Central and peripheral pathology of kuru: pathological analysis of a recent case and comparison with other forms of human prion disease. Philosophical transactions of the Royal College of London Series B: Biological Sciences, Vol. 363(1510), pp. 3755-63.

[86]   Liberski, PP and Brown, P. (2009) Kuru: its ramifications after fifty years. Experimental Gerontology, Vol. 44(1-2), pp. 63-9.

[87]   Choi, EM,(2009) Prion proteins in subpopulations of white blood cells from patients with sporadic Creutzfeldt-Jakob disease. Laboratory Investigations: A journal of technical methods and pathology, Vol. 89(6), pp. 624-35.

[88]   Sikorska, B,(2009)Ultrastructural study of florid plaques in variant Creutzfeldt-Jakob disease: a comparison with amyloid plaques in kuru, sporadic Creutzfeldt-Jakob disease and Gerstmann-Sträussler-Scheinker disease., Neuropathology and Applied Neurobiology, Vol. 35(1), pp. 46-59.

[89]   Baldin, E,(2009)A case of fatal familial insomnia in Africa. Journal of Neurology, Vols. DOI 10.1007/s00415-009-5205-4, p. Epublication.

[90]   Prusiner, SB. (1997)Prion diseases and the BSE crisis. Science, Vol. 278, pp. 245–251.

[91]   Mabbott, NA and Bruce, ME. (2004)Complement component C5 is not involved in scrapie pathogenesis. Immunobiology, Vol. 209, pp. 545–549.

[92]   Sim, RB,(2007) C1q binding and complement activation by prions and amyloids. Immunobiology, Vol. 212 (4-5), pp. 355-62.

[93]   Kishore, U and Bonifati, DM. (2007)Role of complement in neurodegeneration and neuroinflammation. Molecular Immunology, Vol. 44(5), pp. 999-1010.

[94]   Flores-Langarica, A,(2009) Scrapie pathogenesis: the role of complement C1q in scrapie agent uptake by conventional dendritic cells. The Journal of Immunology, Vol. 182, pp. 1305-1313.

[95]   Dumestre-Pérard, C,(2007) Activation of classical pathway of complement cascade by soluble oligomers of prion Cellular Microbiology, Vol. 9(12), pp. 2870-9.

[96]   Blanquet-Grossard, F,(2005) Complement protein C1q recognizes a conformationally modified form of the prion protein. Biochemistry, Vol. 44, pp. 4349–4356.

[97]   Patterson, D. (2009)Molecular genetic analysis of Down syndrome. Hum Genet.

[98]   Poirier, J, Danik, M and Blass, JP. (2000) Pathophysiology of Alzheimer's Syndrome. [book auth.] S Gauthier. Clinical diagnosis and management of Alzheimer's disease. pp. 23-24.

[99]   Lott, IT and Lai, F. (1982) Dementia in Down's syndrome: observations from a neurology clinic. Applied research in mental retardation, Vol. 3(3), pp. 233-9.

[100]   Von Bernhardi, R and Ramirez, G. (2001)Microglia astrocyte interaction in Alzheimer's disease: Friends or Foes for the nervous system. Biological Research.

[101]   Stoltzner, SE,(2000) Temporal accrual of complement proteins in amyloid plaques in Down's syndrome with Alzheimer's disease. The American Journal of Pathology, Vol. 156(2), pp. 489-99.

[102]   Johnson, LV,(2002) The Alzheimer's Aβ-peptide is deposited at sites of complement activation in pathologic deposits associated with aging and age-related macular degeneration. Proceedings Of The National Academy Of Science, Vol. 99(18), pp. 11830–11835.

[103]   Zhang, Z,(2000)Presenilins are required for gamma-secretase cleavage of beta-APP and transmembrane cleavage of Notch-1. Nat. Cell. Biol., Vol. 2(7), pp. 463-465.

[104]   Wyss-Coray, T. and Mucke, L. (2002)Inflammation in neurodegenerative disease--a double-edged sword. Neuron, Vol. 35(3), pp. 419-432.

[105]   Vegh, Z,(2003) Maturation dependent expression of C1q binding proteins on the cell surface of human monocyte derived dendritic cells., International Immunopharmacology, Vol. 3, pp. 39-51.

[106]   Tuppos, E and Arias, HR. (2005)The role of inflammation in Alzheimer's disease., International Journal of Biochemistry and Cell Biology, Vol. 87, pp. 289-305.

[107]   Tada, Y,(2004) Differential effects of LPS and TGF-β on the production of IL-6 and IL-12 by Langerhans cells, splenic dendritic cells, and macrophages. Cytokine, Vol. 25(4), pp. 155-161.

[108]   Tacnet- Delorme, P, Chevallier, S and Arlaud, GJ. (2001) Beta-amyloid fibrils activate the C1 complex of complement under physiological conditions: evidence for a binding site for A beta on the C1q globular regions. J Immunol., Vol. 167(11), pp. 6374-81.

[109]   Stevens, B,(2007)The Classical Complement Cascade Mediates CNS Synapse Elimination. Cell, Vol. 131(6), pp. 1164-1178.

[110]   Spillantini, MG(1997)a-Synuclein in Lewy bodies., Nature, Vol. 388, pp. 839-840.

[111]   Shen, Y,(2001)Complement activation by neurofibrillary tangles in Alzheimer's disease. Neurosci Lett., Vol. 305(3), pp. 165-8.

[112]   Sambamurti, K, Greig, NH and Lahiri, DK. (2002) Advances in the cellular and molecular biology of the beta-amyloid protein in Alzheimer's diseaseNeuromolecular Med., Vol. 1(1), pp. 1-31.

[113]   Hirsch, EC and Hunot, S. (2009)Neuroinflammation in Parkinson's disease: a target for neuroprotection? Lancet Neurol, Vol. 8(4), pp. 382-97.

[114]   Kishore, U.,(2003) Modular organization of the carboxyl-terminal, globular head region of human C1q A, B, and C chains. J Immunol, Vol. 171 (2), pp. 812-820.

[115]   Kishore, U,(2004) Structural and functional anatomy of the globular domain of the the complement protein C1q. Immunology letters, Vol. 95, pp. 113-128.

[116]   Kishore, U,(2004) C1q and tumor necrosis factor superfamily: modularity and versatility. Trends in Immunology, Vol. 25(10), pp. 551-561.

[117]   Kimberly, W.T., Esler, W.P. and Ye, W.(2003)Notch and the amyloid precursor protein are cleaved by similar gamma-secretase(s). Biochemistry, Vol. 42(1), pp. 137-144.

[118]   Peuschel, KE. (2000) Pitfalls in prion research. Medical Hypotheses, Vol. 54(5), pp. 698-700.

[119]   P Dandona, H Ghanim, P Mohanty, and A Chaudhuri. (2006) The Metabolic Syndrome: Linking Oxidative Stress and Inflammation to Obesity, Type 2 Diabetes, and the Syndrome;. DRUG DEVELOPMENT RESEARCH, Vol. 67, pp. 619–626.

[120] Morgan, BP and Harris, CL. (1999) Complement Regulatory Proteins. Academic Press.

[121] Mitchell, DA,(2007)Prion protein activates and fixes complement directly via the classical pathway: implications for the mechanism of scrapie agent propagation in lymphoid tissue., Molecular Immunology, Vol. 44(11), pp. 2997-3004.

[122] Maier, M,(2008)Complement C3 deficiency leads to accelerated amyloid beta plaque deposition and neurodegeneration and modulation of the microglia/macrophage phenotype in amyloid precursor protein transgenic mice. Journal of Neuroscience, Vol. 28(25), pp. 6333-41.

[123] MacKenzie, IR. (2000)Anti-inflammatory drugs and Alzheimer-type pathology in aging., Neurology, Vol. 54(3), pp. 732-4.

[124] Mackenzie, I.R. and Munoz, D.G. (1998)Nonsteroidal anti-inflammatory drug use and Alzheimer-type pathology in aging. Neurology, Vol. 50(4), pp. 986-990.

[125] Garred, P,(2006)Mannose-binding lectin and its genetic variants. Genes Immun., Vol. 7(2), pp. 85-94.

[126] Shen, K and Meyer, T. 1999, Dynamic Control of CaMKII Translocation and Localization in Hippocampal Neurons by NMDA Receptor Stimulation. Science, Vol. 284(5411), pp. 162 - 167.

[127] Ohmi, K,(2003) Activated microglia in cortex of mouse models of mucopolysaccharidoses I and IIIB. Proc Natl Acad Sci U S A., Vol. 100(4) pp. 1902-7.

[128] ADAPT Research group. (2007) Naproxen and celecoxib do not prevent AD in early results from a randomized controlled trial. Neurology, Vol. 68., pp. 1800-1808.

[129] Alzheimer, A. (1911)Uber eigenartige Krankheitsfalle des spateren. Ges Neurol Psych, Vol. 4, pp. 356-385.

[130] T. Fujita, M. Matsushita and Y. Endo(2004),, The lectin-complement pathway—its role in innate immunity and evolution, Immunol. Rev. 198 pp. 185–202

[131] P. Eikelenboom and R. Veerhuis, (1996) The role of complement and activated microglia in the pathogenesis of Alzheimer's disease, Neurobiol. Aging 17 (5), pp. 673–680

[132] D.A. Loeffler, D.M. Camp, M.B. Schonberger, D.J. Singer and P.A. LeWitt (2004), Early complement activation increases in the brain in some aged normal subjects, Neurobiol. Aging 25 (8), pp. 1001–1007

[133] V. Geylis and M. Steinitz (2006) Immunotherapy of Alzheimer's disease (AD): From murine models to anti-amyloid beta (Aβ) human monoclonal antibodies Autoimmunity 5 (1) pp. 33-39.

CHAPTER 8

The Roles of Platelet-Activating Factor (PAF) and its Related Signaling and Metabolism in Neurological Diseases

Yutaka Hirashima

Physiological Chemistry, Faculty of Pharmaceutical Sciences, Teikyo University, 1091-1 Sagamiko, Sagamihara, Kanagawa 229-0195, Japan

Abstract: Platelet-activating factor (PAF, 1-*O*-alkyl-2-acetyl-*sn*-glycero- 3-phosphocholine) displays a variety of biological activities in the nervous system. It has been suggested that PAF plays important roles in neuronal physiological function via activation of its specific membranes receptors. Under certain pathological conditions, PAF acts as a potent mediator of leukocyte functions, platelet aggregation, pro-inflammatory signaling and others. Therefore, PAF has been implicated in the pathophysiology of neuronal diseases such as ischemic stroke, hemorrhagic stroke, chronic subdural hematoma after head injury, brain tumor and associated brain edema, dementia due to neurodegenerative diseases such as Parkinson's disease and Prion diseases, and HIV-1-associated dementia. PAF is synthesized in platelet, monocytes/macrophage, neutrophils and endothelial cells in response to physiological and pathological stimuli through the *de novo* and remodeling pathways from cellular membrane phospholipids. PAF is thought to be a pro-inflammatory and pro-thrombotic mediator and also causes direct damage to neuronal cells. At least three types of platelet-activating factor acetylhydrolase (PAF-AH) have been identified in mammals, *i.e.*, intracellular type I and II, and a plasma type. The type I PAF-AH hydrolyzes the *sn*-2 ester bound in PAF-like phospholipids with a marked preference for very short acyl chains, typically acetyl bound. On the other hand, the type II PAF-AH has its substrate specificity similar to the plasma PAF-AH. Both PAF-AHs hydrolyze phospholipids with short to medium length *sn*-2 acyl chains including truncated ones derived from oxidative cleavage of long chain polyunsaturated fatty acyl groups. With respect to atherosclerosis it is not fully understood whether this enzyme plays an anti-atherogenic role or pro-atherogenic role. In this review, the roles of PAF, its signaling and related metabolism including PAH-AHs in a variety of pathological conditions in the central nervous system are discussed.

Keywords: platelet-activating factor (PAF), PAF acetylhydrolase (PAF-AH), PAF receptor, ischemic stroke, atherosclerosis, hemorrhagic stroke, subarachnoid hemorrhage (SAH), cerebral vasospasm, chronic subdural hematoma (CSH), meingioma, Parkinson's disease, Prion diseases, HIV-1-associated dementia (HAD)

INTRODUCTION

Platelet-activating factor (PAF:1-*O*-alkyl-2-acetyl-*sn*-glycero-3-phosphocholine) is a phsopholipid with potent, diverse bioactivities particularly as a inflammation mediator. It has contractile activity on vessel smooth-muscles and induces activation of platelets. Under pathological conditions, PAF becomes a pro-inflammatory and pro-thrombotic mediator and also causes direct damage to neuronal cells [1-4]. It is released by many types of cells including platelet, monocytes/macrophage, neutrophils and endothelial cells [5]. PAF is synthesized in these cells in response to physiological and pathological stimuli through the *de novo* and remodeling pathways from cellular membrane phospholipids. The *de novo* synthesis of PAF involves the direct transfer of a choline moiety to 1-*O*-alkyl-2-acetyl-*sn*-glycerol, In remodeling pathway, phospholipase A2 that hydrolyzes 1-*O*-alkyl-2-arachidonyl-*sn*-glycero-3-phosphocholine or the transfer of the sn-2 acyl moiery to a lysoplasmalogen to produce 1-*O*-alkyl-*sn*-glycero-3-phosphocholine (lysoPAF) initiates PAF synthesis. The second step in the synthesis of PAF is carried by acetyl-CoA:lysoPAF acetyltransferase [6]. Synthesized PAF acts as an intracellular signaling molecule in inflammatory, vascular and other tissue cells. Regulation and dysregulation of signaling by PAF may be involved in the occurrence of diseases. Its actions are mediated via the specific receptor, which belongs to the family of seven transmembrane-spanning G-protein-linked receptors. The degradation of PAF by PAF-acetylhydrolase (PAF-AH) (3.1.1.47) located intracellularlly or extrcellularlly is one of the regulation mechanisms regulating PAF level. PAF-AH is a Ca^{2+} independent enzyme and belongs to group VII phospholipase A2 family. PAF is hydrolyzed and converted to lysoPAF

*Address correspondence to: Physiological Chemistry, Faculty of Pharmaceutical Sciences, Teikyo University, 1091-1 Sagamiko, Sagamihara, Kanagawa 229-0195, Japan; Tel.:81-42-685-3743; Fax.:81-42-685-3745; E-mail: yhira@pharm.teikyo-u.ac.jp

by the catalytic reaction of this enzyme. To date, three isoforms of PAF-AH are identified: plasma PAF-AH, which has another nomenclature of lipoprotein-associated phospholipase A2 (Lp-PLA2) due to its characteristics of association with plasma lipoptroteins, and intracellular type I and II PAF-AHs.

Plasma PAF-AH is a monometric polypeptide with a molecular weight of 45 kDa. Type II PAF-AH has a molecular weight of 40 kDa and shares 41% sequence identity with plasma PAF-AH. However type I PAF-AH shows less structural similarity to them. In contrast with type I PAF-AH, plasma PAF-AH and type II PAF-AH preferentially hydrolyzed oxidized phospholipids produced by oxidative stress [7]. Plasma PAF-AH is suggested to have anti-inflammatory properties by the degradation of PAF, which plays a role in the activation of platelets and leukocytes. The human plasma PAF-AH is produced as a protein composed of 441 amino acid residues and is secreted into the blood after cleavage of a hydrophobic signal peptide at the cleavage position, Ala-17/Val-18 [7, 8]. The macrophages and Kupffer cells are suggested to be main source of this enzyme appeared in circulation [9,10].

On the other hand, plasma PAF-AH has been proposed as an inflammatory and pro-atherogenic marker of cardiovascular disease and stroke. These properties of the enzyme have been ascribed to its ability to hydrolyze oxidized phospholipids, leading to the generation of oxidized free fatty acids [11] and lysphosphatidylcholine (lysoPC). In this review, I focus on the previous and recent findings and discuss the role of PAF and related metabolism such as the contribution of PAF-AHs in the neurological diseases, including ischemic stroke, hemorrhagic stroke, chronic subdural hematoma, meningioma and associated brain edema, dementia due to neurodegenerative diseases, and HIV-1-associated dementia.

ISCHEMIC STROKE

Ischemic stroke can be caused by several different kinds of mechanisms. The most common mechanism is narrowing of arteries in the neck internal carotid artery or brain. If the arteries become too narrow, blood cells collect and form clots. When clots occlude the arteries where they are formed, it is called thrombosis. Thrombosis is mainly attributed to atherosclerosis, which provokes intravascular plaque formation. Cerebral thrombosis can also be divided into an additional two categories that correlate to the location of occlusion within the brain: large-vessel artery thrombosis which is occlusion of internal carotid artery or middle cerebral artery, while small-vessel thrombosis involves one or more of deep perforating artery. This latter type of stroke is also called a lacuner stroke. The clots formed at the narrowed artery dislodge and become trapped in peripheral arteries in the brain. This mechanism of ischemic stroke is called embolism. Another cause of embolism is blood clots in the heart, which can occur as a result of arrhythmia such as atrial fibrillation or abnormalities of the heart valves. Activation of circulating blood cells and dysfunction of endothelial cell also contribute to embolism. Rapid decreased blood flow occurs in pathology of embolism while collateral circulations have gradually developed before occlusion of artery in thrombosis. The damage of neuronal tissues is generally more severe in embolism. Ischemic stroke finally provides brain edema, total necrosis, namely, cerebral infarction, or delayed neural death due to vulnerability of neurons such as hippocampal neurons. PAF and type II PAF-AH may be involved in these events in ischemic stroke.

Atherosclerosis and Thrombogenesis

Atherosclerosis is a main mechanism of stenosis of arteries. The following events including oxidative modification of phospholipids, monocyte migration into the vessel wall, subsequent macrophage activation, and foam cell formation play roles in the pathogenesis of atherosclerosis [12]. These biological effects of PAF and some oxidized phospholipids suggest that plasma PAF-AH might play a protective role in pathogenesis of atherosclerosis by degradating such pro-inflammatory and pro-atherogenic phospholipids. A meaningful approach for evaluating the effect of the enzyme has been the use of recombinant forms of the enzyme or overexpression of plasma PAF-AH gene in experimental models of human diseases. Tjoelker and collaborators established directly that plasma PAF-AH has anti-inflammatory properties *in vivo* [7]. Overexpression of plasma PAF-AH by intravenous adenoviral transfer reduced neointima formation and spontaneous atherosclerosis [13]. Taking into the adverse effect of systematically increased plasma PAF-AH activity, local adenoviral gene transfer of the enzyme was performed. This approach greatly reduced oxidized LDL accumulation, shear stress-induced thrombosis and neointima formation without changing the plasma level of PAF-AH activity or titers of autoantibodies to oxidized LDL [14].

Table 1: PAF and its Related Events in Ischemic Stroke

Diseases		References
Ischemic stroke		
Atherosclerosis	[Plasma PAF-AH as a protective factor against ischemic stroke]	
	Overexpression of plasma PAF-AH by intravenous adenoviral transfer reduced neointima formation and atherosclerosis.	[13]
	Local adenoviral gene transfer of plasma PAF-AH reduced oxidized LDL accumulation, shear stress-induced thrombosis and neointima formation.	[14]
	[Plasam PAF-AH as a risk factor of ischemic stroke]	
	Subjects with both high CRP and high plasma PAF-AH levels had the highest risk of stroke.	[19, 25]
	Risk of ischemic stroke increased 2-fold in patients with the highest plasma PAF-AH levels compared to patients with the lowest levels	[11]
	Plasma PAF-AH was an independent predictor of ischemic stroke among postmenopausal women.	[18]
	Plasma PAF-AH activity and mass were independent risk factors of ischemic not to cardiac heart disease.	[26]
	Upregulation of mRNA of plasma PAF-AH occured in a mouse atherosclerosis model.	[27]
Formation of intravascular and intracardial clots PAF level and PAF receptor in ischemic brain	PAF induced expression of adhesion molecules (ICAM-1, VCAM-1), monocyte chemoatractant protein-1, and colony-stimulating factor.	[28]
	PAF levels are increased in response to ischemia at early stage of reperfusion.	[29, 30]
	PAF receptor mRNA was reduced after ischemia and restored at late stage of reperfusion in the perifocal regions of cerebral infarction .	[35]
	PAF receptor antagonists prevented ischemia-induced CNS damage in animal models.	[41, 42]
	PAF induced dose-dependently neuronal death and, PAF antagonists and anti-PAF IgG attenuated glutamate neuronal death in primary neuron culture.	[4]
	Recombinant plasma type PAF-AH was effective for preventing against NMDA induced hippocampal neuronal apoptosis.	[43]
	Overexpression of plasma PAF-AH in cultured neurons decreased apoptosis.	[45]
	Transgenic mice with overexpression of human type II PAF-AH specific in neurons attenuated brain edema, cerebral infarction and hippocampal apoptosis in a middle cerebral artery occlusion-reperfusionmodel.	[46]

PAF; platelet-activating factor, PAF-AH; PAF-acetylhydrolase, CRP; C-reactive protein, ICAM-1; intracellular adehesion molecule-1, VCAM-1; vascular cell adhesion molecules-1, CNS: central nervous system, NMDA; N-methyl-D-aspartate

Traditional risk factors for atherosclerosis include obesity, type 2 diabetes mellitus, hypertension, the metabolic syndrome, hypertriglyceridemia, a low level of HDL cholesterol and smoking. Recently a large number of markers of inflammation have been investigated whether they are associated with cardiovascular diseases. The most promising marker seems to be C-reactive protein (CRP) [15]. Serum amyloid A, secretory group II phospholiase A2 (sPLA2), apolipoprotein J, and PON-1 are candidate markers for prediction of atherosclerotic diseases [15]. By contrast, there is increasing interest in plasma PAF-AH as a new risk factor of arteriosclerosis. This enzyme can also product lyso-PC and oxidatively modified non-esterified fatty acids which could promote the pathogenesis of atherosclerosis. In human plasma, PAF-AH circulates in association with low (LDL) and high-density lipoprotein (HDL). About 70-80% of the total activity is present with LDL and 20-30% binds to HDL [16]. Plasma PAF-AH has been shown to be expressed more in the necrotic core of advanced atherosclerotic plaque compared with less advanced lesions [17]. Therefore, the enzyme may be a marker or mediator of unstable plaque [18]. Increased circulating plasma PAF-AH levels have been shown to be independent predictors of coronary heart disease in the Atherosclerosis Risk in Communities Study [19], West of Scotland Coronary Prevention Study [20], and the Rotterdam Study [11], as well as in some other studies [21-24]. Although enrolled numbers of subjects have generally been modest, the data on the association between plasma PAF-AH and risk of future stroke have been accumulated. The Atherosclerosis Risk in Communities Study showed the highest risk of stroke in subjects with both high CRP and high plasma PAF-AH levels [19] and stroke prediction was most enhanced when both markers were combined in individuals who were intermediate risk for ischemic stroke by traditional risk factors [25]. In the Rotterdam Study, multivariate-adjusted hazard ratios for 4 groups regarding plasma PAF-AH activity showed that 2-fold increase in risk of ischemic stroke among subjects with the highest quartile of plasma PAF-AH levels compared to the lowest quartile [11]. The recent data from the Hormones and Biomarkers Predicting Stroke (HaBPS) Study, a case-control study of 972 stroke subjects and their matched control subjects nested in the Women's Health Initiative (WHI) Observational Study has identified plasma PAF-AH as an independent predictor of ischemic stroke among postmenopausal women without using hormone therapy, whereas there was no significant association among subjects with using hormone use [18]. Malmo Diet Cancer Study showed elevated plasma PAF-AH activity and mass were independent risk factors related to the incidence of ischemic stroke not to that of cardiac heart disease [26].

Regarding to the atherosclerosis, it is not fully understood whether plasma PAF-AH plays an anti-atherosclerotic role or pro-atherogenic role. Its expression is upregulated at the transcriptional level in a mouse atherosclerosis model [27]. These observations favor the hypothesis that the plasma PAF-AH plays the pro-atherogenic role. On the other hand, some researcher thinks that this mechanism is likely responsible for the up-regulation by mediators of inflammation during atherosclerosis and suggests that increased expression of this enzyme is a physiological response to inflammatory stimuli [16].

Although atrial fibrillation or abnormalities of heart valves are basically important for thrombogenesis, hypercoagulation and activation and aggregation of circulating blood cells also contribute to formation of circulating clots. PAF induces expression of adhesion molecules such as intracellular adhesion molecule-1(ICAM-1) and vascular cell adhesion molecules-1(VCAM-1), chemotaxis stimulants such as monocyte chemoatractant protein-1, and colony-stimulating factor [28]. These substances might play an important role on pathophysiology of embolism by forming intravascular or intracardial clots.

PAF Levels and PAF Receptors in Ischemic Brain

PAF levels are increased in gerbil brains in response to ischemia at early stage of reperfusion. It has been established that PAF plays an important role following cerebral ischemia-reperfusion [29, 30]. After middle cerebral artery occlusion in rat, PAF concentrations of simply ischemic brain and early reperfusion brain were also demonstrated to be significant higher in the perifocal regions of cerebral infarction. It has been shown that the *de novo* pathway should mainly contribute to PAF synthesis for maintaining its basal levels under physiological conditions. The remode- ling pathway should be more involved in the production of PAF during ischemia. During reperfusion, the overproduction of PAF should be the result of the concomitant activation of both pathways [31]. The biological effect of PAF is produced via the PAF receptor. Honda *et al.* cloned and expressed the PAF receptor in pulmonary tissues in guinea pigs [32]. The PAF receptor is a member of the G protein-coupled receptor Family, and is expressed ubiquitously in the brains of all studied animals [33, 34]. PAF receptor-dependent stimulation of numerous signal transduction has been reported in variety of cells. Zhang X *et al.* analyzed this in nervous tissue of perifocal infarct regions after middle cerebral artery occlusion. Reverse transcription-polymerase chain reaction (RT-PCR) revealed that PAF receptor mRNA levels were reduced in the perifocal regions of cerebral infarction after a sharp increase in PAF concentration. PAF receptors may be subject to a downward regulation event. At late stage of reperfusion, the release of PAF declines followed by a gradual restoration of PAF receptor gene expression levels [35]. This PAF receptor downregulation are thought to be performed by both receptor internalization [36] and decreased expression of PAF receptor mRNA. The transcription activity of the PAF receptor is regulated by cyclic adenosine monophosphate (cAMP) and protein- kinase C, which are upstream messengers of NF-κB [36-38]. The increased levels of certain cAMP and PKC subunits in ischemic brain are speculated to responsible to reduce the gene transcription rate of the PAF receptor [35].

The current research suggests that PAF may contribute to the damage of the central nervous system after cerebral ischemia [39, 40]. The role of PAF in ischemic injury in the central nervous system (CNS) has been initially proposed based on the results showing that selective PAF receptor antagonists prevented ischemia-induced CNS damage in animal models [41]. Until now, the neuroprotective effects of a novel PAF receptor antagonist in focal cerebral ischemia in animals has been introduced by neurobehavioral and histological assessments [42]. In neurons in culture, the role of PAF was also evaluated in many researchers. We showed that PAF induced dose-dependently neuronal death and PAF receptor antagonists and anti-PAF IgG attenuated glutamate neuronal death in primary rat neuron culture [4]. Although it is unknown whether oxidized phospholipids or PAF itself is a main target, degradation enzyme of PAF, recombinant plasma type PAFAH, was recognized to be effect for preventing against NMDA induced hippocampal neuronal apoptosis [43]. Although the mechanism of neuronal injury by PAF is still obscure, PAF modulates neuronal calcium ionization through the PAF receptor and also enhances the synthesis and release of IL-6, IL-8, IL-10, TNF-α, and other mediators of the inflammatory response. Overall, PAF is a potent neuronal injury messenger [44].We also transfected the plasma type PAF-AH gene in cultured rat neuron and found a decrease in glutamate-induced injury, mainly, apparent as decreased apoptosis [45]. Transgenic mice with over- expression human type II PAF-AH specific in neurons were generated and the brain edema, cerebral infarction and hippocampal neuronal apoptosis in brain subjected to ischemia-reperfusion after middle cerebral artery occlusion were estimated. The brain edema, infarction volume, and numbers of apoptotic cells were all significantly lower in transgenic mice than in wild-type mice [46]. Intracellular overexpression of PAF-AH may mainly attenuate oxidized phospholipids induced by oxidative stress in the membranes of intracellular organelle, restore energy metabolism and suppress the activation of signal pathway of apoptosis [46]. It was known that PAF activates early response genes [47], therefore intracellular receptor was believed to be the signal linkage to the nucleus. Recently, the intracellular PAF receptors were characterized. One form is confined to the endosomes, and the other is confined to the nuclear membrane [48, 49]. Whether these receptors take part of transcriptional action of PAF is of great interest.

HEMORRHAGIC STROKE

Intracerebral hemorrhage (ICH) and subarachnoid hemorrhage (SAH) are main subtypes of hemorrhagic stroke. ICH, the second most common cause of stroke includes intraparenchymal hemorrhage and intraventricular hemorrhage due to the penetration of intraparenchymal hemorrhage into ventricles. It is caused by the bleeding from perforators in the brain. Intraparenchymal and intraventricular hemorrhage due to the ruptures of perforating arteries are more common in adults. Although, hypertension is the highest risk factor of ICH, but atherosclerosis, type 2 diabetes mellitus, hypertriglyceridemia, a low level of HDL cholesterol and smoking have also been thought to be involved in the pathogenesis of ICH. Few studies have examined the association between PAF and its related metabolism, and risk of ICH. It has been reported that low LDL cholesterol levels are associated with elevated risk of death due to intraparenchymal hemorrhage. These associations were not altered substantially after adjustment for known risk factors. This recent Japanese study included 91,219 subjects with no history of stroke or coronary heart disease for completing a base line risk factor survey [50]. Arteriosclerosis of intracerebral arteries in the basal ganglia, thalamus, and brain stem is characterized by angionecrosis [51, 52]. A hypercholesterolemia diet reduces angionecrosis and prevents occurrence of hemorrhagic stroke among spontaneously hypertensive rats [53]. The LDL levels correlate positively with plasma PAF-AH activity [54]. These results suggest that low PAF-AH activity may play important roles in occurrence of ICH and death due to ICH, although the association with low LDL level and increased the prevalence of ICH or mortality have not been evaluated.

SAH is most often caused by the rupture of cerebral aneurysms. When a cerebral aneurysm ruptures and blood fills the subarachnoid space, the area between the arachnoid membrane and the pia matter surrounding the brain. Symptoms of SAH include a severe headache with a rapid onset and vomiting. The cause of cerebral aneurysms is not well understood. They may develop from birth or in childhood and grow very slowly. Cerebral vasospasm, in which the main trunk arteries of brain constrict and thus restrict blood flow about one week after SAH, is a serious complication of SAH. The precise mechanism underlying cerebral vasospasm is not known. The cascade of events leads to abnormal constriction of the artery begins with oxyhemoglobin, a breakdown product of red blood cells. Oxyhemoglobin generates reactive oxygen species (ROS) such as superoxide. ROS damage cells constructing blood vessel wall including endothelial cells, smooth muscle cells, and adventitial fibroblast. Physiological relaxation and contacting function is considerably disturbed. Key mediators which have been implicated in the functional component of cerebral vasospasm include the vasodilators and vasoconstrictors. The former including nitric oxide (NO) and prostacyclin (PGI2) become underactive, and the latter including endothelin-1 (ET-1) and thromboxane A2 (TXA2) become overactive [55-57]. There are also structural changes due to inflammatory reaction in arteries. The vessel wall is invaded by white blood cells, while the smooth muscle layers become thickened, myoproliferation, and the adventitial and smooth muscle layers can become more stiff or fibrotic [58, 59]. These changes are similar to the atherosclerosis. We previously reported that PAF concentrations in cerebrospinal fluid (CSF) and plasma increased during the period when cerebral vasospasm typically occurs [60, 61]. Furthermore, PAF concentrations in the CSF and plasma of patients with cerebral infarctions caused by cerebral vasospasm were likely to be higher than those in patients without cerebral infarction [60, 61]. Intrathecal PAF injection aggravated vasospasm, and PAF receptor antagonists prevented cerebral vasospasm in a rabbit model of basilar artery constriction and neurological deterioration [62].

Table 2: PAF and its Related Events in Hemorrhagic Stroke, Chronic Subdural Hematoma and Brain Tumor

Diseases		References
Hemorrhagic stroke		
Cerebral vasospasm after SAH	PAF concentrations in the CSF and plasma were increased during the when cerebral vasospasm typically occurs.	[60,61]
	PAF concentrations in the CSF and plasma of patients with cerebral infarctions caused by cerebral vasospasm was higher than those in patients without cerebral infarction.	[60,61]
	Intrathecal PAF injection aggravated vasospasm, and PAF receptor antagonists prevented cerebral vasospasm in a rabbit model the highest risk of stroke.	[62]
	There was a tendency of the higher activity of PAF-AH in CSF of SAH patients without infarction than those of that of patients with cerebral infarction.	[60]
	A PAF receptor antagonist, E5880, was safe and effective in the treatment of patients with cerebral vasospasm due to SAH.	[65]
CSD	PAF stain was observed predominantly around sinusoidal vessels of hematoma capsule.	[67]
	A patient taking Ginkgo biloba including a inhibitor of PAF, Ginkgolide B, developed CSD.	[68]
Brain tumor	PAF receptor transcripts, PAF, phospholipase A2 activity, and PAF-AH activity were detected in meningiomas.	[71]
	PAF content and leukocyte infiltration in meningiomas correlated with edema index of tumors.	[75]

SAH; subarachnoid hemorrhage; CSF; cerebrospinal fluid, CSD; chronic subdural hematoma

These results strongly suggest that PAF contributes to the pathogenesis of vasospasm following SAH. These results suggest that low PAF-AH activity may play important roles in occurrence of ICH and death due to ICH, although the association with low LDL level and increased the prevalence of ICH or mortality have not been evaluated.

Not only the constriction of main trunk arteries but also functions of peripheral small arteries suggested being important for ischemic symptoms in patients with SAH. PAF may play roles as mediators regulating autoregulation of CBF and BBB function [63]. The total PAF-AH activity in CSF of SAH was originated from plasma [15, 64]. There was a higher tendency of the activity of PAF-AH in CSF of patients without cerebral infarction than those of patients with cerebral infarction [60]. Although, the histological changes of arteries in patients with cerebral vasospasm resemble closely those of arteriosclerosis, but the enzyme activity was rather lower in patients with severe cerebral vasospasm in contrast with many epidemiological studies regarding atherosclerosis. Open clinical trial to investigate efficacy and safety of PAF receptor antagonist, E5880, was performed in 71 patients with SAH who underwent surgery for ruptured aneurisms within 3 days. These results suggested that E5880 is safe and effective in the treatment of patients with cerebral vasospasm due to SAH [65].

CHRONIC SUBDURAL HEMATOMA

A chronic subdural hematoma (CSH) is an old collection of blood and blood breakdown products between arachnoid membrane and dura. Hematoma forms capsule in subdural space. A hematoma develops gradually when the tiny bleeding occurs on the surface of brain usually due to head injury. Although the mechanism of development of CSH is still unknown, repeated leakage of blood from sinusoidal vein vessels into the hematoma cavity may play an important role. It has been postulated that inflammatory reactions are an important stimulus for the formation of capsule and the growth of CSH [66], and eosinophil infiltration into the capsule may contribute to bleeding into the hematoma cavity through the secretion of plasminogen-rich granules. Using fluorescent antibody method, staining of PAF was observed predominantly around sinusoidal vessels on the hematoma cavity side. These results suggest that PAF may play a role in the development of CSH [67]. By contrast, a case report, in which a patient taking Ginkgo biloba showed that in CSH, speculated that PAF, Ginkgolide B, a PAF inhibitor may responsible to occurrence of the hematoma [68].

Table 3: PAF and its Related Events in Dementia

Diseases		References
Dementia		
Neurodegenerative diseases	Neurons pre-treated with PAF antagonists are resistant to toxic peptidefound in AD. PAF antagonists reduce the caspase-3 activity and production pf prostaglandin E2.	[82]
	PLA2 inhibitors and PAF inhibitors protect neurons against toxic effect of HuPrP82-146 and SPrP106.	[82]
HAD	The combination of PAF antagonist and MMP and TNF-α inhibitor reduces brain inflammation, astrogliosis and microglia activation in a mouse model of HIV-1 encephalitis.	[86]
	Elevated level of PAF is observed in HIV-1-infected monocytes and CSF of patients with HAD.	[88]
LTP	PAF is a retrograde messenger in LTP, and upregulates memory formation	[92,93,94]
	Both incidence and size of LTP are attenuated in PAF receptor-knockout mice and PAF receptor antagonists reduce LTP in wild type mice.	[95]
the others	PMS777 with both functions of PAF receptor antagonist and acetylcholinesterase inhibitor reverses scopolamine-induced memory retrieval deficits in mice.	[91]

AD: Alzheimer's disease, PLA2; phospholipase A2, MMP; metalloproteinase, TNF-α; tumor necrosis factor-α, HIV-1; human immunodeficiency virus type 1, HAD; HIV-associated dementia, LTP; long-term potentiation

BRAIN TUMOR

A brain tumor is an abnormal growth of cells within the brain or inside the skull, which can divided into malignant tumor and benign tumor. Metastatic tumor from other organs also is included in brain tumor. PAF is a lipid mediator that stimulates the *in vitro* growth of various human tumor cell lines and that enhanced the effect of vascular endothelial growth factor that play a role during angiogenesis of human cancer. The group of Denizot Y has vigorously investigated the role of PAF in various tumors [69-71]. They also examined the presence of PAF receptor transcripts, the level of PAF, the phospholipase A2 activity, and the PAF-AH activity in human meningiomas [71]. Meningiomas are the most common tumor of the central nervous system, arising from arachnoid cap of the arachnoid villi in the meninges [72]. These tumors are usually benign in nature; however, they can be malignant [73]. Peritumoral edema is characteristic in this tumor. PAF receptor transcripts, PAF, phospholipase A2 activity, and PAF-AH activity was detected in the tumor, although their levels did not correlated with clinicopathological parameters such as brain tumor grade, edema size, necrosis, mitotic index, degree of neovasculaization and chronic inflammatory response [71]. However, this study concluded that PAF might act on tumor growth by altering the local angiogenic and/or cytokine networks. There are two major types of brain edema, cytotoxic (intracellular) edema and vasogenic (extracellular) edema [74]. Brain tumors are surrounded by vasogenic edema which results from sprouting of new, immature blood vessels with an incomplete blood brain barrier. Among the mediators involved is vascular endothelial growth factor, also known as vascular permeability factors, is far more potent than histamine in inducing capillary permeability [74]. We obtained positive data that PAF content and leukocyte infiltration in meningiomas correlated with edema index of tumors [75].

DEMENTIA CAUSED BY NEURODEGENERATIVE DISEASES AND HIV-1-ASSOCIATED DEMENTIA

Dementia can be caused by a variety of diseases, known as neurodegenerative diseases resulting from protein aggregation in the brain [76]. These diseases include Alzheimer's disease (AD), dementia with Lewy bodies, Huntington's disease and Parkinson's disease [76]. The causes and progression in most of them are not well understood. Infectious disease affecting the central nervous system may lead to dementia. These infections can be caused by different agents such as: abnormal protein in prion diseases, bacteria in syphilis and borrelia, parasites in toxoplasmosis, cryptococcosis and neurocysticercosis [77], however viral agents are the leading cause of infection related dementia. Recently, the role of PAF in pathogenesis of AD and prion diseases has been suggested. AD is the most common form of dementia. Cognitive impairment such as recent memory disturbance and disorientation is recognized in early stages. As the disease advances symptoms aggravate and new symptoms including confusion, irritability and aggression, dysarthria, and long-term memory disturbance are also added. Gradually, bodily functions and activities of daily living are lost, ultimately leading to death. AD arise following the degeneration and subsequent loss of neurons. Both senile plaques and neurofibrillary tangles are histological characteristic findings of AD [78]. Plaques are dense, mostly insoluble deposits of amyloid-β peptides and cellular material outside and around neurons.

Neurofibrillary tangles are aggregates of the microtubule-associated protein tau which has become hyperphosphorylated and accumulate inside the cells themselves. Amyloid-β peptides are derived from the cleavage of the amyloid precursor protein by γ-secretases [79]. Reduction in the activity of the cholinergic neurons is a well-known feature of AD [80]. Acetylcholinesterase inhibitors are employed to reduce the rate at which acetylcholine(ACh) is broken down, thereby increasing Ach level in the brain and combating the loss of Ach due to the degradation of cholinergic neurons [81]. Standard techniques to study the mechanisms of neuronal death *in vitro* include addition of peptides derived from amyloid-β peptides to neuronal cultures. Neurons pre-treated with PAF antagonists were resistant to toxic peptide found in AD, amyloid-β 1-42. Moreover, PAF antagonists reduced the caspase-3 activity, a marker of apoptosis, and production of prostaglandin E2 that is closely involved in neuronal degeneration in AD [82]. PL A2 inhibitors showed same results. The activation of PLA2 leads to the synthesis of PAF in neurons by the way of the remodeling pathway [6]. The role of PAF in neurodegeneration was confirmed in this study [82]. Prion diseases or transmissible spongiform encephalopathyies (TSEs) are included in a family of rare progressive neurodegenerative disorders that affect both humans and animals such as Creutzfeldt-Jakob disease and bovine spongiform encephalopathy. They are distinguished by long incubation periods, characteristic spongiform changes associated with neuronal loss. The causative agent of TSEs is believed to be a prion. A prion is an abnormal transmissible agent that is able to induce abnormal folding of normal cellular prion proteins in the brain, leading to brain damage. In order to study the mechanism of neuronal loss in vitro, the addition of peptides derived from the prion protein (PrP) to neuronal cultures is performed. PLA2 inhibitors and PAF inhibitors also protected neurons against toxic effect of HuPrP82-146 and sPrP106. The role of PAF in neurodegeneration in prion diseases was also confirmed [82]. Among the viruses infecting the brain, human immunodeficiency virus type 1 (HIV-1) is the most common cause dementia. Although highly active antiretroviral therapy (HAART) became available, neuronal cell degradation remains a problem that is frequently found in the brains of HIV-1-infected patients [83]. In the absence of opportunistic infections, major symptoms including memory disturbance, disorientation, mental deterioration, weakness of extremities, gait disturbance are appearing [84, 85]. The terms AIDS dementia complex (ADC), and HIV-associated dementia (HAD) are used to described these neurological and psychiatric symptoms [84, 85]. The HIV-1 associated neuropathology is characterized by the infiltration of macrophages into the central nervous system; the formation of microglial nodules; multinucleated giant cells due to virus-induced fusion of microglia and/or macrophages; astrocyte activation and gliosis; neuronal loss in hippocampus and basal ganglia. Using a mouse model of HIV-1 encephalitis in which HIV-1-infected human monocyte-derived macrophage (MDM) was inoculated into basal ganglia, the effects of combination of PAF antagonist and metalloproteinase (MMP) and tumor necrosis factor-α (TNF-α) inhibitor were evaluated. The drugs markedly reduced brain inflammation, astrogliosis and microglia activation [86]. This observation helps to confirm the hypothesis that HAD is a metabolic encephalopathy associated with neuro- toxin production in infected macrophage [87]. The result reported also demonstrated that anti-inflammatory drugs (PAF and/or MMP/TNF-α inhibitors) can diminish macrophage neurotoxin secretions and the intensity of HAD. Elevated levels of PAF have been observed in HIV-1-infected monocytes and cerebrospinal fluid of patients with HAD [88]. PAF is probably the key element in TNF-α-induced neurotoxicity, since TNF-α-triggered neuronal apoptosis could be decreased by PAF-AH or PAF receptor antagonists [89]. Li J *et al.* reported a novel tetrahydrofuran-derived bis-interacting ligand, PMS777, for PAF receptor antagonism and acetylcholinesterase inhibition [90]. It would be of great interest in the therapeutic potential of dementia in AD, prion diseases and HAD [90]. Practically, PMS777 dose-dependently inhibited PAF-induced platelet aggregation and reversed scopolamine-induced memory retrieval deficits in mice [91]. It alleviated PAF-induced cell apoptosis in SH-SY5Y cells with inhibiting intracellular Ca2+ overload, downregulation of anti-apoptotic bcl-2 mRNA, stimulation of pro-apoptotic bax mRNA expression and activation of caspase-3 pathway [90]. However, effects of PAF antagonists on cognitive and memory functions showed inconsistency. PAF is a retrograde messenger in long-term potentiation (LTP) [92], and upregulates memory formation, both in the hippocampus [93] and in the stratum [94]. Both incidence and size of LTP were attenuated in PAF receptor-knockout mice and PAF receptor antagonists also reduced LTD in wild type mice [95].

CONCLUSIONS

There are numerous evidences that PAF plays important roles not only in physiological condition, but also in pathological conditions in CNS. PAF acts as a potent mediator for pro-inflammatory and pro-thrombotic signalings via activation of its specific membrane receptors, PAF receptors. By contrast, it remains obscure whether plasma PAF-AH plays a pro-atherogenic role or anti-atherogenic role in the onset of atherosclerotic diseases such as ischemic strokes due to thrombosis. Effects of PAF receptor antagonists on cognitive and memory function are likewise inconsistent among many concerned studies. The solution of the inconsistency could promote rPAF-AH or pharmacological agents such as PAF receptor antagonists or PAF-AH inhibitors according to circumstances to use in the clinical setting of various neurological diseases.

ACKNOWLEDGEMENT

This work was supported in part by a grant from the Japanese Ministry of Education, Culture, Sports, Science, and Technology, No. 20591696.

REFERENCES

[1]   Shaw, J.O.; Pickard, R.N.; Ferrigni, K.S.; McManus, L.M., Hanahan, D.J. *J. Immunol.*, **1981**, *127*,1250.

[2]   Libby, P.; Hansson, G.K. *Lab. Invest.*, **1991**, *64*, 5.

[3]   Prehn, J.H.; Krieglstein, J. *J. Neurosci. Res.*, **1993**, *34*, 179.

[4]   Nogami, K.; Hirashima, Y.; Endo, S., Takaku, A. *Brain Res.*, **1997**, *754*, 72.

[5]   Bazan, N.G., *J. Lipid Res.*, **2003**, *44*, 2221.

[6]   Zimmerman, G.A.; McIntyre, T.M.; Prescott, S.M.; Stafforini, D.M. Cri. Care Med., **2002**, *30*, S294.

[7]   Tjoelker, L.W.; Wilder, C.; Eberhardt, C.; Stafforini, D.M.; Dietsch, G.; Schimpf, B.; Hooper, S.; Trong, H.L.; Cousens, L.S.; Zimmerman, G.A.; Yamada, Y.; McIntyre, T.M.; Prescott, S.M.; Gray, P.W. *Nature*, **1995**, *374*, 549.

[8]   Tjoelker, L.W.; Eberhardt, C.; Unger, J.; Trong, H.L. ; Zimmerman, G.A.; McIntyre, T.M.; Stafforini, D.M.; Prescott, S.M.; Gray, P.W. *J. Biol. Chem.* **1995**, *270*, 25481.

[9]   Asano, K.; Okamoto, S.; Fukunaga, K.; Shiomi, T.; Mori, T.; Iwata, M.; Ikeda, Y.; Yamaguchi, K.; *Res. Commun.* **1999**, *261*, 511.

[10]   Stafforini, D.M.; Elstad M.R.: McIntyre, T.M.; Zimmerman, G.A. *J. Biol. Chem.* **1990**, *265*, 9682.

[11]   Oei, H-H.S.; van der Meer, I.M.; Hofman, A.; Koudstaal, P.; Stijnen, T., Breterler M.M.B.; Witteman J.C.M. *Circulation* **2005**, *111*, 570.

[12]   Witztum, J.L.; Steinberg, D. *J. Clin. Invest.* **1991**, *88,* 1785.

[13]   Quarck, R.; De Geest, B.; Stengel, D.; Mertens, A.; Lox, M.; Theilmeier..; Michiel, C.; Raes, M.; Bult, H.; Collen, D.; Van Veldhoven, P.; Minio, E.; Holvoet, P. Circulation **2001**, *103*, 2495.

[14]   Arakawa H.; Qian, J.Y.; Baatar, D.; Karasawa, K.; Asada, Y.; Sasaguri, Y.; Miller, E.R.; Witzum, J.L.; Ueno, H. *Circulation* **2005**, *111,* 3302.

[15]   Karasawa, K. *B. B. A.,* **2006**, *1761*, 1359.

[16]   Stafforini, D.M. *Cardiovasc. Drugs Ther.* **2009**, *23*, 73.

[17]   Kolodgeie, F.D.; Burke, A.P.; Skorija, K.S; Ladich, E.; Kutys, R.; Makuria, A.T.; Virmani, R. *Artherioscler. Thromb. Vasc.Biol.* **2006**, *26,* 2523.

[18]   Wassertheil-Smoller, S.; Kooperberg, C.; McGinn, A.P.; Kaplan, R.C.; Hsia, J.; Hendrix, S.L.; Mnson, J.E.; Berger, J.S.; Kuller, L.H.; Allison, M.A.; Baird, A.E. Hypertemsion **2008**, *51*, 1115.

[19]   Ballantyne, C.M.; Hoogeveen, R.C.; Bang, H.; Coresh, J.; Folom, A.R.; Chambless, L.E.; Myerson, M.: Wu, K.K.; Sharrett, A.R.; Boerwinklw, E. *Arch. Intern. Med.* **2005**, *165*, 2479.

[20]   Packard, C.J.; O'Reilly, D.S.;Caslake, M.J.; McMahon, A.D.; Ford, I.; Cooney, J.; Macphee, C.H.; Suckling, K.E.; Krishna, M.; Wilkinson, F.E.; Rumley, A.; Lowe, G.D. *N. Engl. J. Med.* **2000**, *343*, 1148.

[21]   O'Donoghue, M.; Morrow, D.A.; Sabayine, M.S.; Murphy, S.A.; McCabe, C.H.; Cannon, C.P.; Braunwald, E.. *Circulation* **2006**, *113*, 1745.

[22]   Persson, M.; Hedblad, B.; Nelson, J.J.; Berglund, G. *Artherioscler. Thromb. Vasc.Biol.* **2007**, *27,* 1411.

[23]   Koenig, W.; Khuseyinova, N.; Lowel, H.; Trischler, G.; Meisinger, C. *Circulation* **2004**, *110*, 1903.

[24]   Garza, C.A.; Montori, V.M.; McConnell, J.P.; Somers, V.K.; Kullo, I.J.; Lopez-Jimenez, F, *Mayo Clin. Proc.* **2007**, *82*, 159.

[25]   Nambi, V.; Hoogeveen, R.C.; Chambless, L.; Hu, Y.; Bang, H.; Coresh, J.; Ni, H.; Boerwinkle, E.; Mosley, T.; Sharrett, R.; Folsom, A.R.; Ballantyne C.M. *Stroke* **2009**, *40*, 376.

[26]   Persson, M.; Berglund, G.; Nelson, J.J.; Hedblad, B. *Atherosclerosis* **2008**, *200*, 191.

[27]   Singh, U.; Zhong, S.; Xiong, M., Li, T.B.; Sniderman, A.; Teng, B.B. *Clin. Sci.* **2004**, *106*, 421.

[28]   Collins, T.; Cybulsky, M.I. *J. Clin. Invest.* **2001**, *107*, 255.

[29]   Prescott, S.M.; Zimmerman, G.A.; Stafforini, D.M.; McIntyre, T.M. *Annu. Rev. Biochem.* **2000**, *69,* 419.

[30]   Stafforini, D.M.; McIntyre, T.M.; Zimmerman, G.A.; Prescott, S.M. *Crit. Rev. Clin. Lab. Sci.* **2003**, *40*, 643.

[31]   Francescangeli, E.; Boila, A.; Goracci, G. *Neurochem. Res.* **2000**, *25*, 705.

[32]   Honda, Z.; Nakamura, M.; Miki, I.; Minami, M.; Watanabe, T.; Seyama, Y.; Okado, H.; Toh, H.; Ito, K.; Miyamoto, T.; Shimizu, T. *Nature* **1991**, *349*, 342.

[33]   Bito, H.; Nakamura, M., Honda, Z.; Izumi, T.; Iwatsubo, T.; Seyama, Y., Ogura, A.; Kudo, Y.; Shimizu, T. *Neuron*, **1992**, *9*, 285.

[34]   Mori, M.; Aihara, M.; Kume, K.; Hamanoue, M.; Kohsaka, S.; Shimizu, T. *J. Neurosci.* **1996**, *16*, 3590.

[35]   Zhang, X.; Pan, X.-L.; Liu, X.-T.; Wang, S.; Wang, L.-J. *Neurochem. Res.* **2007**, *32*, 451.

[36]   Dupre, D.J.; Chen, Z.; Le Gouill, C. ; Theriault, C. ; Parent, J.L. ; Rola-Pleszczynski, M. ; Stankova, J. *J. Biol. Chem.* **2003**, *278*, 48228.

[37]   Nakao, A.; Watababe, T.; Bitoh, H.; Imaki, H.; Suzuki, T.; Asano, K., Taniguchi, S.; Nosaka, K.; Shimizu, T.; Kurokawa, K. *Am. J. Physiol.,* **1997**, 273, F445.

[38]   Zhuang, Q.; Bastien, Y.; Mazer, B.D. *J. Immunol.* **2000**, *165*, 2423.

[39]   Xia, S.H.; Fang, D.J. *Chin. Med. J.* **2007**, *120*, 922.

[40]   Umemura, A.; Yamada, K.; Mabe, H.; Nagai, H. *J. Stroke Cerebrovasc. Dis.* **1997**, *6,* 394.

[41]   Spinnewyn, B.; Blavet, N.; Clostre, F.; Bazan, N.G..; Braquet, P. *Prostaglandins,* **1987**, *34*, 337.

[42]   Belayev, L.; Khoutorova, L.; Atkins, K.; Cherqui, A. ; Alvarez-Builla, J. ; Bazan, N.G.. *Brain Res.,* **2009**, *1253*, 184.

[43]   Ogden, F.; DeCoster, M.A.; Bazan, N.G.. *J. Neurosci. Res.,* **1998**, *53*, 677.

[44]   Bazan, N.G.. *Mol. Neurobiol.,* **2005**, *32*, 89.

[45]   Hirashima, Y.; Ueno, H.; Karasawa, K.; Yokoyama, K.; Setaka, M.; Takaku, A. *Brain Res.,* **2000**, *885*, 128.

[46]   Umemura, K.; Kato, I.; Hirashima, Y.; Ishii, Y.; Inoue, T.; Aoki, J.; Kono, N.; Oya, T.; Hayashi, N., Hamada, H.; Endo, S.; Oda, M.; Arai, H.; Kinouchi, H.; Hiraga, K. *Stroke,* **2007**, *38*, 1063.

[47]   Bazan, N.G.; Squinto, S.P.; Braquet, P.; Panetta, T.; Marcheselli, V.L. *Lipids,* **1991**, *26*, 1236.

[48]   Ihida, K.; Predescu, D.; Czekay, R.-P.; Palade, G.E. *J. Cell Sci.,* **1999**, *112*, 285.

[49]   Marrache, A.M.; Gobeil, F.Jr.; Bernier, S.G..; Stankova, J.; Rola-Pleszczynski, M.; Choufani, S.; Bkaily, G.; Bourdeau, A.; Sirois, M.G.; Vazquez-Tello, A.; Fan, L.; Joyal, J.S.; Filep, J.G.., Varma, D.R.; Ribeiro-Da-Silva, A.; Chemtob, S. *J. Immunol.,* **2002**, *169*, 6474.

[50]   Noda, H.; Iso, H.; Irie, F.; Sairenchi, T. ; Ohtaka, E. ; Doi, M. ; Izumi, Y. ; Ohta, H. *Circulation,* **2009**, *119*, 2136.

[51]   Ooneda, G. ; Yoshida, Y. ; Suzuki, K.; Sekiguchi, T. ; *Virchoes Arch. Pathol. Anat.* **1973**, *361*, 31.

[52]   Konishi, M.; Iso, H.; Komachi, Y.; Iida, M. ; Shimamoto, T. ; Jacobs, D.R.Jr.; Terao, A.; Baba, S.; Sankai, T.; Ito, M. *Stroke*, **1993**, *24*, 954.

[53]   Takayama, Y. *J. Jpn. Coll. Angiol.* **1975**, *15*, 455.

[54]   Tsimihodimos, V.; Karabina, S.A.; Tambaki, A.P..; Bairaktari, GE..; Miltiadous, G..; Goudevenos, J.A.; Cariolou, M.A.; Chapman, M.J.; Tselepis, A.D.; Elisaf, M.; *J. Lipid Res.*, **2002**, 256.

[55]   Topcuoglu, M. A.; Singhal, A.B. *Expert Opin. Drug Saf.*, **2006**, *5*, 57.

[56]   Nishizawa, S.; Laher, I. *Trends Cardiovasc. Med.*, **2005**, *15*, 24.

[57]   Macdonald R.L. *Neurosurg. Rev.*, **2006**, *29*, 179.

[58]   Hughes, J.T.; Schianchi, P.M. *J. Neurosurg.*, **1978**, *48*, 515.

[59]   Yamashima, T.; Yamamoto, S. *J. Neurosurg.*, **1983**, 58, 843.

[60]   Hirashima, Y.; Endo, S.; Ohmori, T.; Kato, R.; Takaku, A. *J. Neurosurg.*, **1994**, *80*, 31.

[61]   Hirashima, Y.; Nakamura, S.; Endo, S. ; Kuwayama, N. ; Naruse, Y. ; Takaku, A. *Neurochem Res.*, **1997**, *10*, 1249.

[62]   Hirashima, Y.; Endo, S.; Otsuji, T.; Karasawa, K.; Nojima, S.; Takaku, A. *J. Neurosurg.*, **1993**, *78*, 592.

[63]   Kato, K. ; Clark, G.D. ; Bazan, N.G..; Zorumski, C.F. *Nature*, **1994**, *367*, 175.

[64]   Karasawa, K.; Harada, A.; Satoh, N.; Inoue, K.; Setaka, M. *Prog. Lipid Res.*, **2003**, *42*, 93.

[65]   Hirashima, Y.; Endo, S.; Nukui, H.; Kobayashi, N.; Takaku, A. *Neurol. Med. Chir. (Tokyo)*, **2001**, *41*, 165.

[66]   Labadle, E.L.; Glover, D. *J. Neurosurg.* **1976**, *45*, 382.

[67]   Hirashima, Y.; Endo, Sh.; Kato, R.: Ohmori, T.; Nagahori, T.; Nishijima, M.; Takaku, A. *Acta. Neurochir. (Wien)*, **1994**, *129*, 20.

[68]   Miller, L.G.; Freeman, B. *J. Herb Pharmacother.*, **2002**, *2*, 57.

[69]   Denizot, Y.; Chianea, T.; Labrousse, F.; Truffinet, V.; Delage, M.; Mathonnet, M. *Eur. J. Endocrinol.* **2005**, *153*, 31.

[70]   Mathonnet, M.; Descottes, B.; Valleix, D. ; Truffinet, V. ; Labrousse, F. ; Denizot, Y. *World J. Gastroenterol.* **2006**, *12*, 2773.

[71]   Denizot, Y.; De Armas, R.; Caire, F. ; Pommepuy, I. ; Truffinet, V.; Labrousse, F. *Neuropathol. Appl. Neurobiol.* **2006**, *32*, 674.

[72]   Buetow, M.P.; Buetow, P.C.; Smirniotopoulos, J.G.: *Radiographics*, **1991**, *11*, 1087.

[73]   Goldsmith, B.J.; Wara, W.M.:, Wilson, C.B.; Larson, D.A. *J. Neurosurg.*, **1994**, *80*, 195.

[74]   Kempski, O. *Semin. Neurol.*, **2001**, *3*, 303.

[75]   Hirashima, Y. ; Hayashi, N. ; Fukuda, O. ; Ito, H. ; Endo, S.; Takaku, A. *J. Neurosurg.* **1998**, *88*, 304.

[76]   Forman, M.S.; Trojanowski, J.Q.; Lee, V.M. *Nat. Med.* **2004**, *10*, 1055.

[77]   Almeida, O.P.; Lautenschlager, N.T. *Int. Psychogeiatr.*, **2005**, *17 (Supp 1)*, S65.

[78]   Tiraboschi, P.; Hansen, L.A.; Thal, L.J.; Corey-Bloom, J. *Neurology*, **2004**, *62*, 1984.

[79]   Esler, W.P.; Wolfe, M.S. *Science,* **2001**, *293*, 1449.

[80]   Geula, C.; Mesulam, M.M. *Alzheimer Dis. Assoc. Disord.*, **1995**, *9 (Suppl 2)*, 23.

[81]   Stahl, S.M. *J. Clin. Psychiartry.* **2000**, *61*, 813.

[82]   Bate, C.; Salmona, M.; Williams, A. *NeuroReport*, **2004**, *15*, 509.

[83]   Ghafouri, M.; Amini, S.; Khalili, K. ; Sawaya, B.E. *Retrovirology*, **2006**, *3,* 28

[84]   Janssen, R.S. ; Mwanyanwu, O.C. ; Selik, R.M.; Stehr-Green, J.K. *Neurology*, **1992**, *42*, 1472.

[85]   Reger, M.; Welsh, R.; Razani, J.; Martin, D.J.; Boone, K.B. *J. Int. Neuropsychol. Soc.* **2002**, *8*, 410.

[86]   Persidsky, Y.; Limoges, J.; Rasmussen, J.; Zheng, J.; Gearing, A., Gendelman, H.E. *J. Neuroimmunol.* **2001**, *114*, 57.

[87]   Gendelman, H.E.; Persidsky, Y.; Ghorpade, A.; Limoges, J.; Stins, M.; Fiala, M.; Morrisett, R. *AIDS*, **1997**, *11*, S35.

[88]   Gelbard, H.A.; Nottet, H.S.; Swindells, S.; Jett, M.; Dzenko, K.A.; Genis, P.; White, R.; Wang, L.; Choi, Y.B.; Zhang, D.; Lipton, S.A.; Tourtellotte, W.W.; Epstein, L.G..; Gendelman, H.E. *J. Virol.* **1994**, *68*, 4628.

[89]   Perry, S.W.; Hamilton, J.A., Tjoelker, L.W.; Dbaibo, G.; Dzenko, K.A.; Epstein, L.G.; Hannum, Y.; Whittaker, J.S.; Dewhurst, S.; Gelbard, H.A. *J. Biol. Chem.* **1998**, 273, 17660.

[90]   Li, J.; Shao, B.; Zhu, L.; Cui, Y. ; Dong, C.-Z. ; Ezoulin, J.-M. M. ; Gao, X. ; Ren, Q.; Heymans, F.; Chen, H.-Z. *Cell Mol. Neurobiol.* **2008**, *28*, 125.

[91]   Li, J.; Huang, H. ; Ezoulin, J.-M. M. ; Gao, X. ; Massicot, F. ; Dong, C.-Z.; Heymans, F.; Chen, H.-Z. *Int. J. Neuropsychoph.* **2007**, *10*, 21.

[92]   Wieraszko, A.; Li, G.; Kornecki, E.; Hogan, M.V.; Ehrlich, Y.H. *Neuron*, **1993**, 10, 553.

[93]   Izquierdo, I.; Fin, C.; Schmitz, P.K. ; DaSIlva, R.C. ; Jerusalinsky, D.; Quillfeldt, J.A.; Ferreira, M.B.G.; Medina, J.H.; Bazan, N.G.. *Proc. Natl. Acad. Sci. USA*, **1995**, *92*, 5047.

[94] Packard, M.G.; Teather, L.; Bazan, N.G.. *Neurobiol. Learn Mem.* **1996**, *66*, 176.

[95] Chen, C.; Magee, J.C.; Marcheselli, V.L.; Hardy, M..; Bazan, N.G.. *J. Neurophysiol.* **2001**, *85*, 384.

CHAPTER 9

Novel Mechanism for Oxidative Stress in Neurodevelopmental Pathophysiology and Course of Schizophrenia

Anvita Kale[1], Sadhana R. Joshi[1] and Sahebarao P. Mahadik[2,3]*

[1]*Interactive Research School for Health Affairs, Bharati Vidyapeeth, Pune, 411063, India and* [2]*Department of Psychiatry and Health Behavior, Medical College of Georgia, Augusta GA, 30912, USA;* [3]*Medical Research, VA Medical Center, Augusta, GA 30904, USA*

Abstract: The role of oxidative stress and oxidative cellular deficits have been considered for a long time in the etiopathogenesis as well as in course and treatment outcome of schizophrenia. A large number of reviews and monograms have been published primarily on altered levels of indices of oxidative stress and oxidative cell injury in mostly chronic medicated patients and few in drug naïve early psychotic patients. However, since schizophre-nia is now generally considered to have neurodevelopmental deficits that most likely start in utero it is important to have a specific mechanism that can trigger the oxidative stress at the critical developmental time. This lack of information has limited the success in developing effective treatment strategies in amelioration of oxidative stress-mediated cellular damage and improved neurodevelopment and consequent clinical outcome. Furthermore, chronic use of the major antipsychotics worsens the oxidative cellular damage and contributes to the poor clinical outcome by increased negative symptoms and the cognitive and motor deficits. We have here presented a hy-pothesis that states as, the altered metabolism of key maternal nutrients such as folic acid and B_{12} and omega-3 fatty acids synergistically will trigger the oxidative stress which will alter the early foetal neurodevelopment and alter later in life the cognitive deficits and psychosis key behavioural symptomatology of schizophrenia. The al-tered metabolism of these maternal nutrients will contribute to the altered one carbon metabolism leading to in-creased homocysteine and reduced expression of antioxidant enzyme genes and neurotrophic factors by altered chromatin methylation (epigenesis) that contribute to increased oxidative stress. This novel mechanism will also explain the role of a large number of genetic (e.g., altered expression of antioxidant enzymes and neurotrophins) and environmental (e.g., maternal use of alcohol and drug abuse, smoking, under- and mal-nutrition, and socio-economic stressors) risk factors reported for etiopathogenesis and course of schizophrenia since all of these risk factors trigger the oxidative stress and cellular oxidative damage such as reported in schizophrenia with and with-out treatment with antipsychotics. Finally, the mechanism based on altered micronutrients may provide effective specific nutritional supplementation strategies for prevention of oxidative stress-mediated pathologies associated with onset of schizophrenia as well as its improved clinical outcome by conventional treatments.

Keywords: Oxidative stress; Schizophrenia; oxidative cellular deficits; folic acid; B_{12}; omega-3fatty acid

INTRODUCTION

Based on the extensive studies on the altered markers of oxidative stress and oxidative cellular injury in schizophre-nia [1], a role for oxidative stress has been implicated for a long time in etiopathogenesis, pathophysiology and psy-chopathology of schizophrenia. Furthermore, based on the effects of antipsychotic treatments on the oxidative stress in animals and psychotic subjects, a role of oxidative stress on the clinical course and treatment outcome of schizo-phrenia has also been implicated [2]. However, no specific molecular mechanism for increased oxidative stress and subsequent oxidative cellular damage has ever been proposed. Rather, since a large number of both genetic and en-vironmental factors implicated in the etiopathogenesis of schizophrenia induce the oxidative stress in experimental studies all of these factors have been considered to induce oxidative stress in schizophrenia. Furthermore, since schizophrenia involves the neurodevelopmental deficits there is no information when and how oxidative stress trig-gered by an etiopathogenetic factor(s) affects the early neurodevelopment and contributes to onset of psychopa-thology and its course and treatment throughout life. These issues indicate that oxidative stress is most likely trig-gered during fetal growth and continues through adolescence thereby contributing to neurodevelopmental deficits and onset of illness. This further indicates that oxidative stress may sustain throughout the life of a schizophrenic patient playing a role in clinical course and treatment outcome.

*Address correspondence to: Medical Research, VA Medical Center, 1 Freedom Way, Augusta, GA 30904, USA. Phone: 706-733-0188 ext. 2490; Email: smahadik@mail.mcg.edu

Akhlaq A. Farooqui & Tahira Farooqui (Eds.)

We have proposed here a novel molecular mechanism that can trigger the oxidative stress during fetal life and continue rest of the life of an individual (Fig. 1). This mechanism primarily incorporates the long time ago proposed altered one carbon metabolism in schizophrenia [3] and includes the role of omega-3 fatty acid in one carbon metabolism and their combined consequence on the epigenesis and markers of oxidative stress. As it is discussed later, this mechanism is consistent with the proposed role of maternal intake and metabolism of folic acid, vitamin B_{12} and omega-3 fatty acids in one carbon metabolism and induction of oxidative stress during pregnancy by altered metabolites, primarily increased homocysteine and reduced glutathione, altered expression of antioxidant enzymes and neurotrophic factors (see Fig. 1). This chapter will also discuss how this mechanism is more appropriate to explain proposed role of a large number of environmental and genetic factors in etiopathogenesis and progression of schizophrenia. Finally, mechanism also provides potentially more effective early intervention strategies with vital nutritional factors to control oxidative stress and prevent oxidative cellular injury and neurodevelopmental deficits and a serious mental illness.

PRESENT STATUS OF OXIDATIVE STRESS AND SCHIZOPHRENIA

A. Schizophrenia

Schizophrenia is a devastating psychiatric disorder with a broad range of behavioral and biologic manifestations [4] that affects almost 1% population in every culture around the world [5]. Core features of schizophrenia include: auditory hallucinations, changes in thought construction and form; and bizarre delusions [6]. Its symptomatology becomes evident in adolescence and most of the subjects often progressively deteriorate throughout the rest of their lives. The etiopathogenic mechanisms in schizophrenia are unknown although various theories have been proposed that involve a large number of genetic and environmental factors such as socioeconomic, infectious, immunological, nutrition and pregnancy and birth complications which can contribute to abnormal brain development and emergence of symptomatology in later life. As early as the 20th century, Kraepelin and others suggested that some cases of schizophrenia probably result from insults that cause cerebral maldevelopment [7]. Now it is well established that schizophrenia has neurodevelopmental origin due to an interaction between genetic and environmental factors that adversely affect brain development during critical early periods and behavior later in life [6, 8]. Neurodevelopmental deficit hypothesis is very well supported by differential changes in brain regional size, shape, reduced neural cell number and arborization with resultant altered synaptic transmission of several neurotransmitters [9]. Normalization of neurotransmission by drugs with neurotransmitter receptor specific drugs has been the focus of treatment strategies for the last over 60 years [10]. Unfortunately, this strategy has been found to have very limited success. It is now generally accepted that treatment of schizophrenia must involve early intervention by agents that will prevent the neurodevelopmental deficits and trigger the neuroplasticity to ameliorate the neuropathology and thereby improve psychopathology. If oxidative stress is the primary mechanism of neurodevelopmental deficits, one direct treatment strategy will be the use of antioxidants [1, 11]. Several reported studies with use of antioxidants were done in medicated chronic patients and have been indicated to be improperly done [1, 12]. However, this approach is also limited since the mechanisms of oxidative stress are not known and the oxidative stress-mediated cellular injury is a complex process.

B. Oxidative Stress

Oxidative stress is a state when there is an imbalance between the generation of reactive oxygen species and antioxidant defense capacity of the body leading to excessive free radical production [13]. This modifies the normal intracellular balance between oxidant substances produced during aerobic metabolism and antioxidant system processes which perform the function of neutralization, putting a series of protective mechanisms, of both an enzymatic and non-enzymatic nature, in action [14]. Numerous physiological and pathological processes such as ageing, excessive caloric intake, infections, inflammatory disorders, environmental toxins, pharmacological treatments, emotional or psychological stress, ionizing radiation, cigarette smoke and alcohol increase the concentration of oxidizing substances (ROS) or, more commonly, free radicals. These are chemical species which are highly reactive owing to the presence of free unpaired electrons. An increase in free radicals compromises the delicate homeostatic mechanisms which involve neurotransmitters, hormones, oxidizing substances and numerous other mediators.

Oxidative stress leads invariably to oxidative cell damage, namely, DNA breaks and protein inactivation resulting in altered gene expression, peroxidative loss of membrane phospholipids causing abnormal cell growth and differentiation and/or even apoptotic cell death. Starting from conception, the time and magnitude of oxidative stress and the pre-natal and post-natal environment (anti- or pro oxidant) influence the degree and magnitude of neurodevelopmental deficits [1].

Oxidative stress is ubiquitous in every day life of a living being, more critical in humans. Most psychiatric disorders are associated with increased oxidative stress [15]. The brain tissues contain large amounts of polyunsaturated fatty acids and catecholamines, which are thought to be target molecules for free radical-induced peroxidation and neural cell damage [16]. A large number of reviews and chapters have also been published detailing the altered markers of oxidative stress and oxidative cellular injury, and their possible implications to the etiopathogenesis, pathophysiology and psychopathology of schizophrenia [1, 4, 13, 17]. In biological systems, ROS can cause cellular damage/dysfunction by oxidation of lipids, proteins and DNA. Recently more and more converging evidence indicates that oxidative mechanisms may play a role in schizophrenia [18]. The details of these processes and mechanisms have been published in a large number of reviews and chapters [1].

C. Indices of Oxidative Stress and Oxidative Cell Injury

Oxidative status in clinical populations are typically assessed in other ways, such as the measurement of oxidative defenses, particularly key enzymes such as catalase, peroxidase and glutamate cysteine ligase, paralleled by assessment of the consequences of oxidative stress such as plasma lipid peroxides [19]. Malondialdehyde (MDA), an oxidative stress biological marker, is one of the most frequently used markers of lipid peroxidation in schizophrenia research, however data regarding MDA levels in schizophrenia are controversial [20]. Elevated levels of MDA have been shown in plasma, erythrocytes, leucocytes and platelets of schizophrenia patients [21] while levels of TBARS in CSF have been reported as lower in patients compared with controls [22]. The extent of lipid peroxidation has been found to correlate positively with the severity of symptoms in nevermedicated patients, [23] and inversely with the levels of membrane-essential poly un-saturated fatty acids such as arachidonic acid [24]. Free radicals, primarily plasma nitric oxide (NO) was found to be higher [25] or unchanged [26] in chronic patients but lower in deficit patients [26]. Excess NO production leads to changes of neuron structure and function involving neuronal membrane damage [27] and increased indices of lipid peroxidation [21].

Endogenous antioxidant defense generally responds quickly to oxidative stress and prevents oxidative cellular injury. Analyses of RBC antioxidant enzymes (superoxide dismutase, glutathione peroxidase and catalase) have, on the whole, shown evidence of dysregulation in schizophrenia [21]. Reduced levels of the one or more of these antioxidant enzymes are reported in patients with schizophrenia compared to controls [21]. Altered superoxide dismutase (SOD) and increased lipid peroxidation, measured by the thiobarbituric acid reactive substances (TBARS), are increased in schizophrenia [28]. Altered levels of SOD in the brain in patients with schizophrenia suggests a specific neuroanatomical distribution pattern of oxidative stress processes possibly related to the pathophysiology of schizophrenia [29].

Superoxide dismutase has been consistently found increased [30] in drug-free patients or those with TD (tardive dyskinesia) [30], unchanged [21] or decreased [21] in blood from chronic schizophrenic patients. Glutathione peroxidase (GPx) activity was reported increased in drug-free patients [30] while unchanged [31] or decreased [21] in chronic patients. Activity of glutathione peroxidase (GPx), a key antioxidant enzyme, was found to be decreased in red blood cells [21, 32] and plasma [33] of some, but not all schizophrenic patients [21, 34]. Furthermore, plasma GPx levels were significantly and positively correlated with psychosis rating scores in schizophrenic patients [31]. These findings have been neglected for almost three quarters of a century, but have recently attracted renewed attention [35]. These altered levels have been associated with certain clinical symptoms and therapeutic features. Negative symptoms have been associated with low levels of GPx [4] while positive symptoms have been shown to positively correlate with SOD activity [4]. Inverse relationship between blood GPx and structural measures of brain atrophy has been documented, suggesting a link between oxidative dysregulation and progressive structural changes [36]. A few studies have examined levels of non enzymatic antioxidants such as plasma antioxidant proteins (albumin, bilirubine, uric acid), glutathione, alpha-tocopherol (vitamin E), ascorbic acid (vitamin C), flavonoids, the phenol compounds and minerals like zinc and trace elements [21]. Plasma proteins like albumin, uric acid and bilirubine in blood are reported lower in haloperidol-managed [37] and first episode schizophrenic patients [24].

D. Antipsychotic Treatments and Oxidative Stress

The changes in oxidative stress indices have been shown to be related to severity of psychosis [13, 30] and age [37]. The effects of antipsychotic treatments of schizophrenia on the oxidative stress and its implications on the course and clinical outcome have been extensively reported [1]. Recent evidence suggests that some atypical antipsychotic drugs may protect against oxidative stress and consequent neurodegeneration by mechanisms that remain unclear [38]. Typical and atypical antipsychotics significantly differ in their neurotransmitter receptor affinity profiles, and their efficacy and side effects in schizophrenic patients. Typical antipsychotics have been found to increase oxidative (i.e. free radical-mediated) cellular injury in rats.

Since schizophrenia also involves oxidative injury, understanding of differential effects of these antipsychotics on expression of antioxidant enzymes and oxidative injury may be very critical. There is also evidence that psychotic disorders might impair antioxidant defense and increase lipid peroxidation, as antipsychotic treatment itself increases oxidative stress and induces irreversible neuropathological changes in animal models [39]. Some studies [21, 25, 31, 37, 40] indicate that antipsychotic drugs have no significant regulatory action on the antioxidant defense system. However, more studies revealed that antioxidant enzyme activities are associated with the treatment of schizophrenic patients with different neuroleptics [2, 24, 30]. Plasma levels of membrane lipid peroxidation products were found to be elevated significantly in patients treated with first generation antipsychotics (FGAs), compared to most second generation antipsychotics (SGAs) after a 3-week longitudinal study of antipsychotic treatment. SGAs (Olanzapine) therapy for 2 months has been found to partially ameliorate adverse effects on the antioxidant defense mechanism in schizophrenia [41].

Studies have shown that chronic administration of haloperidol (HAL) a typical antipsychotic, but none of the studied atypicals risperidone (RIS) or clozapine (CLZ) or olanzapine (OLZ), induce oxidative stress by persistent changes in the levels of antioxidant enzymes and cause membrane lipid peroxidation [2]. Others studies have shown that atypical antipsychotic drugs slightly up-regulated the expression of copper/zinc superoxide dismutase mRNA, whereas haloperidol strongly increased the expression of copper/zinc superoxide dismutase mRNA [42]. Studies indicate that olanzapine increased SOD enzyme activity in PC12 cells suggesting a neuroprotective action [43] by recovering stress-induced decreases in BDNF mRNA expression [44]. Further, short-term (<45 days) treatment studies in rats have reported increased oxidative stress and oxidative (i.e., oxygen free radical-mediated) neural cell injury with typical antipsychotics such as haloperidol, but not with the atypicals such as clozapine, olanzapine or risperidone [2, 45].

E. Oxidative Stress and Clinical Course and Outcome

Schizophrenia is associated with a broad range of neurodevelopmental, structural and behavioral abnormalities that often progress with or without treatment [1]. Increasing evidence indicates that oxidative injury contributes to the pathophysiology and may also account for deteriorating course and poor outcome in schizophrenia [46]. Deregulated lipid metabolism may be of particular importance for CNS injuries and disorders, as the brain has the highest lipid concentration and play vital role in neurotransmitter signal transduction [47]. The mechanism of ROS mediated cellular damage and the changes in the indices of oxidative cell damage in schizophrenia strongly support the broad range of diffuse neurodevelopmental abnormalities [48] and postmortem neuropathology [49] observed in schizophrenia. Children with schizophrenia have a broad range of structural brain abnormalities [50] and post-mortem neuropathological changes consistent with effects of ROS mediated cellular damage. These changes are associated with clinical features, like impaired psychomotor and neuropsychological development ('neurointegrative defect') in children genetically at risk for schizophrenia [51], pre-morbid dysfunction [52] and minor physical anomalies [53]. These studies have particularly indicated the differential reductions in volumes of several brain regions, disorganized neuronal networks and increased ventricular size. All of these changes likely predate (i.e. begin during neurogenesis) the onset of illness, and these regions may progressively deteriorate by way of proposed increased 'pruning' (excessive removal of nerve endings and processes) but not by classical degeneration [54].

It is also important to indicate that oxidative stress and associated damage is generally systemic and affects the whole body, with brain as the most susceptible region. Even without anti-psychotic treatment that may exacerbate the risk for metabolic disturbance, schizophrenics have a higher prevalence of obesity, diabetes [39], cardiovascular disease [55], dementias, including Parkinson's, and even reduced life span [56]. Thus it is important to prevent the oxidative stress and associated pathologies in schizophrenia. The novel mechanism proposed here is timely to deal with this devastating illness.

NOVEL MECHANISM OF OXIDATIVE STRESS

A. Maternal Malnutrition and Altered One Carbon Metabolism

Maternal malnutrition, particularly altered intake/metabolism of folic acid and vitamin B_{12} is the primary cause of reported altered one carbon metabolism in the etiopathophysiology of schizophrenia. We have recently provided evidence that omega-3 fatty acid also plays a vital role in one carbon metabolism. The altered one carbon metabolism and the changes in associated metabolic process will lead to increased markers of oxidative stress (see Fig. 1). This oxidative stress will then contribute to abnormal foetal development.

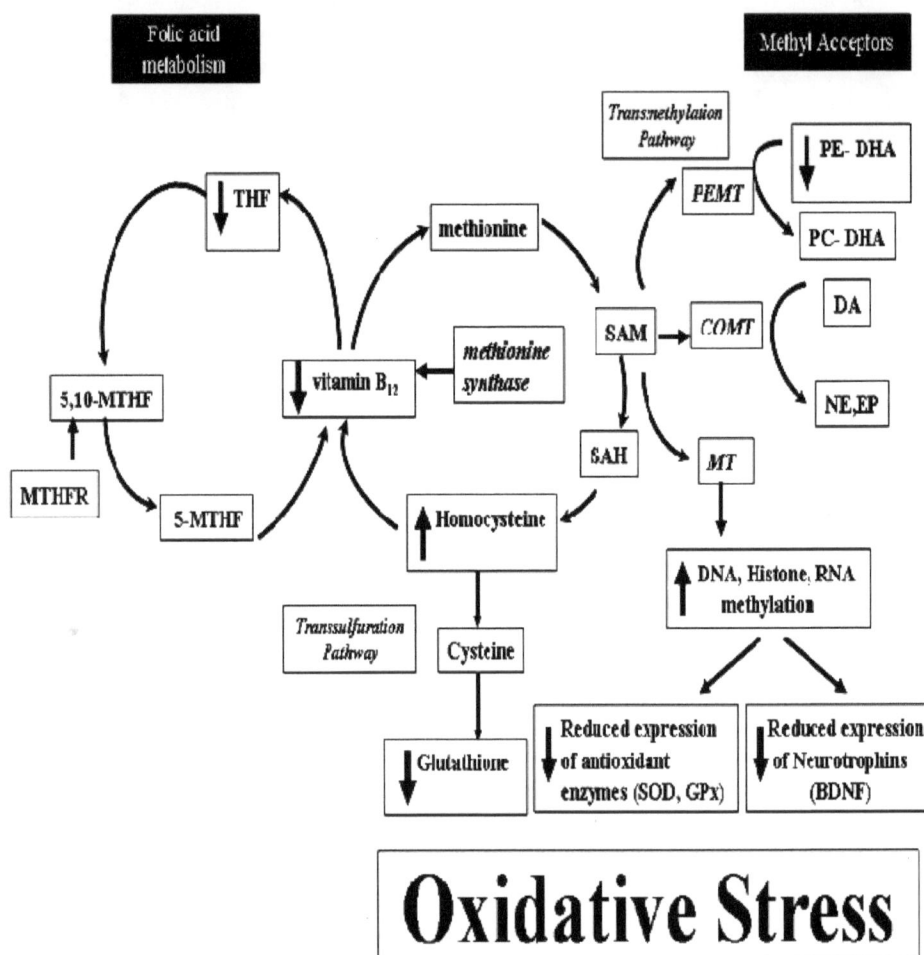

Figure 1: Novel Mechanism of oxidative stress in the neurodevelopmental pathology of schizophrenia. This figure shows the details of maternal nutrient, particularly folic acid, vitamin B_{12} and DHA-mediated molecular mechanism of oxidative stress. The altered intake and/or metabolism of these nutrients alter the one carbon metabolism with ultimate consequences on changes in key indices of oxidative stress: increased homocysteine, and decreased expression of antioxidant defense enzymes and neurotrophic factors. The oxidative stress will cause oxidative cell deficits during fetal and adolescent with onset of psychopathology and will continue throughout life with impact on course and outcome of schizophrenia. This mechanism also indicates its potential for nutritional intervention to prevent the oxidative injury related deficits early in the development and onset of illness later. Arrows indicate the direction of changes in key processes in schizophrenia.

a. Folic Acid and Vitamin B_{12}

During gestation, folic acid is critically important for DNA synthesis and cell proliferation [57]. This demand for folate must be met by adequate maternal dietary intake. Adequate dietary intake of folate is required during critical brain development for neurogenesis in the brain and spinal cord [58]. A severe deficiency can result in neural tube defects. Reduced levels of circulating folate during pregnancy are known to be associated with increased risks of preterm delivery, infant low birth weight and fetal growth retardation [59]. There is mounting evidence that inadequate maternal folate status during pregnancy may lead to low infant birth weight, thereby conferring risk of developmental and long-term adverse health outcomes [60].

The developing brain between 24 and 42 wk of gestation is particularly vulnerable to nutritional insults because of the rapid trajectory of several neurologic processes, including synapse formation and myelination [61]. The young brain is remarkably plastic and therefore more amenable to repair after nutrient repletion. However, the brain's vulnerability to nutritional insults likely outweighs its plasticity, which explains why early nutritional insults result in brain dysfunction not only while the nutrient is in deficit, but also after repletion. Folate deficiency can impede the synthesis and repair of DNA and might thereby increase the risk of de novo mutations and also can impede the production of methyl donors and the methylation of DNA and might thereby affect the expression of genes that regulate neurodevelopmental processes [62]. Folate deficiency can also impede the conversion of homocysteine (Hcy) to methionine and might thereby lead to accumulation of Hcy with adverse effects on fetal brain development.

Studies have also suggested that low maternal folate and high homocysteine levels increase the risk for developing schizophrenia [63]. Prenatal folate deficiency could plausibly influence the risk of offspring schizophrenia [64]. Reduced levels of serum folate but not the red blood cell folate are reported in both schizophrenic and depressive disorder patients [65]. In contrast increased red blood cell folate concentrations have also been reported in schizophrenics [66]. Goff *et al.* (2004) [67] reported lower folate in chronic outpatients compared to published serum folate levels of control samples. In contrast [68] did not find any difference in either folic acid or vitamin B_{12} in an ethnically homogeneous female population with different psychiatric disorders. A meta-analysis of 10 folate studies found lower folate in the majority of patients only in three studies and indicated several methodological shortcomings, including not adequately matched patients and controls or using controls from published studies for comparison [69].

b. Omega-3 Fatty Acids

Essential fatty acids (EFAs) play critical role in brain development. Humans do not have the ability to synthesize these fatty acids de novo and thus are largely dependent upon dietary sources [70]. Docosahexaenoic acid (DHA), an omega-3 fatty acid, is the primary structural fatty acid in the brain, comprising 25%–30% of the structural fatty acids in the gray matter [70, 71]. Varying degrees of altered composition of membrane phospholipid fatty acids have been reported in periphery and postmortem brains from patients at the onset of psychosis as well as on or off drug treatment [72]. The earlier reports including our studies in patients with shortest duration of psychosis (< 5 days) showing significant reductions in both DHA and AA levels in first episode drug naïve schizophrenic patients [10, 15, 24] have led to suggest that abnormalities in AA and DHA probably contribute to the developmental neuropathology before the onset of psychosis, and progression and clinical outcome later in the life [72]. Dietary supplementation of omega-3 fatty acids with improvements also supports these deficits [1, 72, 73]. Recently, a predominant reduction in DHA has been reported in orbito-frontal cortex [72] or RBC [74] of male > female schizophrenic patients with or without drug treatment. The reported reductions in AA mostly in RBC may be related to the differences in the psychopathophysiology and antipsychotic treatments [1, 75]. These studies have led to the "phospholipid" as well as "essential fatty acid" hypotheses of schizophrenia [72]. The mechanism of altered composition of fatty acids in schizophrenia has been suggested to be multifactorial [1, 72, 75, 76]. Our study has also shown a significant reduction in only membrane DHA levels in first episode psychotic patients. On the other hand levels of CSF DHA were significantly increased in these patients as compared to controls suggesting either selective release of DHA from membranes phospholipids and/or its defective synthesis from ALA and EPA since its levels was also increased or defective incorporation into neural phospholipids [77]. Particularly, significantly reduced levels of phosphatidylethanolamine (PE) reported in drugnaïve first episode patients may be result of reduced DHA and that can affect the methylation to PC (see Fig. **1**).

It is well established that long-chain polyunsaturated fatty acids (LCPUFA) have the potential to influence maternal health during pregnancy as well as fetal and child health. Docosahexaenoic acid and arachidonic acid are important LCPUFA for fetal and infant growth and development. In the cerebral cortex and retina, docosahexaenoic acid constitutes a large percentage of the phospholipid fatty acids [78]; while arachidonic acid, is also essential for normal growth and development [79]. These fatty acids, which are essential components of membrane phospholipids, are deposited in relatively large amounts in the central nervous system during brain growth. A few studies suggest that inadequate maternal docosahexaenoic acid status may be associated with suboptimal functional outcomes in infants [80].

Maternal supplementation with cod liver oil, which contains very long-chain omega-3 fatty acids, during pregnancy has been associated with higher IQ at age 4 years [81]. Further, studies indicate that umbilical plasma levels of DHA correlate with increased IQ at 4 years age and umbilical eicosapentaenoic acid was associated with improved mental processing skills in childhood. Umbilical vessel essential fatty acid levels (EFAs), including DHA, have been correlated with decreased neonatal neurological abnormalities [82]. In two additional studies, however, long-chain polyunsaturated fatty acid status at birth was not associated with cognitive function at age 4 and 7 [83]. Nonetheless, further investigation of maternal DHA and other EFAs during pregnancy in relation to adult outcomes, including risk of schizophrenia, may prove fruitful.

B. Increase Homocysteine

Folate and B_{12} deficiency is known to result in increased homocysteine concentrations increasing the generation of reactive oxygen species and contribute to excitotoxicity and mitochondrial dysfunction which may lead to apoptosis [84]. Homocysteine is known to increase oxidative stress-mediated neural apoptosis although the mechanism is not known [84]. A high homocysteine concentration is said to decrease glutathione peroxidase activity and reduce tissue concentrations of glutathione since homocysteine is a precursor of glutathione in neurons and renders them more vulnerable to oxidative attack [85]. Studies have also shown that homocysteine induces oxidative stress in brain of rats by reducing antioxidant defenses and increasing lipid peroxidation [14, 86]. Homocysteine is also a potent glutamatergic agonist and causes excitotoxic neuronal pathology or by auto-oxidation to homocystine and other disulphides releasing O2 and H2O2 [87]. Further, increased oxidative stress-mediated cell injury including apoptosis is well accepted as one of the key pathogenetic mechanism of schizophrenia [1, 87]. Hcy has now been implicated in increased oxidative stress, DNA damage, the triggering of apoptosis and excitotoxicity, all important mechanisms in neurodegeneration [84, 85]. In mild and moderate Hcy, Hcy might primarily influence the epigenetic regulation of gene expression through the interference of transferring methyl-group metabolism. However, at high Hcy concentrations, the impacts might be more injurious through oxidative stress, apoptosis and inflammation [88].

The brain may be particularly vulnerable to high levels of Hcy in the blood because it lacks two major metabolic pathways for its elimination: betaine re-methylation and trans-sulfuration [89]. Some likely mechanisms for the pathophysiological effects of Hcy on the brain are the following. Homocysteine and the vitamins involved in 1-carbon metabolism, folate and vitamin B_{12}, have been associated with a diversity of diseases, including cardiovascular disease [90], birth defects and pregnancy complications [90, 91], Alzheimer's disease, Parkinson's disease, impaired cognitive functioning, and also schizophrenia [90, 92]. Several studies have investigated the 677C→T polymorphism of the methylene- tetrahydrofolate reductase (MTHFR) gene as a risk factor for above diseases, as the C-to-T transition causes reduced enzyme activity, and elevated plasma total homocysteine (tHcy) levels under conditions of impaired folate status [93]. Homocysteine, a potent neurotoxin also exists transitionally at the intersection of the transmethylation and the trans-sulfuration pathways, which regulate its elimination [92]. Such neurotoxic effects may be due to the direct interaction of Hcy with plasma membrane components or to the intracellular accumulation of S-adenosyl-homocysteine (SAH). This latter metabolite inhibits the methylation of catechol substrates resulting in the generation of oxy-radicals and other chemically reactive products that are cytotoxic.

Our recent study showed increased levels of homocysteine in first episode psychotic patients [92]. Elevated homocysteine levels are also reported in young male or newly admitted schizophrenic patients [94]. There are a few inconsistent reports on the levels of homocysteine in schizophrenia [69]. Increased homocysteine levels have also been reported without change in folic acid and vitamin B_{12} levels in chronic schizophrenic patients [95]. However, Muntjewerff *et al.* (2003) [66] found a disturbed folate metabolism in schizophrenics independent of homocysteine while Goff et al (2004) [67] did not find a change in homocysteine levels even though patients had significantly lower folic acid levels compared to published values for normal controls.

c. Exitotoxicity

Hcy functions as an excitatory amino acid by activating group 1 metabotropic glutamate receptors (mGluR) [96] and N-methyl-d-aspartate (NMDA) receptors [97]. Moderately elevated plasma homocysteine levels were found to be associated with human brain atrophy [98], with *in vitro* neurotoxicity via NMDA receptors dependent on glycine levels [97]. Consistent with these actions, the systemic administration of Hcy or homocysteic acid can trigger seizures in animals [99], and patients with homocystinuria suffer from epileptic seizures [100]. However, Hcy is also a partial agonist at the glycine receptor, resulting in reduced NMDA receptor activity. This dual action therefore prevents the overstimulation of the NMDA receptor by Hcy under physiological conditions. In conditions of elevated Hcy levels and disturbance of the blood– brain barrier, such as during a stroke, the neurones are exposed to high levels of Hcy and glycine and overstimulation of NMDA occurs, leading to excitotoxic damage [97].

Evidence for the association between homocysteine and neurotransmitters is found in studies that directly measure neurotransmitter metabolites and in studies demonstrating the antidepressant effects of folate and S-adenosylmethionine, a cofactor and an intermediate metabolite of the methionine-homocysteine pathway. Direct evidence of the association between homocysteine and neurotransmitter levels comes from studies of Parkinson's disease. Parkinson's patients receiving L-dopa, which requires the donation of a methyl group from Sadenosylmethionine to be metabolized, had higher levels of homocysteine than patients not taking L-dopa [101]. Use of L-dopa could create a methyl sink, thus preventing the remethylation of homocysteine to methionine with the result of high homocysteine levels. Another potential mechanism for the homocysteine effect on transmitters is by inhibition of the enzyme necessary to catalyze the methylation reactions between the catecholamines and S-adenosylmethionine (SAMe) [102]. Elevated homocysteine is a risk factor in cardiovascular diseases and neurodegeneration. Among the putative mechanisms of homocysteine-evoked neurotoxicity, disturbances in methylation processes and NMDA receptor-mediated excitotoxicity have been suggested [103]. Homocysteine also over stimulates N-methyl-D-asparate (NMDA) receptors, potentiates glutamate accumulation and amyloid-beta aggregation and neurotoxicity, and induces DNA breakage and lipid peroxidation [104]. Mouse models of AD and Parkinson's disease as well as wild type mice subjected to folate deficiency show elevated Hcy and place neurons at risk of degeneration.

f. Apoptosis

Apoptosis is a highly regulated form of cell death that is often termed as cellular suicide. Apoptosis is pervasive during early development of the central nervous system (CNS)–over half of all developing neurons die by apoptosis [105]–and it also serves to eliminate injured or diseased neurons throughout life. Apoptosis is characterized by cell shrinkage, membrane blebbing, chromatin condensation, DNA fragmentation, and cellular disintegration with phagocytosis [106]. Apoptosis occurs without inflammation and typically requires the formation of new gene products to proceed. The mechanism of apoptosis is focused on regulating the activation of cysteine-dependent aspartate-directed proteases known as caspase proteins [107]. The first study to examine the potential role of apoptosis in schizophrenia measured levels of the anti-apoptotic regulatory protein Bcl-2 in postmortem brain tissue [87]. While many factors have been found to promote apoptosis, several pro-apoptotic stimuli have also been suggested to play a role in the pathophysiology of schizophrenia including glutamatergic excitotoxicity, excess synaptic calcium flux, oxidative stress, and reduced neurotrophin levels. Oxidative stress is known to cause neuronal apoptosis and has been hypothesized to contribute to the pathophysiology of schizophrenia [1]. Further elevated homocysteine may reduce the levels of the antioxidant enzyme glutathione peroxidase making neurons vulnerable to oxidative stress [84]. Hcy promotes apoptosis through DNA breakage because of impaired transmethylation. High homocysteine levels (up to 250 μM in vitro) were found to be involved in neuronal DNA damage and apoptosis [85, 97, 108].

Cultured neurones, in their attempt to repair DNA damage, eventually are depleted of ATP reserves and the caspase pathway is activated [108]. Furthermore, the oxidative stress caused by Hcy increases the activity level of NF-kappaB, a redox-sensitive transcription factor that is important for the control of apoptosis mediation through reactive oxidation species [109]. This results in an influx of Ca^{2+} leading to secondary excitotoxicity [110].

Increased cytosolic Ca^{2+} damages mitochondrion that will result in reduction of ATP, further accumulation of Ca^{2+}, leakage of cytochrome c, activation of caspase-3 pathway and apoptosis [111]. Hcy also enhances the stress response of the endoplasmic reticulum (ER) by enhancing the expression of both chaperone (e.g. grp78, grp94, gadd153) and non-chaperone (e.g. HERP) proteins [112]. The ER stress response is closely linked to the extent of cell injury. Role of apoptosis has been implicated in neurodevelopmental pathology as well as progressive deterioration of schizophrenia [87, 113].

Several possibilities exist for integrating apoptotic mechanisms into the neurodevelopmental hypothesis of schizophrenia. First, the synaptic pruning hypothesis suggests that abnormalities in normal synaptic pruning and related neuromaturational processes such as cortical myelination contribute to schizophrenia [8, 53, 114]. Reductions in synaptic density have been demonstrated in human prefrontal cortex during normal adolescence and early adulthood [113], overlapping the age of highest incidence of first-episode psychosis. While the physiological basis for normal synaptic pruning is unknown, a potential role for apoptosis is suggested by evidence that apoptosis can be activated locally in synapses and terminal dendrites. The possible roles of apoptosis in the pruning hypothesis of schizophrenia include a contribution to a genetically mediated acceleration of normal synaptic pruning or to a pathological potentiation of normal pruning by one or more pro-apoptotic stimuli. A second potential overlap between the neurodevelopmental hypo- thesis of schizophrenia and apoptosis could occur in very early development. A fetal or perinatal insult could transiently alter normal developmental apoptosis to yield enduring cytoarchitectural deficits and altered synaptic connectivity leading to deficits in brain circuitry. Interestingly, a number of pro-apoptotic stimuli such as ischemia, hypoxia, and pro-inflammatory cytokines [115] have been implicated as early life insults that increase the risk of developing schizophrenia [116]. These studies indicate that several apoptotic regulatory proteins are altered in schizophrenia and that the vulnerability to apoptotic activation may be increased as evidenced by high Bax/Bcl-2 ratio and low Bcl-2 levels. It remains possible that in the CNS, exposure to oxidative stress is either time-limited or that effective compensatory mechanisms eventually limit the effect of chronic stress, thereby creating a transient pro-apoptotic environment in the early stages of schizophrenia.

C. Decreased Glutathione

Glutathione (GSH) is the brain's dominant cellular free radical scavenger, [35] and is a tripeptide composed of glutamate, cysteine and glycine. It shuttles between reduced monomeric (GSH) and oxidised dimeric forms (GSSG) in the scavenging process. Glutathione plays a vital role in NMDA receptor-mediated neurotransmission, which is involved in the pathophysiology of schizophrenia [117]. Oxidative stress and reduced brain glutathione (GSH) levels have been reported in psychiatric illnesses including schizophrenia and bipolar disorder. However the role of GSH in cognitive impairment in the illness remains unclear [118]. Accumulating evidence suggests that oxidative stress associated with impaired metabolism of the antioxidant glutathione (GSH) plays a key role in the pathogenesis of schizophrenia [119]. It was observed that the GPx-1 activity was inhibited under severe hyperhomocysteinemia (50-500 microM Hcy) conditions, especially at low glutathione concentrations [120]. The role of glutathione (GSH) in schizophrenia was initially examined as early as 1934, [121] where decreased circulating glutathione and increased lactate were reported. Reduced levels of GSH have been associated with a number of neurological diseases, presumably as a result of increased oxidative stress [122]. Oxidative stress and reduced brain glutathione (GSH) levels have been reported in psychiatric illnesses including schizophrenia and bipolar disorder. However the role of GSH in cognitive impairment in the illness remains unclear. GSH depletion resulted in disruption of short-term spatial recognition memory in a Y-maze test. In conclusion, GSH depletion induces cognitive impairment, which may be relevant to the role of GSH in psychiatric illnesses [118]. A study using postmortem brain samples demonstrated decreased levels of GSH, oxidized GSH (GSSG), GSH-Px, and GSH reductase in the caudate region of brains from schizophrenic patients, suggesting impaired GSH metabolism in schizophrenic brains [123]. GSH plays a major role in the modulation of redox-sensitive sites on the N-methyl-D-aspartate (NMDA) receptors [124], which are implicated in the pathophysiology of schizophrenia [125]. It has been reported that GSH-deficient mice showed enhanced dopamine neurotransmission, altered serotonin function, and augmented locomotor responses to low doses of the NMDA receptor antagonist phencyclidine, suggesting that the GSH deficiency produced alterations in monoaminergic function and behavior in mice relevant to schizophrenia [126]. Cultured skin fibroblasts from schizophrenia patients and control subjects were challenged with oxidative stress, and parameters of the rate-limiting enzyme for the GSH synthesis, the glutamate cysteine ligase (GCL), were measured. Taken together, the study provides genetic and functional evidence that an impaired capacity to synthesize GSH under conditions of oxidative stress is a vulnerability factor for schizophrenia [127]. Schizophrenia is associated with a cerebral glutathione deficit, which may leave the brain susceptible to oxidants. Low brain glutathione and ascorbic acid levels have been shown to be associated with a perturbation of the dopaminergic system actively participate in the development of some cognitive deficits affecting schizophrenic patients [128].

D. Trans-methylation and Trans-sulfuration

Folate, also known as Vitamin B9, is of utmost importance as it provides, via the end product of its metabolism, 5-methyl tetrahydrofolate (5-MTHF), a methyl group that is required for the reconversion of Hcy back to methionine. Also vitamin B_{12} is an indispensable cofactor in the transmethylation reaction in the brain. During remethylation of

Hcy to methionine, a methyl group is transferred to the MS cofactor, cob(I)alamin, which is then activated by forming methylcobalamin. The activation of cob(I)alamin to methylcobalamin requires the methyl donor SAM. However, once the first molecule of methylcobalamin is formed and used for the reconversion of Hcy to methionine, subsequent molecules could be regenerated by using 5-MTHF as methyl donor to serve the same purpose. The absence of the alternative betaine remethylation pathway in the central nervous system (CNS) greatly reduces methylation capacity. Therefore, folate deprivation inhibits transmethylation reactions by reducing SAM and therefore to further potentiates Hcy accumulation in the CNS. Considering that the action of Vitamin B_{12} plays a role in Hcy metabolism that is similar to that of folate [101], and because folate and vitamin B_{12} deficiencies retard methionine regeneration, SAM levels are also reduced as a consequence of lack of folate or deficiency in vitamin B_{12} action [101]. Furthermore, the endogenously-generated methionine is catabolized to produce S-adenosylmethionine (SAM), a methyl donor that mediates the enzymatic reaction utilizing an endogenous antioxidant and downstream metabolite of Hcy metabolism via the trans-sulfuration pathway called glutathione (GSH), under the catalysis of glutathione S-transferase (GST) [100]. Therefore, utilization of GSH by GST would promote Hcy elimination via the trans-sulfuration pathway. This clearly highlights the important role that folate plays in the elimination of Hcy via both the transmethylation and the trans-sulfuration pathways.

Altered one-carbon metabolism, in addition to increasing the levels of homocysteine, is known to alter the levels of methylation of catecholamines, phospholipids and chromatin (histone and DNA), leading to epigenetic regulation of vital developmental genes in schizophrenia [129]. Homocysteine is metabolized to methionine after activation to S-adenosylmethionine, which is known to act as a methyl donator. Furthermore, homocysteine itself influences global and gene promoter-specific deoxyribonucleic acid (DNA) methylation [130].

Acute homocysteine treatment has been shown to cause misregulation of different gene-specific promoters and changes of corresponding messenger ribonucleic acid (mRNA) levels [131]. There is growing evidence that altered promoter-DNA methylation permits genome plasticity and adaption to a variety of environmental factors. This may provide the molecular base for a dynamic interaction between the environment and gene expression, regulated by modified promoter methylation. It has been suggested that these molecular mechanisms play a substantial role in the pathogenesis of different psychiatric disorders associated with hyperhomocysteinemia [132], including schizophrenia, eating disorders, and addiction [130, 132, 133]. These epigenetic alterations may directly influence mono-aminergic neurotransmission by modifying promoter methylation of candidate genes such as COMT and 5-HTTLPR [133, 134]. Alterations in the epigenetic modulation of gene expression have been implicated in several developmental disorders, cancer, and recently, in a variety of mental retardation and complex psychiatric disorders. DNA methylation and histone deacetylation are two major epigenetic modifications that contribute to the stability of gene expression states. Perturbing DNA methylation, or disrupting the downstream response to DNA methylation-methyl-CpG-binding domain proteins (MBDs) and histone deacetylases (HDACs) - by genetic or pharmacological means, has revealed a critical requirement for epigenetic regulation in brain development, learning, and mature nervous system stability, and has identified the first distinct gene sets that are epigenetically regulated within the nervous system. Epigenetically modifying chromatin structure in response to different stimuli appears to be an ideal mechanism to generate continuous cellular diversity and coordinate shifts in gene expression at successive stages of brain development - all the way from deciding which kind of a neuron to generate; through to how many synapses a neuron can support [135].

E. Altered Expression of Antioxidant Enzymes

Hyperhomocysteinemia-induced oxidative stress may occur as a result of decreased expression and/or activity of key antioxidant enzymes as well as increased enzymatic generation of superoxide anion (the precursor for multiple reactive oxygen and reactive nitrogen species) [136]. Recently a role for DNA methylation in controlling SOD2 expression is emerging. Hypermethylation of CpG islands at SOD2 decreases Mn-SOD expression in immortalized fibroblasts [137] and multiple myeloma cells [138]. This is facilitated by DNA methylation inhibiting transcription factor binding at the SOD2 promoter [139]. Furthermore, a causal link between DNA methylation and SOD2 gene silencing is also being developed through the use of pharmacological inhibitors of DNA methyltransferase activity [138]. Hypermethylation of CpG islands decreases the expression of many tumor suppressor genes in cancer, and SOD2 appears to be another likely candidate for transcriptional control by DNA methylation in part because it contains a CpG rich promoter.

More recently, additional studies have shown that methylation of the SOD2 promoter decreases the expression of Mn-SOD in multiple myeloma [138]. Aberrant methylation of the CREB binding site in the SOD2 promoter may participate in loss of Mn-SOD expression in cancer [140]. These studies probably indicate the possible mechanisms underlying the reported altered antioxidant enzymes at various stages of illness (discussed earlier).

F. Altered Expression of Neurotrophins

The important role of neurotrophins in the pathophysiology and pharmacotherapy of schizophrenia is gaining interest [141]. Neurotrophins, including nerve growth factor (NGF) and brain-derived neurotrophic factor (BDNF), and their respective receptors serve important roles in neurodevelopment [142], while, in the mature nervous system, they promote neuronal survival and modulate synaptic plasticity of dopaminergic, serotonergic, and interestingly, cholinergic neurons [143]. Impaired function of several neurotrophic factors, particularly nerve growth factor (NGF), brain derived neurotrophic factor (BDNF), neurotrophin-3 (NT-3), vascular endothelial growth factor (VEGF) and platelet derived growth factor (PDGF) has been implicated [141, 144] in the neurodevelopmental pathogenesis and the neuropathophysiology of course and treatment outcome of schizophrenia [145]. Further, alteration in the expression of neurotrophic factors in post-mortem brain tissues of schizophrenic patients was indicated to be a cause for neural maldevelopment and disturbed neural plasticity [146].

Neurotrophic factors BDNF and NT-3 are critical to neuronal survival and maturation as well as to promoting synapse formation [142], and their withdrawals are among the best studied triggers of neuronal apoptosis [115]. Recent studies have documented reduced levels of BDNF mRNA and protein in prefrontal cortex in schizophrenia [147] as well as concomitant reductions in the BDNF TrkB mRNA receptor [147]. While the basis for lower BDNF levels in schizophrenia is uncertain, the evidence for fewer cortical glial cells in schizophrenia may be a contributing factor, given that glia represents an important source of neurotrophins for CNS neurons. Reduced neurotrophic support represents another potential source of pro-apoptotic stress in schizophrenia.

Circulating BDNF levels is suggested to reflect the BDNF levels in the brain [148]. Decreased BDNF levels in the brain and the serum of patients with psychotic disorders have been reported [149]. Further correlations of schizophrenia with polymorphisms in the BDNF gene and changes in BDNF mRNA levels have been reported [150]. In addition, altered expression of brain-derived neurotrophic factor (BDNF) has been implicated in these illnesses [151]. BDNF gene expression is known to be dynamically regulated during develop- ment, but the regulatory controls of normal differential expression are not well understood. Methylation of CpG dinucleotides within gene promoters is emerging as an important epigenetic control mechanism of transcription, and the BDNF complex promoter contains several CpG dinucleotides. These studies demonstrate that DNA methylation of this regulatory region may be an important mechanism controlling differential expression of BDNF during forebrain development [152]. DNA methylation plays critical roles in gene silencing through chromatin remodeling. Changes in DNA methylation perturb neuronal function, and mutations in a methyl-CpG-binding protein, MeCP2, are shown to be associated with Rett syndrome.

NOVEL MECHANISM OF OXIDATIVE STRESS AND SCHIZOPHRENIA

Pregnancy and Birth Complications and Fetal Developments

The successful progression of pregnancy depends on the complex interactions between numerous biological molecules within the uterine microenvironment [153]. This involves an interaction of intracellular and extracellular factors including micronutrients, hormones, adhesion molecules, growth factors and immunomodulators that determine the fetal growth outcome. Low birth weight is a key determinant of neonatal mortality, morbidity, subsequent growth and developmental retardation and also early onset of adulthood diseases [154]. There is a vast literature on risk of pregnancy complications and bad birth outcome in schizophrenia [9, 155].

With increased use of convenience and fast foods, women can be over fed, but under-nourished in our modern society which can lead to nutrient deficiencies which can have an impact on the gestation and pregnancy outcome [156]. The role of macronutrients in birth outcome is well studied but role of micronutrients is poorly understood. Adequate amounts of micronutrients are essential for immediate as well as long term well being of the embryo, fetus and neonate and deficiencies compromise pregnancy outcome [157]. Reduced levels of circulating folate during pregnancy are known to be associated with increased risks of preterm delivery, infant low birth weight and fetal growth retardation [58]. Our studies in pre-eclampsia and women delivering preterm have shown that increased oxidative stress leading to reduced LCPUFA especially DHA [158] and increased homocysteine concentrations (unpublished data). Altered one carbon metabolism is suggested to be at the heart of intrauterine programming increasing the risk for non communicable diseases in later life [62, 159]. Epigenetic regulation, involving DNA methylation, resulting in nutritional programming, is demonstrated in animal models [160]. Recent reports suggest an epigenetic molecular mechanism potentially underlying lifelong and trans-generational perpetuation of changes in gene expression and behavior incited by early abusive behaviors [161].

The intrauterine environment plays a key role in shaping the prenatal nervous system and produces structural defects of the hippocampus, a critical area for learning and memory functions. Fetal exposure to unfavorable intrauterine conditions is suggested to compromise cognitive function in adult life [162]. Nutrients like omega-3 fatty acids are known to repair neuropathology and trigger neuroplasticity through regulation of a variety of growth factors such as NGF and BDNF [163] which play vital role in brain development during fetal and early postnatal life [164]. Recent studies have shown that children born preterm or to mothers who develop preeclampsia during pregnancy are associated with increased risk for developing childhood behavioral disorders like ADHD [165] and further develop schizophrenia in later life [166] which may be due to delayed brain maturation and functional impairment [167]. Our recent studies have shown that altered one carbon metabolism, reduced membrane DHA and increased homocysteine levels exist at the onset of psychosis [92].

These studies support the novel molecular mechanism proposed for oxidative stress and oxidative cellular injury early in the development that may lead to onset of illness in adult life.

PREVENTION OF OXIDATIVE STRESS

A. Antioxidants & Omega-3 Fatty Acids

Furthermore, evidence suggests that oxidative cell injury can be prevented and corrected by a combination of antioxidants and omega-3 fatty acid supplementation. Since increased oxidative stress mediated EPUFA peroxidative degradation as well as defective phospholipids- EPUFA metabolism exist in schizophrenia, the use of a combination of EPUFAs and antioxidants (e.g. vitamin E and C) for supplementation may be preferable [1, 10]. Some studies have already reported beneficial effects of these supplements. Dietary supplementation of either antioxidants or omega-3 fatty acids was found to improve symptoms of schizophrenia [10]. Oral supplementation of vitamin C with atypical antipsychotic reverses ascorbic acid levels, reduces oxidative stress, and improves psychiatric score hence both the drugs in combination can be used in the treatment of schizophrenia [10]. Recently, several extensive reviews and chapters have been published on this topic [1].

B. Folic Acid and Vitamin B_{12} in better outcome of schizophrenia

Genetic and clinical data suggest that folate and homocysteine may play a role in the pathogenesis of psychiatric disorders [168]. Animal studies have shown that chronic homocysteine administration is known to reduce the antioxidant capacity of the brain and this state could be prevented by the administration of folic acid [14]. Studies have also shown that folic acid prevents oxidative stress induced by Hcy administration. This B-vitamin has intrinsic antioxidant properties, acting as a Fenton-modulator [169] and/or scavenging ROS, such as OH- [170] and also [171] prevented the generation of O_2 caused by Hcy in macrophages. A few studies have examined the supplementation of B vitamins in schizophrenic patients having DHA marginality, moderate hyperhomo- cysteinemia or a combination of these and concluded that supplementation with B vitamins to a sub group of schizophrenics with mild hyperhomocysteinemia seem to correct possible relation of poor EFA and B-vitamin status with some of their psychiatric symptoms [172]. Also a study by Frankenburg, 2007 [3] suggests that physicians often check folate and cobalamin levels in patients with schizophrenia and depression [3]. Our study showing reduced levels of folic acid, Vitamin B12 and increased homocysteine in first episode psychotic patients suggest that balanced dietary supplementation of folate-vitamin B12-DHA may ameliorate the neuropathology and thereby symptomatology associated with altered one-carbon metabolism in some patients with schizophrenia [92]. Since schizophrenia has a neurodevelopmental origin and also emerging evidence suggesting that use of micronutrient containing prenatal vitamins before and during pregnancy is associated with reductions in the risk of congenital defects, preterm delivery, low infant birth-weight, and preeclampsia [158, 173] there is a need to examine if supplementation of micronutrients prenatally could ameliorate the risk for adult psychiatric disorders like schizophrenia. Further, there are no studies which have examined the effect of supplementation of micronutrients (folic acid and vitamin B_{12}) and omega 3 fatty acids during pregnancy for the prevention of schizophrenia.

FUTURE DIRECTIONS IN SCHIZOPHRENIA RESEARCH

The novel mechanism of oxidative stress presented here indicates that the more common global malnutrition of folic acid, vitamin B_{12} and omega-3 fatty acid can have very complex biological consequences through their key role in altered one carbon metabolism during fetal life in predisposing the fetus to adult schizophrenia. Although there is some evidence to support these mechanisms future studies in animal models of schizophrenia and also studies in patients will have to be carried out to validate this mechanism and associated pathologies. Furthermore, any hypothesis that is based on malnutrition is potentially very advantageous for early intervention to prevent/restore the abnormal development before it is too late. It is also important to examine the effect of supplementation of combination of micronutrients (folic acid, vitamin B_{12}), and omega 3 fatty acids (DHA) during pregnancy to examine the effects on birth outcome and further assess the behavioral and cognitive functions in their children as well as in adult patients on psychopathology. This could also be examined using animal models of schizophrenia.

REFERENCES

[1]   Mahadik, S.P.; Pillai, A.; Joshi, S.; Foster, A. *Int Rev Psychiatry*, **2006**, *18*, 119; Das, U.N. *Eur J Clin Nutr*, **2004**, *58*, 195; Farooqui, A.A.; Horrocks, L.A. *Prostaglandins Leukot Essent Fatty Acids*, **2004**, *71*, 161; Mahadik, S.P.; Evans, D. *Drugs of Today*, **1997**, *33*, 5; Mahadik, S.P.; Evans, D.; Lal, H. *Progr Neuropsychopharmacol Biol Psychiatry*, **2001**, *25*, 263; Peet, M.; Laugharne, J.D.E.; Mellor, J.; Ramchand, C.N. *Prostagland Leukotr Essent Fatty Acids*, **1996**, *55*, 71; Peet, M.; Stokes, C. *Drugs*, **2005**, *65*, 1051.

[2]   Parikh, V.; Khan, M.M.; Mahadik, S.P. *J Psychiatr Res*, **2003**, *37*, 43; Pillai, A.; Parikh, V.; Terry, A.V. Jr.; Mahadik, S.P. *J Psychiatr Res*, **2007**, *41*, 372.

[3]   Osmond, H.; Smythies, *J. Journal of Mental Science* **1952**, *98*, 309; Smythies, J.R. *Guy's Gazet*, **1966**, *14*, 2; Frankenburg, F.R. *Harv Rev Psychiatry*, **2007**, *15*, 146.

[4]   Fendri, C.; Mechri, A.; Khiari, G.; Othman, A.; Kerkeni, A.; Gaha, L. *Encephale*, **2006**, *32*, 244.

[5]   Craig, T.J.; Siegel, C.; Hopper, K.; Lin, S.; Sartorius, N. *Br J Psychiatry*, **1997**, *170*, 229; Jablensky, A.; Sartorius, N.; Ernberg, G.; Anker, M.; Korten, A.; Cooper, J.E.; Day, R.; Bertelsen, A. *Psychol Med Monogr Suppl*, **1992**, *20*,1; Regier, D.A.; Narrow, W.E.; Rae, D.S.; Manderscheid, R.W.; Locke, B.Z.; Goodwin, F.K. *Arch Gen Psychiatry*, **1993**, *50*, 85.

[6]   McGrath, J.J. *Am J Epidemiol*, **2003**, *158*, 4301.

[7]   Southard, E.E. *American Journal of Insanity*, **1915**, *71*, 603; Kraeplin, E. (ed). *Livingstone Edinburgh*, **1919**.

[8]   Susser, E.B.; Brown, A.; Matte, T.D. *Can J Psychiatry*, **1999**, *44*, 326; Weinberger, D.R. *Archives of General Psychiatry*, **1987**, *44*, 660; Murray, R.M.; Lewis, S.W. *British Medical Journal*, **1987**, *295*, 681; Murray, R.M.; O'Callaghan, E.; Castle, D.J.; Lewis, S.W. *Schizophr Bull*, **1992**, *18*, 319; Bloom, F.E. *Archives of General Psychiatry*, **1993**, *50*, 224; Xu, M.Q.; Sun, W.S.; Liu, B.X.; Feng, G.Y.; Yu, L.; Yang, L.; He, G.; Sham, P.; Susser, E.; St Clair, D.; He, L. *Schizophr Bull*, **2009**, *35*, 568; Brown, A.S.; Vinogradov, S.; Kremen, W.S.; Poole, J.H.; Deicken, R.F.; Penner, J.D.; McKeague, I.W.; Kochetkova, A.; Kern, D.; Schaefer, C.A. *Am J Psychiatry*, **2009**, *166*, 683; Corroon, B. E. *TSMJ Volume 6*, **2005**; Fatemi, S.H.; Folsom, T.D. *Schizophr Bull*, **2009**, *35*, 528.

[9]   Olney, J.W.; Farber, N.B. *Arch Gen Psychiatry*, **1995**, *52*, 998; Reynolds, G.P.; Harte, M.K. *Biochem Soc Trans*, **2007**, *35*, 433.

[10]   Tuppurainen, H.; Kuikka, J.T.; Viinamäki, H.; Husso, M.; Tiihonen, J. *Psychiatry Clin Neurosci*, **2009**, In Press; Remington, G. *Prog Brain Res*, **2008**, *172*, 117; Lieberman, J.A.; Bymaster, F.P.; Meltzer, H.Y.; Deutch, A.Y.; Duncan, G.E.; Marx, C.E.; Aprille, J.R.; Dwyer, D.S.; Li, X.M.; Mahadik, S.P.; Duman, R.S.; Porter, J.H.; Modica-Napolitano, J.S.; Newton, S.S.; Csernansky, J.G. *Pharmacol Rev*, **2008**, *60*, 358; Olijslagers, J.; Werkman, T.; McCreary, A.; Kruse, C.; Wadman, W. *Curr Neuropharmacol*, **2006**, *4*, 59; Huang, X.F.; Tan, Y.Y.; Huang, X.; Wang, Q. *Neurosci Res*, **2007**, *59*, 314; Heresco-Levy, U.*Expert Opin Emerg Drugs*, **2005**, *10*, 827; de Bartolomeis, A.; Fiore, G.; Iasevoli, F. *Curr Pharm Des*, **2005**, *11*, 3561; Lipina, T.; Labrie, V.; Weiner, I.; Roder, J. *Psychopharmacology (Berl)*, **2005**, *179*,54; Centonze, D.; Usiello, A.; Costa, C.; Picconi, B.; Erbs, E.; Bernardi, G.; Borrelli, E.; Calabresi, P. *J Neurosci*, **2004**, *24*,82; van der Heijden, F.M.; Tuinier, S.; Fekkes, D.; Sijben, A.E.; Kahn, R.S.; Verhoeven, W.M. *Eur Neuropsychopharmacol*, **2004**, *14*,259.

[11]   Arvindakshan, M.; Ghate, M.; Ranjekar, P.K.; Evans, D.R.; Mahadik, S.P. *Schizophr Res*, **2003b**, *62*, 195; Sivrioglu, E.Y.; Kirli, S.; Sipahioglu, D.; Gursoy, B.; Sarandöl, E. *Prog Neuropsychopharmacol Biol Psychiatry,* **2007**, *31*, 1493; Dakhale, G.N.; Khanzode, S.D.; Khanzode, S.S.; Saoji, A. *Psychopharmacology (Berl),* **2005**, *182,* 494.

[12]   Mahadik, S.P.; Gowda, S. *Drugs Today,* **1996**, *32*, 1; Mahadik, S.P.; Scheffer, R.E. *Prostaglandins Leukot. Essent. Fat. Acids*, **1996**, *55*, 45; Mahadik, S. P.; Evans, D.R. *Psychiatr Clin North Am,* **2003**, *26,* 85.

[13]   Zhang, X.Y.; Chen, D.C.; Xiu, M.H.; Wang, F.; Qi, L.Y.; Sun, H.Q.; Chen, S.; He, S.C.; Wu, G.Y.; Haile, C.N.; Kosten, T.A.; Lu, L.; Kosten, T.R. *Schizophr Res,* **2009**, In Press.

[14]   Halliwell, B. *Drugs Aging,* **2001**, *18,* 685 ; Halliwell, B.; Whiteman, M. *Br. J. Pharmacol,* **2004**, *142,* 231; Halliwell, B.; Gutteridge, J.M.C. *Oxford University Press, New York,* **2007.**

[15]   Tsaluchidu, S.; Cocchi, M.; Tonello, L.; Puri, B.K. *BMC Psychiatry,* **2008**, *8,* S5.

[16]   Ogawa, N. *Rinsho Shinkeigaku,* **1994**, 34, 1266.

[17]   Kunz, M.; Gama, C.S.; Andreazza, A.C.; Salvador, M.; C1eresér, K.M.; Gomes, F.A.; Belmonte-de-Abreu, P.S.; Berk, M.; Kapczinski, F. *Prog Neuropsychopharmacol Biol Psychiatry,* **2008**, *32*, 1677.

[18]   Akyol, O.; Canatan, H.; Yilmaz, H.R.; Yuce, H.; Ozyurt, H.; Sogut, S.; Gulec, M.; Elyas, H. *Clin Chim Acta,* **2004**, *345,* 151; Lohr, J.B.; Browning, J.A. *Psychopharmacol Bull,* **1995**, *31*, 159; Yao, J.K.; Reddy, R.D.; van Kammen, D.P. *CNS Drugs*, **2001**, *15*, 287.

[19]   Khan, M.M.; Evans, D.R.; Gunna, V.; Scheffer, R.E.; Parikh, V.V.; Mahadik, S.P. *Schizophr Res*, **2002**, *58*, 1.

[20]   Morera, A.L.; Intxausti, A.; Abreu-Gonzalez, P. *World J Biol Psychiatry,* **2008**, *5*, 1.

[21]   Li, H.C.; Chen, Q.Z.; Ma, Y.; Zhou, J.F. *J Zhejiang Univ Sci B,* **2006**, *7,*981; Ranjekar, P.K.; Hinge, A.; Hegde, M.V.; Ghate, M.; Kale, A.; Sitasawad, S.; Wagh, U.V.; Debsikdar, V.B.; Mahadik, S.P. *Psychiatry Res,* **2003**, *121,*109; Akyol, O.; Herken, H.; Uz, E.; Fadillioglu, E.; Unal, S.; Sogut. S.; Ozyurt, H.; Savaş, H.A. *Pro Neuro-Psychopharmacol Biol Psychiatry,* **2002**, *26,* 995; Herken, H.; Uz, E.; Ozyurt, H.; Söğüt, S.; Virit, O.; Akyol, O. *Mol Psychiatry,* **2001**, *6,* 66; Srivastava, N.; Barthwal, M.; Dalal, P.; Agarwal, A.K.; Nag, D.; Srimal, R.C.; Dikshit, M. *Psychopharmacology,* **2001**,*158*,140; Altuntas, I.; Aksoy, H.; Coskun, I.; Cayköylü, A.; Akçay, F. *Clin Chem Lab Med,* **2000**, *38*, 1277; Dietrich-Muszalska, A.; Olas, B.; Rabe-Jablonska, J. *Platelets,* **2005**, *16*, 386; Zhang, X,Y.; Tan, Y.L.; Cao, L.Y.; Wu, G.Y.; Xu, Q.; Shen, Y.; Zhou, D.F. *Schizophr Res,* **2006**, *81*, 291; Abdalla, D.S.; Monteiro, H.P.; Oliveira, J.A.; Bechara, E.J. *Clin Chem,* **1986**, *32*, 805.

[22]   Skinner, A.O.; Mahadik, S.P.; Garver, D.L. *Schizophr Res*, **2005**, *76*, 83.

[23]   Arvindakshan, M.; Sitasawad, S.; Debsikdar, V.; Ghate, M.; Evans, E.; Horrobin, D.F.; Bennett, C.; Ranjekar, P.K.; Mahadik, S.P. *Biol Psychiatry*, **2003a**, *53*, 56.

[24]   Reddy, R.D.; Keshavan, M.S.; Yao, J.K. *Schizophrenia Bulletin*, **2004**, *30*, 901.

[25]   Taneli, F.; Pirildar, S.; Akdeniz, F.; Uyanik, B.S.; Ari, Z. *Arch Med Res,* **2004**, *35*, 401; Zoroglu, S.S.; Herken, H.; Yürekli, M.; Uz, E.; Tutkun, H.; Savaş, H.A.; Bagci, C.; Ozen, M.E.; Cengiz, B.; Cakmak, E.A.; Dogru, M.I.; Akyol, O. *J Psychiatr Res*, **2002**, *36*, 309.

[26]   Suzuki, E.; Nakaki, T.; Nakamura, M.; Miyaoka, H. *J Psychiatry Neurosci,* **2003**, *28*, 288.

[27]   Yao, J.K.; Leonard, S.; Reddy, R.D. *Schizophr Res*, **2000**, *42*, 7.

[28]   Gama, C.S.; Salvador, M.; Andreazza, A.C.; Lobato, M.I.; Berk, M.; Kapczinski, F.; Belmonte-de-Abreu, P.S. *Neurosci Lett,* **2008**, *433*, 270.

[29   Michel, T.M.; Thome, J.; Martin, D.; Nara, K.; Zwerina, S.; Tatschner, T.; Weijers, H.G.; Koutsilieri, E. *J Neural Transm,* **2004**, *111*, 1191.

[30]   Vaiva, G.; Thomas, P.; Leroux, J.M.; Cottencin, O.; Dutoit, D.; Erb, F.; Goudemand, M. *Therapie,* **1994**, *49*, 343; Yao, J.K.; Reddy, R.; McElhinny, L.G.; van, Kammen, D.P. *J Psychiatr Res,* **1998b**, *32*, 385; Zhang, Y. ; Gu, Y. ; Lucas, M.J. ; Wang, Y. *J Soc Gynecol Investig,* **2003**, *10*, 5.

[31]   Yao, J.K.; Reddy, R.D.; van Kammen, D.P. *Biol Psychiatry,* **1999**, *45*, 1512.

[32]   Ben Othmen, L.; Mechri, A.; Fendri, C.; Bost, M.; Chazot, G.; Gaha, L.; Kerkeni, A. *Prog Neuropsychopharmacol Biol Psychiatry,* **2008**, *32*, 155.

[33]   Zhang, X.Y.; Tan, Y.L.; Zhou, D.F.; Cao, L.Y.; Wu, G.Y.; Haile, C.N.; Kosten, T.A.; Kosten, T.R. *J Clin Psychiatry,* **2007**, *68*, 754.

[34]   Reddy, R.; Sahebarao, M.P.; Mukherjee, S.; Murthy, J.N. *Biol Psychiatry,* **1991**, *30*, 409.

[35]   Ng, F.; Berk, M.; Dean, O.; Bush, A.I. *Int J Neuropsychopharmacol,* **2008**, *11*, 851; Berk, M.; Ng, F.; Dean, O.; Dodd, S.; Bush, A.I. *Trends Pharmacol Sci,* **2008**, *29*, 346; Berk, M. *Acta Neuropsychiatrica,* **2007**, *19*, 259; Do, K.Q.; Trabesinger, A.H.; Kirsten-Krüger, M.; Lauer, C.J.; Dydak, U.; Hell, D.; Holsboer, F.; Boesiger, P.; Cuénod, M. *Eur J Neurosci,* **2000**,*12*, 3721.

[36]   Buckman, T.D.; Kling, A.S.; Eiduson, S.; Sutphin, M.S.; Steinberg, A. *Biol Psychiatry*, **1987**, *22*, 1349.

[37] Yao, J.K.; Reddy, R.; van Kammen, D.P. *Psychiatry Res,* **2000b**, *97*, 137.

[38] Magliaro, B.C.; Saldanha, C.J. *Brain Res,* **2009**, In Press.

[39] Mahadik, S.P.; Mukherjee, S.; Scheffer, R.; Correnti, E.E.; Mahadik, J.S. *Biol Psychiatry,* **1998**, 43, 674; Cadet, J.L.; Perumal, A.S. *Biol Psychiatry,* **1990**, *28*, 738.

[40] Yao, J.K.; Reddy, R.; McElhinny, L.G.; van Kammen, D.P. *Schizophr Res,* **1998a**, *32*, 1; Yao, J.K.; Reddy, R.; McElhinny, L.G.; van Kammen, D.P. *J Psychiatr Res,* **1998b**, *32*, 385.

[41] Al-Chalabi, B.M.; Thanoon, I.A.; Ahmed, F.A. *Neuropsychobiology,* **2009**, *59*, 8.

[42] Qing, H.; Xu, H.; Wei, Z.; Gibson, K.; Li, X.M. *Eur J Neurosci,* **2003**, *17*,1563.

[43] Wei, Z.; Bai, O.; Richardson, J.S.; Mousseau, D.D.; Li, X. M. *J Neurosci Res,* **2003**, *73*, 364.

[44] Park, S.W.; Lee, C.H.; Lee, J.G.; Lee, S.J.; Kim, N.R.; Choi, S.M.; Kim, Y.H. J *Psychiatr Res,* **2009**, *43*, 274-281.

[45] Kropp, S.; Kern, V.; Lange, K.; Degner, D.; Hajak, G.; Kornhuber, J.; Rüther, E.; Emrich, H.M.; Schneider, U.; Bleich, S. *J Neuropsychiatry Clin Neurosci,* **2005**, *17*, 227.

[46] Young, J.; McKinney, S.B.; Ross, B.M.; Wahle, K.W.; Boyle, S.P. *Prostaglandins Leukot Essent Fatty Acids,* **2007**, *76*, 73.

[47] Adibhatla, R.M.; Hatcher, J.F. *Subcell Biochem,* **2008**, *49*, 241.

[48] Degreef, G.; Ashtari, M.; Bogerts, B.; Bilder, R.M.; Jody, D.N.; Alvir, J.M.; Lieberman, J.A. *Arch Gen Psychiatry,* **1992**, *49*, 531; DeLisi, L.E.; Hoff, A.L.; Kushner, M.; Degreef, G. *Psychiatry Res,* **1993**, *50*, 193; Lawrie, S.M.; Abukmeil, S.S. *Br J Psychiatry,* **1998**, *172*, 110; Lieberman, J.; Bogerts, B.; Degreef, G.; Ashtari, M.; Lantos, G.; Alvir, J. *Am J Psychiatry,* **1992**, *149*, 784.

[49] Arnold, S.E.; Hyman, B.T.; Van Hoesen, G.W.; Damasio, A.R. *Arch Gen Psychiatry,* **1991**, *48*, 625; Heckers, S. *Schizophr Bull,* **1997**, *23*, 403; Selemon, L.D.; Rajkowska, G.; Goldman-Rakic, P.S. *Arch Gen Psychiatry,* **1995**, *52*, 805.

[50] Lawrie, S.M.; Whalley, H.; Kestelman, J.N.; Abukmeil, S.S.; Byrne, M.; Hodges, A.; Rimmington, J.E.; Best, J.J.; Owens, D.G.; Johnstone, E.C. *Lancet,* **1999**, *353*, 30.

[51] Fish, B.; Marcus, J.; Hans, S.L.; Auerbach, J.G.; Perdue, S. *Arch Gen Psychiatry,* **1992**, *49*, 221.

[52] Cannon, M.; Jones, P.; Gilvarry, C.; Rifkin, L.; McKenzie, K.; Foerster, A.; Murray, R.M. *Am J Psychiatry,* **1997**, *154*, 1544.

[53] Gupta, S.; Rajaprabhakaran, R.; Arndt, S.; Flaum, M.; Andreasen, N.C. *Schizophr Res,* **1995**, *16*, 189; Lane, A.; Kinsella, A.; Murphy, P.; Byrne, M.; Keenan, J.; Colgan, K.; Cassidy, B.; Sheppard, N.; Horgan, R.; Waddington, J.L.; Larkin, C.; O'Callaghan, E. *Psychol Med,* **1997**, *27*, 1155.

[54] Keshavan, M.S.; Anderson, S.; Pettegrew, J.W. *J Psychiatr Res,* **1994**, *28*, 239.

[55] Hanson, D.R.; Gottesman, II. *BMC Med Genet,* **2005**, *11*, 6.

[56] Horrobin, D.F.; Manku, M.S.; Hillman, H.; Iain, A.; Glen, A.I.M. *Biol Psychiatry,* **1991**, *30*, 795; Horrobin, D.F. *Prostaglandin Leukotr Essent Fatty Acids,* **1996**, *55*, 3; Mahadik, S.P.; Sitasawad, V.; Mulchandani, M. *Phospholipid Spectrum Disorders in Psychiatry. Marius Press, Lancashire, UK,* **1999**, pp. 99.

[57] Chitambar, C.R.; Antony, A.C. *Modern nutrition in health and disease eds. In: Shils, M.E.; Shike. M.; Ross, A.C.; Caballero, C.; Cousins, R.J.* 10th ed. Philadelphia, PA: Lippincott Williams & Wilkins, **2005**, 1436.

[58] Zeisel, S.H. *Am J Clin Nutr,* **2009**, *89*, 673S.

[59] Refsum, H. *Br J Nutr,* **2001**, *85*, S109.

[60] Molloy, A.M.; Kirke. P.N.; Brody. L.C.; Scott. J.M.; Mills. J.L. *Food Nutr Bull,* **2008**, *29*, S101.

[61] Georgieff, M.K.; Rao, R. *In: Nelson CA, Luciana M, eds. Handbook in developmental cognitive neuroscience. Cambridge, MA: MIT Press,* **2001**, 491; Dobbing, J.; *In: Dobbing J, ed. Brain, behavior and iron in the infant diet. London, United Kingdom: Springer,* **1990**, 1; Rao, R.; Georgieff, M.K. *In: Nelson CA, ed. Minnesota Symposium on Child Psychology, Vol 31. Hillsdale, NJ: Erlbaum Associates,* **2000**, 1; Kretchmer, N.; Beard, J.L.; Carlson, S. *Am J Clin Nutr,* **1996**, *63*, 997S; Thompson. R.A.; Nelson, C.A. *Am Psychol,* **2001**, *56*, 5.

[62] Waterland, R.A.; Jirtle, R.L. *Nutrition,* **2004**, *20*, 63.

[63] Picker, J.D.; Coyle, J.T. *Harvard Reviews in Psychiatry,* **2005**, *13*, 197.

[64] Blount, B.C.; Mack, M.M.; Wehr, C.M.; MacGregor, J.T.; Hiatt, R.A.; Wang, G.; Wickramasinghe. S.N.; Everson, R.B.; Ames. B.N. *Proc Natl Acad Sci U S A,* **1997**, *94*, 3290; Ames, B.N. *Mutat Res,* **2001**, *475*, 7; Friso. S.; Choi, S.W. *J Nutr,* **2002**, *132*, 2382S; Fenech. M. Food Chem Toxicol, **2002**, *40*, 1113; Beetstra, S.; Thomas, P.; Salisbury, C.; Turner, J.; Fenech, M. *Mutat Res,* **2005**, *578*, 317.

[65] Kay, S.; Fiszbein, A.; Opler, L.A. *Schizophrenia Bulletin,* **1987**, *13,* 261; Herran, A.; Garcia-Unzueta, M.T.; Amado, J.A.; Lopez-Cordovilla, J.J.; Diez-Manrique, J.F.; Vazquez-Barquero, J.L. *Psychiatry Clinical Neuroscience,* **1999**, *53*, 531.

[66] Muntjewerff, J.W.; van der Put, N.; Eskes, T.; Ellenbroek, B.; Steegers, E.; Blom, H.; Zitman, F. *Psychiatry Research,* **2003**, *121*, 1.

[67]   Goff, D.C.; Bottiglieri, T.; Arning, E.; Shih, V.; Freudenreich, O.; Evins, A.E.; Henderson, D.C.; Baer,L.; Coyle, J. *American Journal of Psychiatry*, **2004**, *161*, 1705.

[68]   Reif, A.; Pfuhlmann, B.; Lesch, K.P. *Prog Neuropsychopharmacol Biological Psychiatry,* **2005**, *29*, 1162.

[69]   Muntjewerff, J.W.; Blom, H.J. *Progress in Neuropsychopharmacology and Biological Psychiatry*, **2005**, *29*, 1133.

[70]   Salem, N.; Kim, H.Y.; Yergey, J. In: Simopoulos, A.P.; Kifer, R.R.; Martin, R.E.; eds. *Health Effects of Polyunsaturated Fatty Acids in Seafoods. London, UK: Academic Press*, **1986**, 263.

[71]   O'Brien, J.S.; Sampson, E.L. *J Lipid Res*, **1965**, *6*, 545.

[72]   Horrobin, D.F. *Schizophr Res*, **1998**, *30*,193; Peet, M.; Glen, I.; Horrobin, D.; *(eds.), Phospholipid Spectrum Disorders in Psychiatry and Neurology, Carnforth, Lancashire, UK, Marius Press*, **2003**; McNamara, R.K.; Jandacek, R.; Rider, T.; Tso, P.; Hahn, C.G.; Richtand, N.M.; Stanford, K.E. *Schizophr Res*, **2007**, *91*, 37. Keshavan, M.S.; Mallinger, A.G.; Pettegrew, J.W.; Dippold, C. *Psychiatry Res*, **1993**, *49*, 9. Mahadik, S.P.; Yao, J.K., *In: Lieberman, J.A., Stroup, TS, Perkins DO (eds). Textbook of Schizophrenia. American Psychiatric Publishing: Washington, DC*, **2006**, 117.

[73]   Peet, M.; Horrobin, D.F. *J. Psychiatr. Res*, **2002**, *36*, 7.

[74]   Assies, J.; Lieverse, R.; Vreken, P.; Wanders, R.J.A.; Dingemans, P.M.J.A.; Linszen, D.H. *Biol Psychiatry*, **2001,** *49*, 510.

[75]   Mahadik, S.P.; Sitasawad, V.; Mulchandani, M. Membrane peroxidation and the neuropathology of Schizophrenia, in: Peet, M.; Glen, A.L.; Horrobin, D.F.; (Eds.), Phospholipid Spectrum Disorders in Psychiatry, Marius Press, Lancashire, UK, **1999**, 99; Yao, J.K.; Sistilli, C.G.; van Kammen, D.P. Prostaglandins Leukot Essent FattyAcids, **2003**, 69, 429.

[76]   Fenton, W.S.; Dickerson, F.; Boronow, J.; Hibbeln, J.R.; Knable, M. *Am. J. Psychiatry*, **2001**, *158*, 2071; Yao, J.K.; Reddy, R.D.; van Kammen, D.P. *CNS Drugs*, **2001**,*15*, 287.

[77]   Kale, A.; Joshi, S.; Naphade, N.; Sapkale, S.; Raju, M.S.V.K.; Pillai, A.;Nasrallah, H.; Mahadik, S. *Schizophrenia Research*, **2008**, *98*, 295.

[78]   Clandinin, M.T. ; Chappell, J.E. ; Leong, S. ; Heim, T. ; Swyer, P.R.; Chance, G.W. *Early Hum Dev*, **1980**, *4*, 121; Martinez, M. *J Pediatr*, **1992**, *120*, 129.

[79]   Carlson, S.E.; Cooke, R.J.; Werkman, S.H.; Tolley, E.A. *Lipids*, **1992**, *27*, 901.

[80]   Rioux, F.M.; Lindmark, G.; Hernell, O. *Acta Paediatr,* **2006**, *95*, 137.

[81]   Helland, I.B.; Smith, L.; Saarem, K.; Saugstad, O.D.; Drevon, C.A. *Pediatrics*, **2003**,*111*, e39.

[82]   Dijck-Brouwer, D.A.; Hadders-Algra, M.; Bouwstra, H.; Decsi, T.; Boehm, G.; Martini, I.A.; Boersma, E.R.; Muskiet, F.A. *Prostaglandins Leukot Essent Fatty Acids*, **2005**, *72*, 21.

[83]   Ghys, A.; Bakker, E.; Hornstra, G.; van den Hout, M. *Early Hum Dev*, **2002**, *69*,83; Bakker, E.C.; Ghys, A.J.; Kester, A.D.; Vles, J.S.; Dubas, J.S.; Blanco, C.E.; Hornstra, G. *Eur J Clin Nutr*, **2003**, *57*, 89.

[84]   Kronenberg, G.; Colla, M.; Endres, M. *Curr Mol Med*, **2009**, *9*, 315.

[85]   Mattson, M.P.; Shea, T.B. *Trends in Neuroscience*, **2003**, *26*, 137; Chen, H.; Zhang, M.S.; Schwarzschild, A.M.; Heran, A.M.; Logroscino, G.; Willet, C.W.; Ascherio, A. *American Journal of Epidemiology*, **2004**, *160*, 368.

[86]   Wyse, A.T.S.; Zugno, A.I.; Streck, E.L.; Matte´, C.; Calcagnotto, T.; Wannmacher, C.M.D.; Wajner, M. *Neurochem. Res,* **2002**, *27*, 1685; Streck, E.L.; Vieira, P.S.; Wannmacher, C.M.D.; Dutra-Filho, C.S.; Wajner, M.; Wyse, A.T.S. *Metab. Brain Dis,* **2003**,*18*, 147; Matte´, C.; Monteiro, S.C.; Calcagnotto, T.; Bavaresco, C.S.; Netto, C.A.; Wyse, A.T.S. *Int. J. Dev. Neurosci,* **2004**, *22*,185; Matte´, C.; Scherer, E.B.S.; Stefanello, F.M.; Barschak, A.G.; Vargas, C.R.; Netto, C.A.; *Wyse, A.T.S. Int. J. Dev. Neurosci,* **2007**, *25, 545.*

[87]   Boldyrev, A.A.; Johnson, P. *Journal of Alzheimer's Disease*, **2007**, *11*, 219; Kim, W.K.; Pae, Y.S. *Neurosci. Let*, **1996**, *216*, 117; Dayal, S.; Arning, E.; Bottiglieri, T.; Boger, R.H.; Sigmund, C.D.; Faraci, F.M.; Lentz, S.R. *Stroke*, **2004**, *35*, 1957; Boldyrev, A.A.; Carpenter, D.O.; Johnson, P. *J. Neurochem*, **2005**, *95*, 913; Faraci, F.M.; Lentz, S.R. Stroke, **2004**, *35*, 345; Ho, P.I.; Ortiz, D.; Rogers, E.; Shea, T.B. *J. Neurosci. Res*, **2002**, *70*, 694; Ho, P.I.; Ashline, D.; Dhitavat, S.; Ortiz, D.; Collins, S.C.; Shea, T.B.; Rogers, E. *Neurobiol. Dis*, **2003**, *14*, 32; Jara-Prado, A.; Ortega-Vazquez, A.; Martinez-Ruano, L.; Rios, C.; Santamarı´a, A. *Neurotoxicol. Res*, **2003**, *5*, 237; Zieminska, E.; Lazarewicz, J.W. Acta Neurobiol. Exp, **2006**, *66*, 301; Zhang, F.; Slungaard, A.; Vercellotti, G.M.; Iadecola, C. *Am. J. Physiol*, **1998**, *274*, 1704.

[88]   Jarskog; L.F.; Gilmore, J.H.; Selinger, E.S.; Lieberman, J.A. *Biol Psychiatry*, **2000**, *48*, 641.

[89]   Jiang, Y.; Sun, T.; Xiong, J.; Cao, J.; Li, G.; Wang, S. *Acta Biochim Biophys Sin (Shanghai)*, **2007**, *39*, 657.

[90]   Finkelstein, J.D. *Eur J Pediatr*, **1998**, *157*, S40.

[91]   Herrmann, W. *Clin Lab,* **2006**, *52*, 367; Refsum, H.; Ueland, P.M.; Nygård, O.; Vollset, S.E. *Annu Rev Med,* **1998**, *49*, 31. Diaz-Arrastia, R. *Arch. Neurol*, **2000**, *57*, 1422; Mattson, M.P.; Kruman, I.I.; Duan, W. *Ageing Res. Rev*, **2002**, *1*, 95; Mattson, M.P.; Haberman, F. *Curr. Med. Chem*, **2003**, *10*, 1923; Regland, B. *Prog. Neuropsychopharmacol. Biol. Psychiatry*, **2005**, *29*, 1124; Sachdev, P. *Rev. Bras. Psiquitr*, **2004,** *26*, 49.

[92]   Scholl, T.O.; Johnson, W.G. *Am J Clin Nutr*, **2000**, *71*, 1295S.

[93]   Clarke, R.; Smith, A.D.; Jobst, K.A.; Refsum, H.; Sutton, L.; Ueland, P.M. *Arch. Neurol,* **1998,** *55,* 1449; Selhub, J.; Miller, J.W. *Am. J. Clin. Nutr,* **1992,** *55,* 131; Kale, A.; Naphade, N.; Sapkale, S.; Raju, M.S.V.K.; Pillai, A.; Joshi, S.; Mahadik, S. *Psychiatry Research,* **2009,** In Press.

[94]   Ueland, P.M.; Hustad, S.; Schneede, J.; Refsum, H.; Vollset, S.E. *Trends Pharmacol Sci,* **2001,** *22,* 195.

[95]   Levine, J.; Stahl, Z.; Sela, B.A.; Gavendo, S.; Ruderman, V.; Belmaker, R.H. *American Journal of Psychiatry,* **2002,** *159,* 1790; Applebaum, J.; Shimon, H.; Sela, B.A.; Belmaker, R.H.; Levine, J. *Journal of Psychiatric Research,* **2004,** *38,* 413.

[96]   Haidemenos, A.; Kontis, D.; Gazi, A.; Kallai, E.; Allin, M.; Lucia, B. *Progress in Neuro-psychopharmacol Biological Psychiatry,* **2007,** *31,* 1289.

[97]   Lazarewicz, J. ; Urbanska, E.M. *J. Neurosci. Res,* **2005,** *79,* 375.

[98]   Lipton, S.A.; Kim, W.K.; Choi, Y.B.; Kumar, S.; D'Emilia, D.M.; Rayudu, P.V.; Arnelle, D.R.; Stamler, J.S. *Proc. Natl. Acad. Sci. U. S. A,* **1997,** *94,* 5923.

[99]   Sachdev, P.; Loo, C.; Mitchell, P.; Malhi, G. *Psychiatry and Clinical Neuroscience,* **2005,** *59,* 354.

[100]   Kubová, H.; Folbergrová, J.; Mares, P. *Epilepsia,* **1995,** *36,* 750; Folbergrová, J.; Haugvicová, R.; Mares, P. *Exp Neurol,* **2000,***161,* 336.

[101]   Mudd, S.H.; Skovby, F.; Levy, H.L.; Pettigrew, K.D.; Wilcken, B.; Pyeritz, R.E.; Andria, G.; Boers, G.H.; Bromberg, I.L.; Cerone, R.; *et al. Am J Hum Genet,* **1985,** *37,* 1.

[102]   Miller, J.W.; Selhub, J.; Nadeau, M.R.; Thomas, C.A.; Feldman, R.G.; Wolf, P.A. *Neurology,* **2003,** *60,* 1125.

[103]   Zhu, B.T. *Curr Drug Metab,* **2002,** *3,* 321.

[104]   Zieminska, E.; Lazarewicz, J.W. *Acta Neurobiol Exp (Wars),* **2006,** *66,* 301.

[105]   McKeever, M.P.; Weir, D.G.; Molloy, A.; Scott, J.M. *Clin Sci (Lond),* **1991,** *81,* 551; Ichinohe, A.; Kanaumi, T.; Takashima, S.; Enokido, Y.; Nagai, Y.; Kimura, H. *Biochem Biophys Res Commun,* **2005,** *338,* 1547.

[106]   Burek, M.J.; Oppenheim, R.W. *Brain Pathol,* **1996,** *6,* 427.

[107]   Bredesen, D.E. *Ann. Neurol,* **1995,** *38,* 839.

[108]   Friedlander, R.M. *N. Engl. J. Med,* **2003,** *348,* 1365.

[109]   Kruman, I.; Culmsee, C.; Chan, S.L.; Kruman, Y.; Guo, Z.; Penix, L.; Mattson, M.P. *J. Neurosci,* **2000,** *20,* 6920.

[110]   Chern, C.L.; Huang, R.F.; Chen, Y.H.; Cheng, J.T.; Liu, T.Z. *Biomed Pharmacother,* **2001,** *55,* 434.

[111]   Greene, J.G.; Greenamyre, J.T. *J Neurochem,* **1995,** *64,* 430.

[112]   Huang, N.K.; Lin, Y.W.; Huang, C.L.; Messing, R.O.; Chern, Y. *J Biol Chem,* **2001,** *276,* 13838.

[113]   Outinen, P.A.; Sood, S.K.; Liaw, P.C.; Sarge, K.D.; Maeda, N.; Hirsh, J.; Ribau, J.; Podor, T.J.; Weitz, J.I.; Austin, R.C. *Biochem J,* **1998,** *332,* 213; Kokame, K.; Agarwala, K.L.; Kato, H.; Miyata, T. *J Biol Chem,* **2000,** *275,* 32846.

[114]   Margolis, R.L.; Chuang, D.M.; Post, R.M. *Biol. Psychiatry,* **1994,** *35,* 946; Lewis, D.A.; Lieberman, J.A. *Neuron,* **2000,** *28,* 325; Berger, G.E.; Wood, S.; McGorry, P.D. *Psychopharmacol. Bull,* **2003,** *37,* 79; Jarskog, L.F.; Glantz, L.A.; Gilmore, J.H.; Lieberman, J.A. *Prog Neuropsychopharmacol Biol Psychiatry,* **2005,** *29,* 846; Zhong, L.T.; Kane, D.J.; Bredesen, D.E. Brain Res. *Mol. Brain Res,* **1993a,** *19,* 35; Zhong, L.T.; Sarafian, T.; Kane, D.J.; Charles, A.C.; Mah, S.P.; Edwards, R.H.; Bredesen, D.E. *Proc. Natl. Acad. Sci. U. S. A,* **1993b,** *90,* 4533; Lawrence, M.S.; Ho, D.Y.; Sun, G.H.; Steinberg, G.K.; Sapolsky, R.M.; *J. Neurosci,* **1996,** *16,* 486; Huttenlocher, P.R. *Brain Res,* **1979,** *163,* 195; Huttenlocher, P.R.; Dabholkar, A.S. *J. Comp. Neurol,* **1997,** *387,* 167.

[115]   Feinberg, I. *J. Psychiatr. Res,* **1982,***17,* 319.

[116]   Thompson, C.B. *Science,* **1995,** *267,* 1456.

[117]   Geddes, J.R.; Lawrie, S.M. *Br. J. Psychiatry,* **1995,***167,* 786; Gilmore, J.H.; Jarskog, L.F. *Schizophr. Res,* **1997,** *24,* 365; Urakubo, A.; Jarskog, L.F.; Lieberman, J.A.; Gilmore, J.H. *Schizophr. Res,* **2001,** *47,* 27.

[118]   Matsuzawa, D.; Obata, T.; Shirayama, Y.; Nonaka, H.; Kanazawa, Y.; Yoshitome, E.; Takanashi, J.; Matsuda, T.; Shimizu, E.; Ikehira, H.; Iyo, M.; Hashimoto, K. *PLoS One,* **2008,** *3,* e1944.

[119]   Dean, O.; Bush, A.I.; Berk, M.; Copolov, D.L.; van den Buuse, M. *Behav Brain Res,* **2009,** *198,* 258.

[120]   Owen, M.J.; Craddock, N.; O'Donovan, M.C. *Trends Genet,* **2005,** 21, 518; Harrison, P.J.; Weinberger, D.R. *Mol Psychiatry,* **2005,** *10,* 40.

[121]   Durmaz, A.; Dikmen, N. *J Enzyme Inhib Med Chem,* **2007,** *22,* 733.

[122]   Looney, J.M.; Childs, H.M.*J. Clin Invest,* **1934,** 13, 963.

[123]   Schulz, J.B.; Lindenau, J.; Seyfried, J.; Dichgans, J. *Eur J Biochem,* **2000,** *267,* 4904.

[124]   Yao, J.K.; Leonard, S.; Reddy, R. *Dis Markers,* **2006,** *22,* 83.

[125]   Sucher, N.J. ; Lipton, S.A. *J Neurosci Res,* **1991,** *30,* 582; Kohr, G.; Eckardt, S.; Luddens, H.; Monyer, H.; Seeburg, P.H. *Neuron,* **1994,** *12,* 1031; Varga, V.; Jenei, Z.; Janaky, R.; Saransaari, P.; Oja, S.S. *Neurochem Res,* **1997,** *22,* 1165.

[126]   Javitt, D.C.; Zukin, S.R. Am J Psychiatry, 1991, *148,* 1301; Tamminga, C.A. *Crit Rev Neurobiol,* **1998,** *12,* 21; Goff, D.C.; Coyle, J.T. *Am J Psychiatry,* **2001,** *158,*1367; Hashimoto, K.; Fukushima, T.; Shimizu, E.; Komatsu, N.; Watanabe,

H.; Shinoda, N.; Nakazato, M.; Kumakiri, C.; Okada, S.; Hasegawa, H.; Imai, K.; Iyo, M. *Arch Gen Psychiatry*, **2003a**, *60*, 572; Hashimoto, K.; Okamura, N.; Shimizu, E.; Iyo, M. *Curr Med Chem-CNS Agents*, **2003b**, *60*, 572.

[127] Fukami, G.; Hashimoto, K.; Koike, K.; Okamura, N.; Shimizu, E.; Iyo, M. *Brain Res*, **2004**, *1016*, 90.

[128] Gysin, R.; Kraftsik, R.; Sandell, J.; Bovet, P.; Chappuis, C.; Conus, P.; Deppen, P.; Preisig, M.; Ruiz, V.; Steullet, P.; Tosic, M.; Werge, T.; Cuénod, M.; Do, K.Q. *Proc Natl Acad Sci U S A*, **2007**,*104*, 16621.

[129] Castagné, V.; Rougemont, M.; Cuenod, M.; Do, K.Q. *Neurobiol Dis*, **2004**, 15, 93.

[130] Sharma, R.P. *Schizophrenia Research*, **2005**, *72*, 79.

[131] Bleich S.; Lenz, B.; Ziegenbein, M.; Beutler, S.; Frieling, H.; Kornhuber, J.; Bönsch, D. *Alcohol Clin Exp Res*, **2006**, *30*, 587.

[132] Lenz, B.; Bleich, S.; Beutler, S.;Schlierf, B.; Schwager, K.; Reulbach, U.; Kornhuber, J.; Bönsch, D. *Exp Cell Res*, **2006**, *312*, 4049.

[133] Frieling, H.; Gozner, A.; Römer, K.D.; Lenz, B.; Bönsch, D.; Wilhelm, J.; Hillemacher, T.; de Zwaan, M.; Kornhuber, J.; Bleich, S. *Mol Psychiatry*, **2007**, *12*, 229.

[134] Abdolmaleky, H.M.; Cheng, K.H.; Faraone, S.V.; Wilcox, M.; Glatt, S.J.; Gao, F.; Smith, C.L.; Shafa, R.; Aeali, B.; Carnevale, J.; Pan, H.; Papageorgis, P.; Ponte, J.F.; Sivaraman, V.; Tsuang, M.T.; Thiagalingam, S. *Hum Mol Genet*, **2006**, *15*, 3132.

[135] Mill, J.; Petronis, A. *Mol Psychiatry*, **2007**, *12*, 799.

[136] MacDonald, J.L.; Roskams, A.J. *Prog Neurobiol*, **2009**, *88*, 170.

[137] Faraci, F.M. *Arterioscler Thromb Vasc Biol*, **2003**, *23*, 371; Loscalzo, J. *Biochem Soc Trans*, **2003**, *31*, 1059.

[138] Huang, Y.; He, T.; Domann, F.E. *DNA Cell Biol*, **1999**, *18*, 643.

[139] Hodge, D.R.; Xiao, W.; Peng, B.; Cherry, J.C.; Munroe, D.J.; Farrar, W.L. *Cancer Res*, **2005**, *65*, 6255; Hodge, D.R.; Peng, B.; Pompeia, C.; Thomas, S.; Cho, E.; Clausen, P.A.; Marquez, V.E.; Farrar, W.L. *Cancer Biol Ther*, **2005**, *4*, 585.

[140] Huang, Y.; Peng, J.; Oberley, L.W.; Domann, F.E. Free Radic Biol Med, 1997, *23*, 314.

[141] Hitchler, M.J.; Wikainapakul, K.; Yu, L.; Powers, K.; Attatippaholkun, W.; Domann, F.E. *Epigenetics*, **2006**, *1*, 163.

[142] Buckley, P.F.; Pillai, A.; Evans, D.; Stirewalt, E.; Mahadik, S. *Schizophr Res*, **2007a**, *91*, 1; Buckley, P.F.; Mahadik, S.; Pillai, A.; Terry, A. Jr. *Schizophr Res*, **2007b**, *94*, 1; Shoval, G.; Weizman, A. *Eur Neuropsychopharmacol*, **2005**, *15*, 319.

[143] McAllister, A.K. *Cell Mol. Life Sci*, **2001**, *58*, 1054.

[144] Fisher, C.M. *Can. J. Neurol. Sci*, **1991**, *18*, 18; Dawson, N.M.; Hamid, E.H.; Egan, M.F.; Meredith, G.E. *Synapse*, **2001**, *39*, 70; Knusel, B.; Hefti, F. *J. Neurochem*, **1991**, *57*, 955.

[145] Mathalon, D.M.; Sullivan, E.V.; Lim, K.O.; Pfefferbaum, A. *Arch. Gen. Psychiatry*, **2001**, *58*, 148; Nawa, H.; Futamura, T.; Mizuno, M.; Takahashi, M.; Toyooka, K.; Someya, T. *Nippon Rinsho*, **2003**, *61*, 521; Pillai, A. *Neurosignals*, **2008**, *16*, 183.

[146] Arnold, S.E.; Hyman, B.T.; Van Hoesen, G.W.; Damasio, A.R. *Arch Gen Psychiatry*, **1991**, *48*, 625; Benes, F.M.; Sorensen, I.; Bird, E.D. *Schizophr Bull*, **1991**, *17*, 597; Akbarian, S.; Bunney, W.E. Jr.; Potkin, S.G.; Wigal, S.B.; Hagman, J.O.; Sandman, C.A.; Jones, E.G. *Arch Gen Psychiatry*, **1993**, *50*, 169; Keshavan, M.; Montrose, D.M.; Rajarethinam, R.; Diwadkar, V.; Prasad, K.; Sweeney, J.A. *Schizophr Res*, **2008**, *103*, 114; Lewis, D.A.; Levitt, P. *Ann Rev Neurosci*, **2002**, *25*, 409; Powchik, P.; Davidson, M.; Haroutunian, V.; Gabriel, S.M.; Purohit, D.P.; Perl, D.P.; Harvey, P.D.; Davis, K.L. *Schizophr Bull*, **1998**, *24*, 325.

[147] Durany, N.; Michel, T.; Zöchling, R.; Boissl, K.W.; Cruz-Sánchez, F.F.; Riederer, P.; Thome, J. *Schizophr Res*, **2001**, *52*, 79.

[148] Hashimoto, T.; Volk, D.W.; Buchheit, S.E.; Lewis, D.A. *Society for Neuroscience, Online Abstract Viewer/Itinerary Planner*, **2002**, *703*, 7; Weickert, C.S.; Hyde, T.M.; Lipska, B.K.; Herman, M.M.; Weinberger, D.R.; Kleinman, J.E. *Mol. Psychiatry*, **2003**, *8*, 592.

[149] Reis, H.J.; Nicolato, R.; Barbosa, I.G.; Teixeira do Prado, P.H.; Romano-Silva, M.A.; Teixeira, A.L. *Neurosci Lett*, **2008**, *439*, 157.

[150] Rizo, E.N.; Rontos, I.; Laskos, E.; Arsenis, G.; Michalopoulou, P.G.; Vasilopoulos, D.; Gournellis, R.; Lykouras, L. *Prog Neuropsychopharmacol Biol Psychiatry*, **2008**, *32*, 1308.

[151] Lu, B.; Martinowich, K. *Novartis Found Symp*, **2008**, *289*, 119.

[152] Choy, K.H.; de Visser, Y.; Nichols, N.R.; van den Buuse, M. *Hippocampus*, **2008**, *18*, 655.

[151] Dennis, K.E.; Levitt, P. *Brain Res Mol Brain Res*, **2005**, *140*, 1.

[153] Shankar, R.; Gude, N.; Cullinane, F.; Brennecke, S.; Purcell, A.W.; Moses, E.K. *Reproduction*, **2005**, *129*, 685.

[154] Fall, C.H.D.;Yajnik, C. S.; Rao, S.; Coyaji, K.J. *In Fetal programming: Influences on development and disease in Later Life (O'Brien, P.M.S, Wheeler, T.& Barker,D.J.P.,eds), RCOG Press, London,U.K,* **1999**; Ericsson, J.G. *BMJ*, **2005**, *330*, 1096.

[155] Brown, A.S.; Susser, E.S. *Schizophr Bull*, **2008**, *34*, 1054; Susser, E.; St Clair, D.; He, L. *Ann N Y Acad Sci*, **2008**, *1136*, 185; Susser, E.S. ; Lin, S.P. *Arch Gen Psychiatry*, **1992**, *49*, 983; Susser, E.; Neugebauer, R.; Hoek, H.W.; Brown, A.S.; Lin, S.; Labovitz, D.; Gorman, J.M. *Arch Gen Psychiatry*, **1996**, *53*, 25; St Clair, D.; Xu, M.; Wang, P.; Yu, Y.; Fang, Y.; Zhang, F.; Zheng, X.; Gu, N.; Feng, G.; Sham, P.; He, L. *JAMA*. **2005**, *294*, 557.

[156] Glenville, M. *Curr. Opin. Obstet. Gynecol*, **2006**, *18*, 642.

[157] Lee, B.E.; Hong, Y.C.; Lee, K.H.; Kim, Y.J.; Kim, W.K.; Chang, N.S.; Park, E.A.; Park, H.S.; Hann, H.J. *Euro. J. Clin. Nutr*, **2004**, *58*, 1365.

[158] Mehendale, S.; Kilari, A.; Dangat, K.; Tarlekar, V.; Mahadik, S.; Joshi, S. *Int J Gynaecol Obstet*, **2008**, *100*, 234.

[159] Cooney, C.A.; Dave, A.A.; Wolff, G.L. *J Nutr*, **2002**, *132*, 2393S; Lillycrop, K.A.; Phillips, E.S.; Jackson, A.A.; Hanson, M.A.; Burdge, G.C. *J Nutr*, **2005**, *135*, 1382.

[160] Yajnik, C.S. ; Deshpande, S.S. ; Jackson, A.A. ; Refsum, H. ; Rao, S. ; Fisher, D.J. ; Bhat, D.S. ; Naik, S.S. ; Coyaji, K.J. ; Joglekar, C.V. ; Joshi, N. ; Lubree, H.G. ; Deshpande, V.U. ; Rege, S.S. ; Fall, C.H. *Diabetologia*, **2008**, *51*, 29.

[161] Roth, T.L.; Lubin, F.D.; Funk, A.J.; Sweatt, J.D. *Biol Psychiatry*, **2009**, *65*, 760.

[162] Gomez-Pinilla, F.; Vaynman, S. *Exp Neurol*, **2005**, *192*, 235.

[163] Wu, A.; Ying, Z.; Gomez-Pinilla, F. *Journal of Neurotrauma*, **2004**, *21*, 1457.

[164] Georgieff, M.K. *Am J Clin Nutr*, **2007**, *85*, 614S; Malamitsi-Puchner, A.; Nikolaou, K.E.; Puchner, K.P. *Ann N Y Acad Sci*, **2006**, *1092*, 293.

[165] Shum, W.W.; Da Silva, N.; McKee, M.; Smith, P.J.; Brown, D.; Breton, S. *Cell*, **2008**, *135*, 1108.

[166] Bennedsen, B.E. *Schizophr Res*, **1998**, *33*, 1.

[167] Barkley, R.A. *J Dev Behav Pediatr*, **1990**, *11*, 343.

[168] Monji, A.; Yanagimoto, K.; Maekawa, T.; Sumida, Y.; Yamazaki, K.; Kojima, K. *Journal of Clinical Psychopharmacology*, **2005**, *25*, 3.

[169] Patro, B.S.; Adhikari, S.; Mukherjee, T.; Chattopadhyay, S. *Med. Chem*, **2006**, *2*, 407.

[170] Joshi, R.; Adhikari, S.; Patro, B.S.; Chattopadhyay, S.; Mukherjee, T. *Free Radic. Biol. Med*, **2001**, *30*, 1390.

[171] Au-Yeung, K.K.W.; Yip, J.C.W.; Siow, Y.L.; Karmin, O. *Can. J. Physiol. Pharmacol*, **2006**, *84*, 141.

[172] Kemperman, R.F.; Veurink, M.; van der Wal, T.; Knegtering, H.; Bruggeman, R.; Fokkema, M.R.; Kema, I.P.; Korf, J.; Muskiet, F.A. *Prostaglandins Leukotrienes and Essential Fatty Acids*, **2006**, *74*, 75.

[173] Dangat, D.K.; Mehendale, S.S.; Yadav, R.H.; Kilari, S.A.; Kulkarni, V.A.; Taralekar, S.V.; Joshi, R.S. *Neonatology*, **2009**, In Press; Kilari, S.A.; Mehendale, S.S.; Dangat, D.K.; Yadav, R.H.; Gupta, A.; Taralekar, S.V.; Joshi, R.S. *Journal of perinatal medicine*, **2009**, In Press.

CHAPTER 10

Neuroprotective Actions of Polyunsaturated Fatty Acids with Particular Reference to Alzheimer's Disease

Undurti N Das*

Jawaharlal Nehru Technological University, Kakinada-530 003, India and UND Life Sciences, 13800 Fairhill Road, #321, Shaker Heights, OH 44120, USA

Abstract: Eicosapentaenoic acid (EPA) and docosahexaenoic acid (DHA) are of benefit in Alzheimer's disease by virtue of their anti-inflammatory actions, ability to modulate neural function, including neurotransmission, membrane fluidity, ion channel, enzyme regulation and gene expression. EPA, DHA, and ω-6 arachidonic acid (AA) form precursors to anti-inflammatory compounds: lipoxins, resolvins, protectins and maresins that suppress leukocyte migration and activation, inhibit NF-κB activation, production of pro-inflammatory cytokines tumor necrosis factor-α and interleukin-6, free radical generation, and enhance endothelial nitric oxide generation and augment the healing process. In animal models, the protective action of EPA and DHA against Alzheimer's disease correlated with increased formation of lipoxins, resolvins, protectins and maresins. EPA, DHA and AA stimulate neurite outgrowth by activating syntaxin 3 that is specifically involved in fast calcium-triggered exocytosis of neurotransmitters. SNAP25 (synaptosomal-associated protein of 25 kDa), a syntaxin partner implicated in neurite outgrowth, interacted with syntaxin 3 only in the presence of AA that allowed the formation of the binary syntaxin 3-SNAP 25 complex. AA stimulated syntaxin 3 to form the ternary SNARE complex (soluble N-ethylmaleimide-sensitive factor attachment protein receptor), which is needed for the fusion of plasmalemmal precursor vesicles into the cell surface membrane that leads to membrane fusion that facilitates neurite outgrowth. These results imply that EPA, DHA, and AA when given in optimal amounts are of benefit in the prevention and treatment of Alzheimer's disease. PUFAs enhance the concentrations of neurotrophic factors in the brain that may provide additional protection to neurons. Thus, PUFAs by themselves or their stable synthetic analogues could be of benefit in Alzheimer's disease and other neurodegenerative diseases.

Keywords: Polyunsaturated fatty acids, Alzheimer's disease, lipoxins, protectins, resolvins, maresins, syntaxin, free radicals, nitric oxide, inflammation, cytokines, tumor necrosis factor.

INTRODUCTION

Alzheimer's disease (AD), the most common form of dementia, is a progressive neurodegenerative disorder characterized by amyloid plaques composed of aggregated amyloid beta plaques, neurofibrillary tangles (NFT) composed of hyperphosphorylated tau and synaptic defects resulting in neuritic dystrophy and neuronal death [1].

AD produces loss of memory and other intellectual abilities serious enough to interfere with daily life. It is estimated that about 5 million Americans now have Alzheimer's disease. It is important to note, however, that AD is not a normal part of aging. The duration of Alzheimer's disease can vary from three to twenty years, but many die an average of four to six years after diagnosis. The disease initially presents with mild cognitive impairment such as memory lapses, especially in forgetting familiar words or names or the location of keys, eyeglasses or other everyday objects. As the disease progresses, they are unable to recognize and remember spouse and children, do not respond to the environment, and ultimately lose ability to speak and control movement.

PATHOBIOLOGY OF ALZHEIMER'S DISEASE

Plaques and tangles are prime suspects in damaging neurons of the brain and lead to the onset of Alzheimer's disease. Plaques build up between nerve cells, and contain deposits of the protein fragment *β-amyloid*. Tangles are twisted fibers of another protein called *tau* and form inside dying cells. Most people develop some plaques and tangles as they age, but those with Alzheimer's develop far more. The plaques and tangles form in areas important in

---

*Address correspondence to: Jawaharlal Nehru Technological University Kakinada-530 003, Andhra Pradesh, India

Akhlaq A. Farooqui & Tahira Farooqui (Eds.)

learning and memory and then spread to other regions. It is not clear what role plaques and tangles play in Alzheimer's disease, but are believed to block communication among nerve cells and disrupt activities that cells need to survive and involve mainly nerve cells in areas of the brain that are vital to memory and other mental abilities. The levels of neurotransmitter acetylcholine (Ach) are lower in Alzheimer's disease.

Missense mutations in amyloid precursor protein (APP), presenillin-1 (PS-1) (chromosome 14), presenillin-2 (PS-2) (chromosome 1) genes alter the proteolysis of APP and increase the generation of Aβ42 (amyloid β 42). The accumulation of Aβ42 triggers the inflammatory responses due to microglial activation and release of pro-inflammatory cytokines. Perturbations in the equilibrium between kinases and phosphatases results in hyperphosphorylation of tau protein, leading to neuronal degeneration and loss [2].

Several other genes that are considered to increase susceptibility for AD include: apolipoprotein E (ApoE 4) variant [3], 2-macroglobulin [4], the K-variant of butyryl-cholinesterase [5], and several mitochondrial genes [6]. Other factors that seem to play a role in the aetiopathogenesis of AD include: brain metabolic abnormalities, environmental factors, and age related decrease in neuronal membrane fluidity that could lead to neuronal death, possibly, by increasing the formation of amyloid beta plaques and hyperphosphorylation of tau protein [7].

Mutations in presenilins lead to dominant inheritance of Familial Alzheimer's disease (FAD). These mutations alter the cleavage of γ-secretase of the amyloid precursor protein, resulting in the increased ratio of Aβ42/ Aβ40 and accelerated amyloid plaque pathology in transgenic mouse models [8]. Proteolytic processing of APP by β-secretase, γ-secretase, and caspases generates A-beta peptide and carboxyl-terminal fragments (CTF) of APP, which have a role in the pathogenesis of Alzheimer's disease [9]. Missense mutations in the gene encoding APP, as well as those in the genes encoding PS-1 and PS-2, share the common feature of altering the γ-secre- tase cleavage of APP to increase the production of the amyloidogenic Aβ42, a primary component of amyloid plaques in both familial and sporadic AD.

In a recent bioinformatics study, I observed that three important proteins: presenilin-1 (PS-1), presenilin-2 (PS-2), and amyloid precursor protein (APP) out of 73 proteins that were studied appeared to be the key pathological proteins in the evolution of Alzheimer's disease (unpublished data). Factors that seem to influence the initiation and progression and thus, have a role in the pathophysiology of AD are: i) Aβ42/ Aβ40 ratio and oligomers of these peptides; ii) oxidative stress; iii) proinflammatory cytokines produced by activated glial cells, iv) alterations in cholesterol homeostasis, and v) alterations in cholinergic nervous system [10].

FACTORS THAT INFLUENCE THE GENESIS AND PROGRESSION OF ALZHEIMER'S DISEASE

Familial Alzheimer's disease (FAD) is associated with mutations in APP, PS-1, and PS-2. These substances, along with their normal counterparts, undergo proteolytic processing in the endoplasmic reticulum (ER). The mutated compounds, apart from increasing the ratio of Aβ42 to Aβ40, could down-regulate the calcium buffering activity of the ER. Decrease in the ER calcium pool would cause compensatory increases in other calcium pools, particularly in mitochondria. Increase in mitochondrial calcium levels are associated with enhanced formation of superoxide radical formation, and hence damage to the neurons and their senility [11].

Presenilins act as catalytic subunit of gamma secretase. Presenilins, the causative molecules of FAD, are transmembrane proteins localized predominantly in the ER and Golgi apparatus. Presenilins are thought to be involved in intramembrane proteolysis mediated by their gamma secretase activities. In addition, presenilins interact with FKBP38 (human FK506-binding protein 38) and form macromolecular complexes together with anti-apoptotic Bcl-2, thus it may regulate the apoptotic cell death [12].

Presenilins and their interacting proteins play a major role in the generation of A-beta from the amyloid precursor protein (APP). Three proteins nicastrin, aph-1 and pen-2 interact with presenilins to form a large enzymatic complex known as gamma secretase that cleaves APP to generate Aβ [13].

There are numerous proteases in the brain that could potentially participate in Aβ turnover. Aβ (amyloid-beta) degrader candidates include: cathepsin D and E, gelatinase A and B, trypsin- or chymotrypsin-like endopeptidase, amino peptidase, neprilysin (enkephalinase), serine protease complexed with 2-macroglobulin, and insulin-degrading enzyme [14]. Genetic linkage studies have linked Alzheimer's disease and plasma Aβ42 levels to chromosome 10q, which harbors the IDE (insulin-degrading enzyme) gene. IDE has been observed in human cerebrospinal fluid; and its activity levels and m-RNA are decreased in AD brain tissue and is associated with increased amyloid beta levels [14, 15]. Amyloid beta is the major component of amyloid plaques characterizing Alzhei- mer's disease. Amyloid beta accumulation can be affected by numerous factors including increased rates of its production and/or impaired clearance. Insulin degrading enzyme is responsible for the degradation and clearance of amyloid beta in the brain [16].

Several studies showed that Aβ is toxic to cultured neuronal cells and induces tau phosphorylation [17]. Tau is a microtubule-associated protein that stabilizes neuronal microtubules under normal physiological conditions, how-ever in certain pathological conditions like Alzheimer's disease; tau protein undergoes modifications, mainly through phosphorylation that can result in the generation of aberrant aggregates that are toxic to neurons [18]. Amyloid vaccine (both passive and active immunization against amyloid) arrests and even reverses both plaque pathology and behavioral phenotypes in the transgenic animals [19]. Aβ42 fibrils can significantly accelerate neurofibrillary tangles formation in P301L mice providing further support to the hypothesis that amyloid beta could be a causative pathogenic factor. Mutations in tau give rise to neurofibrillary tangles but not plaques and mutations in APP or in the probable APP proteases give rise to both plaques and tangles indicates that amyloid pathology occurs upstream of tau patho-logy. Although the exact mechanism(s) by which amyloid beta causes neuronal death is not clear, there is evidence to suggest that it could enhance free radical gene-ration and induce inflammation that could result in profound loss in the cholinergic system of brain, including dramatic loss of choline acetyltransferase level, choline uptake, and decrease in acetylcholine (ACh) level which are responsible for cognitive deficits in AD.

OXIDATIVE STRESS AND NEURONAL DEATH

A-beta causes hydrogen peroxide (H_2O_2) accumulation in cultured hippocampal neurons [20] that results in oxidative damage to cellular phospholipid mem-branes suggesting a role for lipid peroxidation in the pathogenesis of AD [21]. The loss of membrane integrity due to Abeta-induced free-radical damage leads to cellular dysfunction, such as inhibition of ion-motive ATPase, loss of calcium homeostasis, inhibition of glial cell Na^+-dependent glutamate uptake system that results in NMDA receptors mediated delayed neurodegeneration, loss of protein transporter function, disruption of signaling pathways, and activation of nuclear transcription factors and apoptotic pathways.

INFLAMMATION AND NEURONAL DEATH

Free radicals including H_2O_2 not only have direct neurotoxic actions but also participate in inflammation. The fact that inflammation plays a significant role in the pathobiology of Alzheimer's disease is supported by the observation that in the early stages of the disease there is activation of microglial cells and reactive astrocytes in neuritic plaques and the appearance of inflammatory markers [22]. Immune activation and/or inflammatory activity have been shown to be significantly elevated in the brains of AD patients compared with age-matched control patients [23]. Continuous neuroinflammatory processes including glial activation is seen in AD [24]. Microglia and astrocytes would be activated, perceiving Abeta oligomers and fibrils as foreign material, since Abeta assemblies are apparently never observed during the development of brain and in the immature nervous system [2].

Beta-Amyloid fibrils have been shown to activate parallel mitogen-activated protein kinase pathways in microglia and THP1 monocytes [25]. Recently, it was reported that microglia from human AD brain exposed to Abeta produced and secreted a wide range of inflammatory mediators, including cytokines, chemokines, growth factors, complements, and reactive oxygen intermediates [26]. Significant dose-dependent increase in the production of prointerleukin-1, interleukin-6, tumor necrosis factor-α, monocyte chemoat-tractant protein-1, macrophage inflam-matory peptide-1, interleukin-8, and macrophage colony-stimulating factor were observed after exposure to pre-aggregated Amyloid beta-42. These evidences emphasize the role of infla-mmation in the pathogenesis of AD [27].

CHOLINERGIC SYSTEM AND ALZHEIMER'S DISEASE

A primary clinical symptom of Alzheimer's dementia is the progressive deterioration in learning and memory ability. There is evidence that suggests that profound loss in the cholinergic system of brain, including dramatic loss of choline acetyltransferase level, choline uptake, and acetylcholine (ACh) level in the neocortex and hippocampus and reduced number of the cholinergic neurons in basal forebrain and nucleus basalis of Meynert occurs that are closely associated with cognitive deficits in AD [28]. Pharmacological interventions that enhance acetylcholine levels or block further fall in ACh levels and thus, improve cholinergic neurotransmission are known to produce improvement in learning and memory in AD [29].

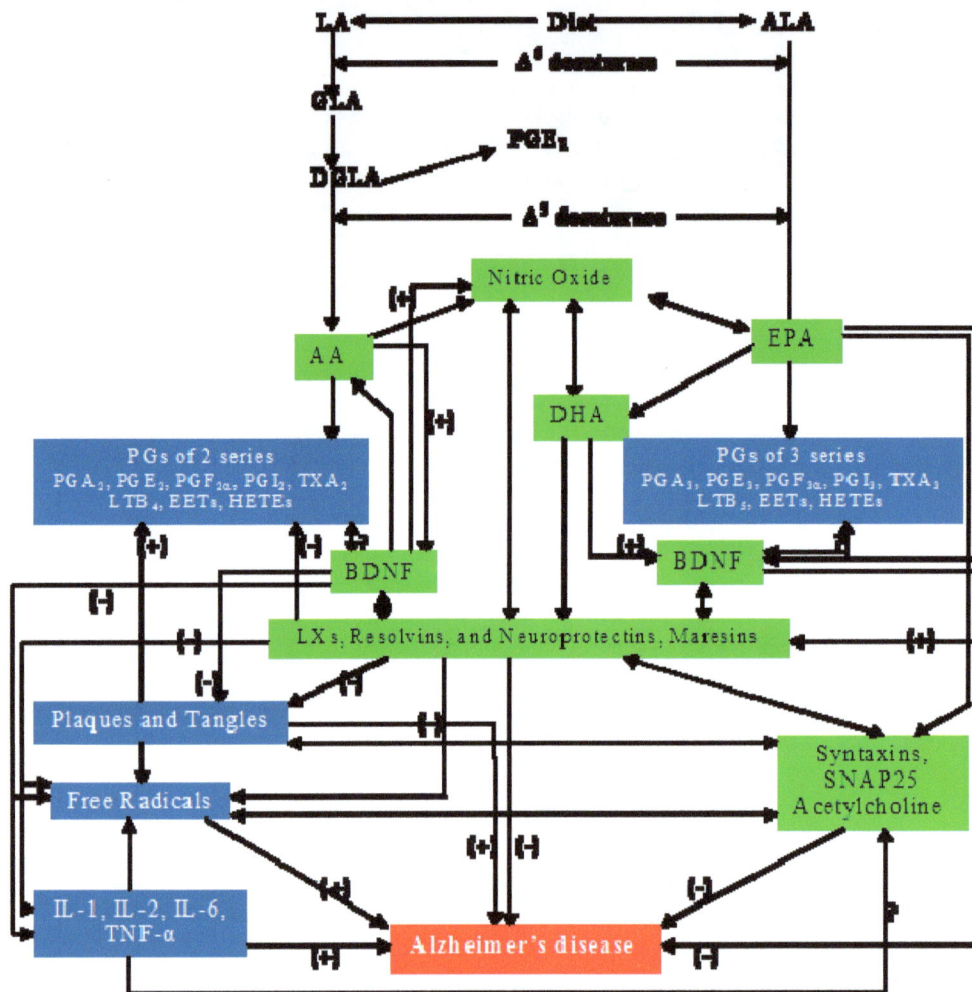

Figure 1: Scheme showing the metabolism of essential fatty acids: LA (linoleic acid) and ALA (α-linolenic acid) and their relationship to BDNF, neurotransmitters such as acetylcholine and cytokines and role in Alzheimer's disease. **(+)** Indicates initiation and/or progression of disease or increase in the synthesis or action.**(-)** Indicates protection from disease and better prognosis or decrease in the synthesis or action. **?** Indicates that possibly, cytokines inhibit syntaxin, SNAP25 and acetylcholine formation or interfere with their action.

In this context it is interesting to note that acetylcholine has anti-inflammatory actions, and hence, a decrease in the levels of ACh may further aggravate the inflammatory process and progression of AD. This "cholinergic anti-inflammatory pathway" mediated by ACh acts by inhibiting the production of TNF, IL-1, MIF, and HMGB1 and suppresses the activation of NF-κB expression [30].

PRO-INFLAMMATORY CYTOKINES IN ALZHEIMER'S DISEASE

Missense mutations in APP, PS-1, and PS-2 genes could alter the proteolysis of APP and increase the generation of Aβ42, whose accumulation as diffuse plaques triggers the inflammatory responses due to microglial activation and release of pro-inflammatory cytokines that ultimately may add to the development of Alzheimer's disease. This is supported by the observation that plasma and cerebrospinal fluid levels of pro-inflammatory cytokines: interleukin-1 (IL-1) and tumor necrosis factor-α (TNF-α) are increased in patients with Alzheimer's disease [31]. Systemic injection of IL-1 decreased extracellular acetylcholine in the hippocampus suggesting that increased concentrations of IL-1 in patients with Alzheimer's disease could be responsible for lowered cerebral acetylcholine levels seen. In addition, IL-1 stimulates the beta-amyloid precursor protein promoter, which is processed out of the larger amyloid precursor protein (APP) that is found in the form of amyloid plaques in the brains of Alzheimer's diseased patients. Furthermore, receptors of IL-1 are on APP mRNA positive cells and its ability to promote APP gene expression suggests that IL-1 plays an important role in Alzheimer's disease [32]. The involvement of inflammatory process in the pathogenesis of Alzheimer's disease is further supported by the observation that inhibition or neutralizing the actions of TNF-α could be of benefit to these patients [33].

NEUROTROPHIC FACTORS IN ALZHEIMER'S DISEASE

For their survival, neurons need the support of various neurotrophic factors such as nerve growth factor (NGF) and brain-derived neurotrophic factor (BDNF). Both NGF and BDNF exert neurotrophic actions on the cholinergic neurons of the basal forebrain nuclei. These neurotrophic factors are synthesized by hippocampal and cortical neurons that are located in the projection field of the basal forebrain cholinergic neurons. Maintenance of the normal levels of NGF- and BDNF-mRNAs seems to be mediated predominantly by NMDA receptors. Synthesis of BDNF and NGF in neurons of the hippocampus is regulated by neuronal activity. The glutamate system is predominantly responsible for upregu- lation and the GABAergic system for down regulation both *in vitro* and in vivo. Experimental evidence exists to indicate that during early postnatal development the activity dependent regulation of NGF and BDNF is mediated by NMDA receptors that are also influenced by the cholinergic system [34]. In postmortem samples of hippocampus from AD and control donors, wherein the levels of BDNF, NGF and NT-3 were examined by in situ hybridization it was noticed that BDNF was decreased. These results suggested the possibility that deceased expression of BDNF may contribute to the progression of cell death in AD [35]. These results coupled with the observation that systemic interleukin-1β (IL-1β) decreases BDNF messenger RNA expression in the rat hippocampal formation indicates that inflammatory cytokines initiate and enhance the progression of AD by decreasing BDNF production, a neurotrophic factor. Thus, it is possible that the accumulation of IL-1 in Alzheimer's disease hippocampus may be responsible for altered hippocampal neuron synaptic plasticity by stimulating beta-amyloid production; and by decreasing BDNF production in the hippocampus of Alzheimer's disease patients [36]. In addition, it was noted that IL-1β renders neurons vulnerable to degeneration by interfering with BDNF-induced neuroprotection. IL-1β compromised the PI3-K/Akt pathway-mediated protection by BDNF and suppressed Akt and MAPK/ERK activation, but not PLC gamma. Activation of CREB, a target of these signaling pathways, was severely depressed by IL-1 β. As the cytokine did not influence TrkB receptor and PLC gamma activation, IL-1 β appears to interfere with BDNF signaling at the docking step conveying activation to the PI3-K/Akt and Ras/MAPK pathways since it was observed that IL-1β suppressed the activation of the respective scaffolding proteins IRS-1 and Shc, an effect that involves ceramide generation. IL-1-induced interference with BDNF neuroprotection and signal transduction was corrected, in part, by ceramide production inhibitors and mimicked by the cell-permeable C2-ceramide, suggesting that IL-1 β interferes with BDNF signaling involving a ceramide-associated mechanism [37].

In contrast to this, combined antidepressant treatment and physical activity have an additive, potentiating effect on BDNF mRNA expression within dentate gyrus and CA 1, CA 3, and CA 4 cellular fields of the rat hippocampus. These results provide support to the concept that physical exercise as a potential enhancer of treatment response to antidepressants [38]. In a study involving 35 older adults without neurological history, it was noted that higher serum BDNF levels were associated with better performance on the Mini-Mental State Examination and short form of the Boston Naming Test. These findings extend work from Alzheimer disease and vascular dementia samples and indicate that higher BDNF levels are associated with better neuropsychological function in healthy older adults [39].

In amyloid-transgenic mice, BDNF gene delivery, when administered after disease onset, reversed synapse loss, partially normalized aberrant gene expression, improved cell signaling and restored learning and memory. These outcomes occurred independently of effects on amyloid plaque load. In aged rats, BDNF infusion reversed cognitive decline, improved age-related perturbations in gene expression and restored cell signaling. In adult rats and primates, BDNF prevented lesion-induced death of entorhinal cortical neurons. In aged primates, BDNF reversed neuronal atrophy and ameliorated age-related cognitive impairment. These and other observations indicate that BDNF exerts substantial protective effects on crucial neuronal circuitry involved in Alzheimer's disease, acting through amyloid-independent mechanisms [40] and suggests that BDNF could be exploited as a potential therapy for Alzheimer's disease [41].

POLYUNSATURATED FATTY ACIDS IN ALZHEIMER'S DISEASE

Several lines of evidence suggest that polyunsaturated fatty acids (PUFAs), especially ω-3 fatty acids, could be useful in the prevention and treatment of Alzheimer's disease. PUFAs are essential for the growth and development of brain [42]. PUFAs; eicosapentaenoic acid (EPA) and docosahexaenoic acid (DHA) of ω-3 series are essential for neurocognitive development and normal brain functioning [42], and DHA improved memory performance in aged mice [43]. A reduction in dietary DHA in an Alzheimer's mouse model showed loss of postsynaptic proteins associated with increased oxidation, increased caspase-cleaved actin, which was localized in dendrites. In contrast, when DHA-restricted mice were given DHA, the fatty acid protected them against dendritic pathology and behavioral deficits and increased anti-apoptotic BAD phosphorylation. These results suggest that DHA is useful in preventing Alzheimer's disease in which synaptic loss is critical [44].

DHA attenuated amyloid-β secretion accompanied by the formation of neuroprotectin D1 (NPD1), a DHA-derived 10,17S-docosatriene (see Fig. 1for the metabolism of essential fatty acids and their products), in addition to giving rise to lipoxins, resolvins and maresins that have neuroprotective actions [42, 45], inhibited interleukin-6 (IL-6) and tumor necrosis factor-α (TNF-α) production that are neurotoxic and increased the synthesis of endothelial nitric oxide (eNO), a neurotransmitter. In Alzheimer's disease, hippocampal DHA and NPD1 were reduced including the expression of enzymes involved in NPD1 synthesis, cytosolic phospholipase A_2 and 15-ipoxygenase [45]. NPD1 repressed amyloid β-induced activation of pro-inflammatory genes and upregulated the antiapoptotic genes encoding Bcl-2, Bcl-xl and Bfl-1 (A1) indicating its (NPD1) anti-inflammatory nature. Soluble amyloid precursor protein-α stimulates NPD1 synthesis from DHA [45] that, in turn, could prevent neuronal death.

Presenilin, a major component of γ-secretase, generates amyloid-β. Overexpression of phospholipase D1 decreases the catalytic activity of γ-secretase [46], and releases PUFAs as evidence by increased formation of prostaglandin E_2 [47]. This suggests that PUFAs could regulate the activity of γ-secretase. PUFAs (especially arachidonic acid and DHA) enhance acetylcholine release in the brain [48] accounting for their beneficial effects in Alzheimer's. Furthermore, EPA and DHA have the ability to enhance NO generation [42], suppress production of pro-inflammatory cytokines [42], and enhance brain acetylcholine levels [48], a neurotransmitter whose levels are decreased in Alzheimer's disease [49].

Alzheimer's disease is an inflammatory condition since elevated levels of cytokines occurs in the plasma and brains of these patients [50]. Acetylcholine, the principal neurotransmitter inhibits the production of pro-inflammatory cytokines: IL-1, IL-2, IL-6, and TNF-α [51], and enhances NO, whereas acetylcholinesterase enhances acetylcholine levels suggesting the close association that exists between inflammation, cholinergic system, and Alzheimer's disease.

DHA stimulated for neurite outgrowth [52] by activating syntaxin 3 that is specifically involved in fast calcium-triggered exocytosis of neurotransmitters. Furthermore, SNAP25 (synaptosomal-associated protein of 25 kDa), a syntaxin partner implicated in neurite outgrowth, interacted with syntaxin 3 only in the presence of AA that allowed the formation of the binary syntaxin 3-SNAP 25 complex. AA stimulated syntaxin 3 to form the ternary SNARE complex (soluble N-ethylmaleimide-sensitive factor attach- ment protein receptor), which is needed for the fusion of plasmalemmal precursor vesicles into the cell surface membrane that leads to membrane fusion that facilitates neurite outgrowth. These results imply that EPA, DHA, and AA when given in optimal amounts could be of significant benefit in the prevention and treatment of Alzheimer's disease.

In addition, DHA promoted neuronal survival by facilitating membrane translocation/activation of Akt. *In vivo* reduction of DHA by dietary depletion increased hippocampal neuronal susceptibility to apoptosis [53]. Retinal pigment epithelium and synaptic membranes contain the highest content of DHA of all cell membranes, and DHA is required for retinal pigment epithelium functional integrity. NPD_1 that is formed from DHA counteracted

H_2O_2/TNF-α-induced apoptosis by up-regulating anti-apoptotic proteins Bcl-2 and Bcl-$_{XL}$ and decreasing pro-apoptotic Bax and Bad proteins. In addition, NPD$_1$ inhibited oxidative stress-induced caspase-3 activation, IL-1β-stimulated expression of cyclooxygenase-2 and thus, prevented oxidative stress-induced apoptosis and inflammation [42, 45].

Furthermore, PUFAs have the ability to modulate the expression, properties, and action dopamine, serotonin, and acetylcholine [54], especially during the perinatal period during which the growth and development of brain is maximum. Thus, ω-3 PUFAs and to some extent AA modulate neural function, including neurotransmission, membrane fluidity, ion channel and enzyme regulation and gene expression, prevent inflammation, and thus, bring about their beneficial actions in Alzheimer's disease. But, it remains to be determined the exact dose of various PUFAs, the ratio of ω-3 and ω-6 that is optimal for neuronal growth and survival, and the duration and time of their institution to obtain their beneficial action in the prevention and treatment of Alzheimer's disease remain to be determined. The beneficial actions of PUFAs in Alzheimer's disease suggest that preparation of stable analogues of lipoxins, resolvins and protectins could be of benefit in its management.

In addition, EPA protected heart against ischemia/reperfusion-induced injury by inhibiting caspase-3 activation and PARP-cleavage and reducing the apoptotic index during reperfusion by increasing ERK phosphorylation and decreasing p38 phosphorylation during reperfusion [55]. DHA promoted neuronal survival by facilitating membrane translocation/ activation of Akt. *In vivo* reduction of DHA by dietary depletion increased hippocampal neuronal susceptibility to apoptosis [53]. Retinal pigment epithelium and synaptic membranes contain the highest content of DHA of all cell membranes, and DHA is required for retinal pigment epithelium functional integrity. NPD$_1$ that is formed from DHA upon its release from the cell membrane by the action of phospholipase A$_2$ counteracted H_2O_2/TNF-α-induced apoptosis by upregulating anti-apoptotic proteins Bcl-2 and Bcl-$_{XL}$ and decreasing pro-apoptotic Bax and Bad proteins. In addition, NPD$_1$ inhibited oxidative stress-induced caspase-3 activation, IL-1β-stimulated expression of cyclooxygenase-2 and thus, prevented oxidative stress-induced apoptosis and inflammation [45].

Essential fatty acid deficiency promotes respiratory uncoupling [56]. Mild respi-ratory uncoupling, a form of mito-chondrial dysfunction, increases oxidative stress and decreases nitric oxide availability or biological action [57]. PUFAs modulate the activities of uncoupling proteins [58], and under certain specific conditions may function as anti-oxidants and prevent free radical-induced damage to cells [59]. PUFAs modulate the expression, properties, and action dopamine, serotonin, and acetyl- choline [54], especially during the perinatal period during which the growth and development of brain is critical. Thus, ω-3 PUFAs modulate neural function, neurotransmission, membrane fluidity, ion channel and enzyme regulation and gene expression, prevent inflammation, and thus, bring about their beneficial actions in Alzheimer's disease.

CROSS-TALK BETWEEN PUFAS AND BDNF

Since both BDNF and PUFAs have a role in Alzheimer's disease, it is logical to expect that these two molecules interact with each other. In animal experiments, it was noted that inhibition of COX enzymes blocks the induction of long-term potentiation (LTP; the major contemporary model of synaptic plasticity), and causes substantial and sustained deficits in spatial learning in the watermaze test. Increases in BDNF and PGE2 following spatial learning and LTP were also blocked. It is interesting to note that 4 days of prior exercise in a running wheel increased endogenous BDNF levels sufficiently to reverse the COX inhibition (that is there is an enhancement of formation of PGs) seen with the use of Cox inhibitors and restored a parallel increase in LTP and learning-related BDNF and PGE2. These results indicate that COX enzymes play a permissive role in synaptic plasticity and spatial learning via a BDNF-associated mechanism [60]. In contrast to these results, it was reported that PGs are responsible for memory deficits in contextual fear conditioning that occur following IL-1β injection into the dorsal hippocampus of Sprague-Dawley rats [61], whereas COX inhibition blocked the disruption in contextual fear conditioning produced by IL-1β and COX inhibition alone also disrupted contextual memory, suggesting an inverted U-shaped relationship between PG levels and memory. This can be interpreted to mean that physiological concentrations of PGs may enhance memory whereas high or low concentrations could cause memory deficits. Furthermore, PGs injected directly into the dorsal hippocampus impaired context memory and significantly reduced post-conditioning levels of BDNF within the hippocampus, suggesting that PGs interact with BDNF to produce their memory-impairing effects.

In a study involving the evaluation of a possible synergistic action between DHA dietary supplementation and voluntary exercise on modulating synaptic plasticity and cognition, it was reported that DHA-enriched diet significantly increased spatial learning ability, and these effects were enhanced by exercise. The DHA-enriched diet increased levels of pro-brain-derived neurotrophic factor (P-BDNF) and mature BDNF, whereas the additional application of exercise boosted the levels of both. In addition, the levels of the activated forms of CREB and synapsin I was incremented by the DHA-enriched diet with greater elevation by the concurrent application of exercise. With a concomitant reduction in hippocampal oxidized protein level in the DHA group whereas a combination of a DHA diet and exercise resulted in a much greater reduction. The levels of activated forms of hippocampal Akt and CaMKII were also increased by the DHA-enriched diet and even a greater elevation was attained by a combination of diet and exercise [62]. Thus, these results indicated that DHA enhances synaptic plasticity and spatial learning ability and memory and that it produces these beneficial actions by enhancing BDNF levels and thus, may reduce the risk of neurological disorders. What is more interesting is the fact that both exercise and BDNF complement each other's actions.

These results coupled with the observation that omega-3 enriched dietary supplements can provide protection against reduced plasticity and impaired learning ability after traumatic brain injury in experimental animals by normalizing BDNF levels and reducing oxidative damage suggests that these fatty acids directly stimulate the production of BDNF in the brain [63]. These and other results emphasize the fact that EPA/DHA/AA are involved in neuronal growth, synaptic plasticity, memory improvement and reduction in oxidative stress in the brain and that they enhance BDNF levels in the brain to bring about these beneficial actions. It is possible that BDNF needs these fatty acids for its own actions and for its stabilization in the brain. The benefits of exercise in memory formation and improvement in synaptic plasticity can also be attributed to increased formation and utilization of PUFAs and enhanced levels of BDNF. Thus, both PUFAs and BDNF act in concert to prevent Alzheimer's disease and are needed to preserve neuronal integrity, maintain synaptic plasticity and memory formation.

CONCLUSIONS

Alzheimer's disease is the most common neurodegenerative disorder, which is known to present in the initial stage with mild cognitive impairment that ultimately could lead to severe dementia. It is characterized by the presence of plaques and tangles that induce damage to the neurons of the brain. Accumulation of excess of *β-amyloid* that constitutes the plaques and *tau* of tangles cause inappropriate accumulation of intracellular Ca^{2+} resulting in neuronal apoptosis. Furthermore, excess plaques and tangles could decrease the levels of acetylcholine, the principle neurotransmitter, and that of other neurotransmitters in the brain and enhance the production of proinflamma- tory cytokines that are neurotoxic leading to disruption of communication, metabolism, and repair processes seen in Alzheimer's disease.

PUFAs, especially EPA, DHA and AA, are not only essential for the growth and development of brain but also improve memory and protect against dendritic pathology and behavioral deficits and increased anti-apoptotic BAD phosphorylation, suggesting that they could be useful in preventing Alzheimer's disease. PUFAs attenuate amyloid-β secretion, decrease the catalytic activity of γ-secretase, inhibit IL-6 and TNF-α production that are neurotoxic, and increase the synthesis of eNO, a neurotransmitter, stimulate neurite outgrowth by activating syntaxin 3 that is specifically involved in fast calcium-triggered exocytosis of neurotransmitters and enhance the production of acetylcholine, events that explain their ability both in the prevention and treatment of Alzheimer's disease. Some of the beneficial actions of PUFAs can be attributed to the formation of lipoxins, resolvins, protectins and maresins that are potent anti-inflammatory and neuroprotective molecules. Furthermore, PUFAs augment the production of BDNF, an important neurotrophic molecule that is essential for growth, differentiation and survival of neurons in the brain. These results suggest that a combination of PUFAs and BDNF could be of significant benefit in the prevention and treatment of Alzheimer's disease. Since AA/EPA/DHA form precursors to potent anti-inflammatory and neuroprotective molecules such as lipoxins, resolvins, protectins and maresins, it is likely that the plasma, cerebrospinal fluid and tissue levels of various PUFAs and their metabolites could serve as markers both to predict and prognosticate the development and progression of Alzheimer's and other neurodegenerative diseases. It is possible to develop synthetic analogues of lipoxins, resolvins, protectins and maresins that are more stable, potent and orally active for the treatment of various degenerative neurological conditions.

In a recent study, it was noted that inflammation of microglia -- an abundant cell type that plays an important supporting role in the brain -- does not appear to be associated with dementia in Alzheimer's disease. This supports the results of recent clinical results that indicate anti-inflammatory drugs are not effective at fighting dementia in patients with Alzheimer's disease.

Instead, it was noted that that the brain's immune system, made up of microglia, is not activated in the brains of Alzheimer's patients but microglia are degenerating. This led to the claim that loss of microglial cells contributes to the loss of neurons, and thus to the development of dementia. It is worthwhile to note that microglial cells, a subset of glial cells, provide physical and nutritional support to neurons. Glial cells, which outnumber neurons 10-to-1, are at the heart of Alzheimer's disease that suggests β-amyloid cause memory loss and dementia. It has been suggested that microglia become "activated" and mount an immune response to β-amyloid and release toxic chemicals that worsen the disease effects. However, high-resolution observations did not reveal evidence that β-amyloid activates, or inflames, human microglia cells nor evidence that inflammation is to blame for brain cell death. It was noted that microglial cells from Alzheimer's patients were not distinctly larger or unusually shaped, which would have been the case had they been inflamed. On the other hand, it was noted that glial cells are dying, and so the neurons lose support. The microglial cells had a tangled, fragmented appearance, similar to neurons in the throes of Alzheimer's disease or - old age proving that microglial cells are subject to aging and may undergo degeneration, and that the loss of these cells precedes the loss of neurons and so the development of Alzheimer's disease. If this true, it is important to find what is leading to the degeneration of microglia and hence, the loss of support to neurons.

ACKNOWLEDGEMENTS

The author was in receipt of Ramalingaswami fellowship of the Department of Biotechnology, New Delhi during the tenure of this study.

REFERENCES

[1]   Hutton, M.; McGowan, E. *Neuron.* **2004**, *43*, 293.
[2]   Selkoe, D.J. *Physiol. Rev.* **2001**, *81*, 741.
[3]   Poierier, J.; Minnich, A.; Davignon, J. *Ann. Med.* **1995**, *27*, 663.
[4]   Blacker, D.; Wilcox, M. A.; Laird, N. M.; Rodes, L.; Horvath, S. M.; Go, R. C.; Perry, R.; Watson, B.; Bassett, S. S.; McInnis, M. G.; Albert, M. S.; Hyman, B. T.; Tanzi, R. E. *Nat. Genet.* **1998**, *19*, 357.
[5]   Sridhar, G. R.; Thota, H.; Allam, A. A.; Babu, C. S.; Prasad, A. S.; Divakar, Ch. *LipidsHealth Dis.* **2006**, *5*, 28.
[6]   Law, A.; Gauthier, S.; Quirion, R. *Brain Res. Brain Res. Rev.* 2001, *35*, 73.
[7]   Iqbal, K.; Grundke-Iqbal, I. *Acta Neuropathol. (Berl).* **2005**, *109*, 25.
[8]   Wang, R.; Wang, B.; He, W.; Zheng, H. *J. Biol. Chem.* **2006**, *281*, 15330.
[9]   Selkoe, D. J. *Nature* **1999**, *399*, A23.
[10]   Rojo L.; Sjoberg, M. K.; Hernandez, P.; Zambrano, C.; Maccioni, R. B. *J. Biomed. Biotechnol.* **2006**, *2006*, 73976.
[11]   Harman, D. *Ann. N Y Acad. Sci.* **2002**, *959*, 384.
[12]   Wang, H. Q.; Nakaya, Y.; Du, Z.; Yamane, T.; Shirane, M.; Kudo, T.; Takeda, M.; Takebayashi, K.; Noda, Y.; Nakayama, K. I.; Nishimura, M. *Hum. Mol. Genet.* **2005**, *14*,   1889.
[13]   Verdile, G.; Gandy, S. E.; Martins, R. N. *Neurochem. Res.* **2007**, *32*, 609.
[14]   Saido, T. *Neurobiol. Aging* **1998**, *19(1 Suppl)*, S69.
[15]   Zhao, L.; Teter, B.; Morihara, T.; Lim, G. P.; Ambegaokar, S. S.; Ubeda, O. J.; Frautschy, S. A.; Cole, G. M. *J. Neurosci.* **2004**, *24*, 11120.
[16]   Edland, S. D. *J. Mol. Neurosci.* **2004**, *23*, 213.
[17]   Takashima, A.; Noguchi, K.; Sato, K.; Hoshino, T.; Imahori, K. *Proc. Natl. Acad. Sci. USA.* **1993**, *90*, 7789.
[18]   Avila, J.; Lucas, J. J.; Perez, M.; Hernandez F. *Physiol. Rev.* **2004**, *84*, 361.
[19]   Morgan, D.; Diamond, D. M.; Gottschall, P. E.; Ugen, K. E.; Dickey, C.; Hardy, J.; Duff, K.; Jantzen, P.; DiCarlo, G.; Wilcock, D.; Connor, K.; Hatcher, J.; Hope, C.; Gordon, M.; Arendash, G W. *Nature.* **2000**, *408*, 982.
[20]   Mattson, M. P.; Lovell, M. A.; Furukawa, K.; Markesbery, W. R. *J. Neurochem.* **1995**, *65*, 1740.
[21]   Koppaka, V.; Axelsen, P. H. *Biochemistry.* **2000**, *39*, 10011.
[22]   Chong, Y. H.; Sung, J. H.; Shin S. A.; Chung, J. H.; Suh, Y. H. *J. Biol. Chem.* **2001**, *276*, 23511.
[23]   Dumery, L.; Bourdel, F.; Soussan, Y.; Fialkowsky, A.; Viale, S.; Nicolas, P.; Reboud-Ravaux, M. *Pathol. Biol. (Paris).* **2001**, *49*, 72.

[24]   Calingasan, N. Y.; Erdely, H A.; Altar, A. C. *Neurobiol. Aging.* **2002,** *23,* 31.

[25]   McDonald, D. R.; Bamberger, M. E.; Combs, C. K.; Landreth, G. E. *J. Neurosci.* **1998,** *18,* 4451.

[26]   Lue, L. F.; Rydel, R.; Brigham, E. F.; Yang, L. B.; Hampel, H.; Murphy, G M.; Brachova, L.; Yan, S. D.; Walker, D. G.; Shen, Y.; Rogers, J. *Glia.* **2001,** *35,* 72.

[27]   Cummings, J. L.; Kaufer, D. *Neurology.* **1996,** *47,* 876.

[28]   Giacobini, E. *Jpn. J. Pharmacol.* **1997,** *74,* 225.

[29]   Giacobini, E. *Pharmacol. Res.* **2004,** *50,* 433.

[30]   Borovikova, L. V.; Ivanova, S.; Zhang, M.; Yang, H.; Botchkina, G. I.; Watkins, L. R.; Wang, H.; Abumrad, N.; Eaton, J. W.; Tracey, K. J. *Nature.* **2000,** *405,* 458; Pavlov, V. A.; Tracey, K. J. *Biochem. Soc. Trans.* **2006,** *34 (Pt 6),* 1037; Wang, H.; Liao, H.; Ochani, M.; Justiniani, M.; Lin, X.; Yang, L.; Al-Abed, Y.; Wang, H.; Metz, C.; Miller, E. J; Tracey, K. J.; Ulloa, L. *Nat. Med.* **2004,** *10,* 1216; Czura, C. J.; Friedman, S. G.; Tracey, K. J. *J. Endotoxin. Res.* **2003,** *9,* 409.

[31]   Cacabelos, R.; Barquero, M.; Garcia, P.; Alvaez, X. A.; Varela de Seijas, E. Methods. Find. Exp. Clin. Pharmacol. 1991, 13, 455; Fillit, H.; Ding, W.H.; Buee, L.; Kalman, J.; Altstiel, L.; Lawlor, B.; Wolf-Klein, G. *Neurosci. Lett.* **1991,** *129,* 318; Chao, C. C.; Hu, S.; Frey, W. H. 2nd.; Ala, T. A.; Tourtellotte, W. W.; Peterson, P. K. *Clin. Diagn. Lab.Immunol.* **1994,** *1,* 109.

[32]   Donnelly, R. J.; Friedhoff, A. J; Beer, B.; Blume, A. J.; Vitek, M. P. *Cell. Mol. Neurobiol.* **1990,** *10,* 485; Blume, A. J.; Vitek, M. P. *Neurobiol. Aging.* **1989,** *10,* 406.

[33]   Tobinick, E.; Gross, H.; Weinberger, A.; Cohen, H. *Med. Gen. Med.* **2006,** *8,* 25; Rosenberg, P. B. *Med. Gen. Med.* **2006;** *8: 24.*

[34]   Berzaghi, M. P.; Cooper, J.; Castren, E.; Zafra, F.; Sofroniew, M.;, Thoenen, H.; Lindholm, D. *J. Neurosci.* **1993,** *13,* 3818.

[35]   Phillips, H. S.; Hains, J. M.; Armanini, M.; Laramee, G. R.; Johnson, S. A.; Winslow, J. W. *Neuron.* **1991,** *7,* 695; Connor, B.; Young, D.; Yan Q.; Faull, R. L.; Synek, B.; Dragunow, M. *Brain. Res. Mol. Brain. Res.* **1997,** *49,* 71; Soontornniyomkii, V.; Wang, G.; Pittman, C. A.; Hamilton, R. L.; Wiley, C. A.; Achim, C. L. *Acta. Neuropathol.* **1999,** *98,* 345.

[36]   Lapchak, P. A.; Araujo, D. M.; Hefti, F. **Neuroscience. 1993,** *53,* 297.

[37]   Tong, L.; Balazs, R.; Soiampornkul, R.; Thangnipon, W.; Cotman, C. W. *Neurobiol. Aging.* **2008,** *29,* 1380.

[38]   Russo-Neustadt, A.; Beard, R. C.; Cotman, C. W. *Neuropsycho- pharmacology.* **1999,** *21,* 679.

[39]   Gustad, J.; Benitez, A.; Smith, J.; Glickman, E.; Spitznagel, M. B.; Alexander, T.; Juvancic-Heltzel, J.; Murray, L. *J. Geriatr. Psychiatry. Neurol.* **2008,** *21,* 166.

[40]   Nagahara, A. H.; Merrill, D. A.; Coppola, G.; Tsukada, S.; Schroeder, B. E.; Shaked, G. M.; Wang, L.; Blesch, A.; Kim, A.; Conner, J. M.; Rockenstein, E.; Chao, M. V.; Koo, E. H.; Geschwind, D.; Masliah, E.; Chiba, A. A.; Tuszynski, M. H. *Nat. Med.* **2009,** *15,* 331.

[41]   Rao, A. A.; Sridhar, G. R.; Srinivas, B.; Das, U. N. *Med. Hypotheses.* **2008,** *70,* 424.

[42]   Das, U. N. *Current. Pharmaceutical. Biotech.* 2006, *7,* 467; Das, U. N. *Biotech. J.* 2006, *1,* 420.

[43]   Gamoh, S.; Hashimoto, M.; Hossain, S.; Masumura, S. *Clin. Exp. Pharmacol. Physiol.* 2001, *28,* 266.

[44]   Lim, G. P.; Calon, F.; Morihara, T.; Yang, F.; Teter, B.; Ubeda, O.; Salem, N. Jr.; Frautschy, S. A.; Cole, G. M. *J. Neurosci.* 2005, *25,* 3032; Hashimoto, M.; Hossain, S.; Shimada, T.; Shido, O. *Clin. Exp. Pharmacol. Physiol.* 2006, *33,* 934.

[45]   Lukiw, W. J.; Cui, J. G.; Marcheselli, V. L.; Bodker, M.; Botkjaer, A.; Gotlinger, K.; Serhan, C. N.; Bazan, N. G. *J. Clin. Invest.* **2005,** *115,* 2774; Marcheselli, V. L.; Hong, S.; Lukiw, W. J.; Tian, X. H.; Gronet, K.; Musto, A.; Hardy, M.; Gimenez, J. M.; Chiang, N.; Serhan, C. N.; Bazan, N. G. *J. Biol. Chem.* **2003,** *278,* 43807; Mukherjee, P. K.; Marcheselli, V. L.; Serhan, C. N.; Bazan, N. G. *Proc. Natl. Acad. Sci. USA.* **2004,** *101,* 8491.

[46]   Cai, D.; Netzer W. J.; Zhong, M.; Lin, Y.; Du, G.; Frohman, M.; Foster, D. A.; Sisodia, S.; Xu, H.; Gorelick, F. S.; Greengard, P. *Proc. Natl. Acad. Sci. USA.* **2006,** *103,*1941.

[47]   Kim, S. Y.; Ahn, B. H.; Min, K. J.; Lee, Y. H.; Joe, E. H.; Min, D. S. *J. Biol. Chem.* **2004,** *279,* 38125.

[48]   Almeida, T.; Cunha, R. A.; Ribeiro, J. A. *Brain. Res.* **1999,** *826,* 104; Aid, S.; Vancassel, S.; Linard, A.; Lavialle, M.; Guesnet, P. *J. Nutr.* **2005,** *135,* 1008.

[49]   Kihara, T.; Shimohama, S. *Acta Neurobiol. Exp (Warsaw).* **2004,** *64,* 99.

[50]   Das, U. N. *Prostaglandins. Leukot. Essent. Fatty. Acids.* **2008,** *78,* 11.

[51]   Das, U. N. *Neuropsychopharmacology.* **2007,** *32,* 2053; Borovikova, L. V.; Ivanova, S.; Zhang, M.; Yang, H.; Botchkina, G. I.; Watkins, L. R.; Wang, H.; Abumrad, N.; Eaton, J. W.; Tracey, K. J. *Nature.* **2000,** *405,* 458.

[52]   Darios, F.; Davletov, B. *Nature.* **2006,** *440,* 813.

[53]   Akbar, M.; Calderon, F.; Wen, Z.; Kim, H. Y. *Proc. Natl. Acad. Sci. USA.* **2005,** *102,* 10858.

[54]   de La Presa Owens, S.; Innis, S. M. *J. Nutr.* **1999,** *129,* 2088.

[55]   Engelbrecht, A. M.; Engelbrecht, P.; Genade, S.; Niesler, C.; Page, C.; Smuts, M.; Lochner, A. *J. Mol. Cell. Cardiol.* **2005,** *39,* 940.

[56]   Klein, P. D.; Johnson, R. M. *J. Biol. Chem.* **1954,** *211,* 103; Hayashida, T.; Portman, O. W. *Proc. Soc. Exp. Biol. Med.* **1960,** *103,* 656.

[57]   Bernal-Mizrachi, C.; Gates, A. C.; Weng, S.; *et al. Nature* **2005,** *435,* 502.

[58]   Cha, S. H.; Fukushima, A.; Sakuma, K.; Kagawa, Y. *J. Nutr.* **2001,** *131,* 2636.

[59]   Suresh, Y.; Das, U. N. *Nutrition.* **2003,** *19,* 93; Suresh, Y.; Das, U.N. *Nutrition.* **2003,** *19,* 213.

[60]   Shaw, K. N.; Commins, S.; O'Mara, S. M. *Eur. J. Neurosci.* **2003,** *17,* 2438.

[61]   Hein, A. M.; Stutzman, D. L.; Band, S. T.; Barrientos, R. M.; Watkins, L. R.; Rudy, J. W.; Maier, S. F. *Neuroscience.* **2007,** *150,* 754.

[62]   Wu, A.; Ying, Z.; Gomez-Pinilla, F. *Neuroscience.* **2008,** *155,* 751.

[63]   Wu, A.; Ying, Z.; Gomez-Pinilla, F. *J. Neurotrauma.* **2004,** *21,* 1457.

CHAPTER 11

Future Perspective: Directions for Future Development on Various Aspects of Neurodegeneration and Neuroprotection in Neurological Disorders

Akhlaq A. Farooqui[1*] and Tahira Farooqui[2]

[1]*Department of Molecular and Cellular Biochemistry,* [2]*Department of Entomology/Center for Molecular Neurobiology, The Ohio State University, Columbus, Ohio 43210, USA*

Abstract: Neuronal cell death in neurodegenerative and neurotraumatic diseases is accompanied by excitotoxicity, oxidative stress, and inflammation. These processes lead to the generation of lipid mediators. The intensity of cross talk among lipid mediators dictates the rate of neurodegeneration in above neurological diseases. In neurodegenerative diseases cell death occurs slowly (months to years), where as in neurotraumatic diseases, neurodegeneration occurs rapidly (minutes to hours) through necrosis at the core of injury, whereas in penumbral region neurons undergo delayed neurodegeneration through apoptosis. The presence of lipid mediators in biological fluids can be used for the diagnosis of above neurological disorders.

Keywords: Neurodegeneration; neuroprotection; neurological disorders; apoptosis; oxidative stress; inflammation; excitotoxicity; lipid mediators

INTRODUCTION

Excitotoxicity, oxidative stress, inflammation, and alterations in immune system are major components associated with the pathogenesis of acute neural trauma (ischemia, spinal cord trauma, and head injury), neurodegenerative diseases (AD; PD; HD; and ALS) [1-6]. In severe acute neural trauma neurodegeneration occurs rapidly (minutes to hours) by necrotic cell death at the core of injury, whereas in penumbral region or surrounding area neurons undergo delayed neurodegeneration through apoptotic cell death [2, 5, 7]. At the same time acute neural trauma triggers neuroprotective processes such as the induction of heat shock proteins, anti-inflammatory, cytocytokines, and production of endogenous antioxidants and anti-inflammatory metabolites, such as resolvins and neuroprotectins [8-14]. Heat shock proteins serve as molecular chaperones involved in the protection of cells from various forms of stress including ischemia. In the brain, heat shock protein synthesis is induced not only after hyperthermia, but also following alterations in the intracellular redox environment [2-5].

Neurodegenerative diseases are accompanied by site-specific premature and slow death of certain neuronal populations in specific area of central and peripheral nervous system [1,15]. The neurodegenerative process is followed by the increased expression of genes that are linked to neurotransmission, neuroplasticity, axonal transport, aerobic metabolism and neuroprotection. This process represents adaptations that induce and facilitate greater neuronal activity. Several genes associated with increased neuronal activity are extremely vulnerable to oxidative stress and neuroinflammation. Changes in expression of such genes can disrupt not only the modulation of neuroplasticity and axonal transport, but promote the abnormal accumulation of peptides, such as β-amyloid that is characteristic of AD, over-expression of α-synuclein and parkin in PD, and abnormalities of huntingtin function in HD. Increased intake of animal fats alters the balance of polyunsaturated fatty acids in the neuronal membrane and makes neuronal membrane more susceptible to oxidative stress [16]. Present day Western diet contains more ω-6 fatty acids than ω-3 fatty acids (ω-6:/ω-3 ratio, 20:1). Consumption of n-6 fatty acids results in formation of arachidonic acid-derived eicosanoids that promote a higher susceptibility to diseases associated with inflammation and oxidative stress [17]. In contrast, consumption of ω-3 fatty acid enriched food (fish and fish oil) generates docosanoids that prevent inflammation and oxidative stress [12]. It is becoming increasingly evident that the present day Western diet promotes the pathogenesis of many chronic diseases such as cardiovascular disease, inflammatory

*Address correspondence to: Department of Molecular and Cellular Biochemistry, The Ohio State University, Columbus, OH 43210, USA; Tel.: (614) 488-0361; Email: farooqui.1@osu.edu

and autoimmune diseases. The consumption of ω-3 fatty acids has numerous beneficial effects on human brain [5, 6]. The beneficial effects of ω-3 may be due not only to its effect on the physicochemical properties of neural membranes, but also due to modulation of neurotransmission, gene expression, activities of enzymes, ion channels, receptors, and immunity [5]. Thus, a diet enriched in ω-3 fatty acids exerts cardioprotective, immunosuppressive, and neuroprotective effects [5, 6].

Like acute neural trauma, neurodegenerative diseases are also accompanied by the induction of neuroprotective processes such as the induction of heat shock proteins, anti-inflammatory cytokines, and production of anti-inflammatory metabolites (resolvins and neuroprotectins) provided by ω-3 fatty acids are present in diet [2, 5, 6, 12]. In neurodegenerative diseases neurons die through apoptotic as well as necrotic cell death [4].

Neuropsychiatric disorders involve abnormalities in cerebral cortex and limbic system (thalamus, hypothalamus, hippocampus, and amygdale) along with loss of cortical gray matter resulting in the impairment of cognitive function, a process closely associated every day problem-solving behavior including the ability to learn and store the memory, to retrieve stored memory for further use and to apply the stored memory to efficiently solve problems [18]. In addition, neuroimaging studies also show a progressive ventricular enlargement in some neuropsychiatric patients. Although, the molecular mechanism associated with pathogenesis of neuropsychiatric disorders is not fully understood, but perinatal exposure to infectious agents and environmental toxins (mercury, fluoride, and aluminum) are closely linked to the pathogenesis of neuropsychiatric disorders [19, 20]. Both environmental and dietary excitotoxins have been reported to exacerbate the pathological and clinical problems by worsening excitotoxicity and by microglial priming. In neuropsychiatric disorders, neurons are lost predominantly by apoptosis. Neuochemical studies indicate that in neuropsychiatric diseases several neurotransmitter systems are simultaneously altered within a single microcircuit and each transmitter system shows circuitry changes in more than one region. Changes in microcircuits and neurotransmitters (synthesis and transport) may not only vary on a region-by-region basis, but also from one disease to another. Both macro and microcircuitry within the specific brain system such as limbic system may serve as 'triggers' for the onset of neuropsychiatric condition [21, 22].

CROSS TALK AMONG EXCITOTOXICITY, OXIDATIVE STRESS, AND NEUROINFLAMMATION AS A POSSIBLE CAUSE OF NEURODEGENERATION

Although, the molecular mechanism associated with the pathogenesis of acute neural trauma, sporadic forms neurodegenerative diseases and neuropsychiatric disorders remain unclear, but excitotoxicity, oxidative stress, inflammatory processes, and alterations in immunological parameters are closely associated with above neurological conditions [2]. Excitotoxicity, inflammation, and oxidative stress are interrelated processes that may induce neurodegeneration independently or synergistically. An upregulation of cross talk among excitotoxicity, oxidative stress, and neuroinflammation through neural membrane-derived lipid mediators may increase the vulnerability of neurons in acute neural trauma, neurodegenerative diseases, and neuropsychiatric disorders [2, 23]. In acute neural trauma, increased intensity of cross talk among excitotoxicity, oxidative stress and neuroinflammation may result in rapid neurodegeneration (minutes to hours) due to the sudden lack of oxygen, rapid decrease in ATP level, and sudden collapse of ion gradients. In contrast, in neurodegenerative diseases, cross talk among excitotoxicity, oxidative stress, and neuroinflammation occurs at a slow rate. Oxygen, nutrients, and ATP are available to neurons so ionic homeostasis is maintained to a limited extent. These parameters result in a neurodegenerative process that takes several years to develop [2]. In neurodegenerative diseases and neuropsychiatric disorders pain perception remains below the threshold of detection and immune system continues to attack brain tissue at the cellular and subcellular levels. This results in lingering of chronic inflammation for years causing continued insult to the brain tissue, ultimately reaching the threshold of detection many years after the onset of the neurodegenerative and neuropsychiatric diseases [24]. As stated in chapter 2 that many internal and external factors (activities of PLA_2, COX isoforms, cytokines and growth factors) modulate the dynamic aspects of chronic inflammation and mild oxidative stress in neurodegenerative diseases [2, 25]. The cross talk among excitotoxicity, oxidative stress, and neuroinflammation may be modulated by the diet, genetic, and environmental factors [26]. The onset of neurodegenerative diseases is often subtle and usually occurs in mid to late life. Progression of neurodegenerative diseases depends not only on genetic and dietary factors, but also on environmental factors [15], leading to progressive cognitive and motor disabilities with devastating consequences. Perhaps in acute neural trauma, neurodegenerative diseases, and neuropsychiatric disorders, the intensity of cross talk among lipid mediators generated during excitotoxicity, oxidative stress, and neuroinflammation may lead to unique manifestations that are characteristic features of AD, PD, HD, and ALS. This cross talk may disturb neuronal lipid, protein, nucleic acid, and carbohydrate metabolism leading to irreversible neuronal damage [2, 5, 6]. It remains controversial whether excitotoxicity, oxidative stress, and neuroinflammation are the cause or consequence of neurodegeneration [27, 28]. However, it is becoming increasingly evident that interactions among lipid mediators of glycerophospholipid-, sphingolipid-, and cholesterol-derived lipid mediators modulate pathogenesis of acute neural trauma, neurodegenerative diseases and neuropsychiatric disorders [2, 5].

SIGNAL TRANSDUCTION AND NEURODEGENERATION

Neurons are more susceptible to oxidative stress and excitotoxicity than glial cells. Oxidative stress contributes to brain damage through direct and indirect effects. During direct effects oxidative stress initially occurs in the specific region of brain. For example, in nucleus basalis and hippocampal area through the accumulation of β-amyloid in the AD brain and in substantia nigra and brain stem in PD brain through the accumulation of α-synuclein.

Subsequent events in AD, PD, HD, and ALS are mediated by glutamate-induced neurotoxicity and increased cytosolic Ca^{2+} influx, resulting in activation of Ca^{2+}-dependent enzymes including NADPH oxidase, cytosolic PLA_2, xanthine oxidase, transglutaminases, and neuronal nitric oxide synthase (NOS). These enzymes generate ROS and RNS, which oxidatively induce the modification of nucleic acid, lipid, carbohydrate, and protein [4]. These processes not only result in nuclear and mitochondrial dysfunction, but also proteasome inhibition, and induction of endoplasmic reticulum stress [29]. Mitochondrial dysfunction produces both release of ROS from the electron transport chain and leakage of Ca^{2+} from mitochondria. In addition, neuronal injury results in localization of protein kinases to mitochondria. These kinases include mitogen activated protein kinases (MAPK) such as extracellular signal regulated protein kinases (ERK) and c-Jun N-terminal kinases (JNK), protein kinase B/Akt, and PTEN-induced kinase 1 (PINK1). Although site(s) of action within mitochondria and specific kinase targets are still unclear, these signaling pathways regulate mitochondrial respiration, transport, fission-fusion, calcium buffering, ROS production, mitochondrial autophagy and apoptotic cell death [30]. It is proposed that interactions among α-synuclein, leucine rich repeat kinase 2 (LRRK2), DJ-1 and parkin play an important role in parkinsonian neurodegeneration.

Oxidative stress mediated alterations in the nuclear metabolism result in p53 activation, and inhibition of proteasome function. Increased levels of p53 in the nucleus induce Bax activation and Bcl-2 inhibition, followed by a release of cytochrome c into the cytosol that truncates procaspase-9. Oxidative stress in endoplasmic reticulum mediates activation of caspase-12 as well as caspase-9 via the TNF-α receptor-associated factor-2/ apoptosis-signaling kinase-1/c-Jun N-terminal kinase pathway. Oxidative stress and the redox state are also implicated in the signaling pathway that involves phosphatidyl-inositol 3-kinase/Akt and downstream signaling, which are associated with neuronal survival. Studies on genetically modified mice or rats that overexpress or are deficient in superoxide dismutase strongly, support of the role of mitochondrial dysfunction and oxidative stress [31, 32].

During indirect effects, oxidative stress activates a number of cellular pathways resulting in the expression of stress-sensitive genes and proteins to cause oxidative injury. Moreover, oxidative stress also activates mechanisms that result in a glia-mediated inflammation that also causes secondary neuronal damage. Thus, oxidative stress also mediates the release of cytokines, chemokines, growth factors, and cell adhesion molecules from astrocytes and microglia inducing neurodegeneration through inflammation and apoptosis [29]. Activated glial cells are histopathological hallmarks of acute neural trauma and neurodegenerative diseases. Even though direct contact of activated glia with neurons per se may not necessarily be toxic, the immune mediators (e.g. NO, ROS, proinflammatory cytokines and chemokines, and ω-6 fatty acid derived lipid mediators) released by activated glial cells may serve as neurotoxins for the induction and propagation of neurodegenerative process [33].

Although, the cross talk among exicitotoxicity, oxidative stress, and neuroinflammation may be a general phenomenon, but in acute neural trauma, neurodegenerative diseases, and neuropsychiatric disorders, intensity of this phenomenon may turn on specific genes that modulate neurodegenerative process in specific neuronal population in specific brain region depending upon the nature of neurodegenerative disease [34, 35]. This proposal is supported by the view that the nature of neuron-neuron connections as well as interactions between neurons and glial cells are essential for determining the selective neuronal vulnerability of neurons and alterations of neuron-glia interactions may be involved in the development of neurodegenerative diseases [36; 37]. Astrocytes support neuronal function in several ways, including synapse formation and plasticity, energetic and redox metabolism, and synaptic homeostasis of neurotransmitters and ions. Microglial cells are resident macrophages and represent the immune system of the brain. They provide defense against neuroinflammation in acute neural trauma, neurodegenerative diseases, and neuropsychiatric disorders. Oligodendrocytes contribute to myelination and are involved in regulation of steroid synthesis which is important for neuroprotection against degeneration. Glia cell mediated inflammatory response is involved in dramatic changes in activity of neuritic plaque-associated astrocytes and micro- glia, and contribute to the link between glial activation and neuronal damage or repair [37].

In AD, neurodegeneration occurs in the nucleus basalis, whereas in PD, neurons in the substantia nigra die. The most severely affected neurons in HD are striatal medium spiny neurons. ALS is characterized by selective death of motor neurons in corticospinal tracts. Unlike neurodegenerative diseases, where neurodegeneration occurs in a relatively homogenous population of neurons in a specific area [5, 6], stroke affects multiple different neuronal phenotypes. For example, an infarct might involve the thalamus, hippocampus, and striate visual cortex, affecting three or more very different neuronal populations including neurons, oligodendrocytes, astrocytes, and endothelial cells [38, 39]. In addition, neurodegeneration in stroke also depends on the nature of ROS. For example the selective vulnerability of the CA1 pyramidal neurons following hypoxic/ ischemic injury depends upon superoxide radicals, rather than hydroxyl radicals [36]. Discovering factors and neurochemical events that modulate the specificity of regional neurodegeneration in acute neural trauma, neurodegenerative diseases, and neuropsychiatric diseases remains a most challenging area of neuroscience research [2, 15]. More studies are urgently required on this important topic.

LIPIDOMICS, PROTEOMICS, AND GENOMICS, AND DIAGNOSIS OF NEUROLOGICAL DISORDERS

Development of lipidomics, proteomics, and genomics technologies in recent years has resulted in the identification and full characterization of *in vivo* markers for oxidative stress and neuroinflammation along with protein fingerprinting in complex mixtures with peptoid microarrays and top-down mass spectrometry for annotation of gene products.. The biomarkers include F_2-isoprostanes, prostaglandins, leukotrienes, lipoxins, hydroxyleicosatetraenoic acids, nitrotyrosine, 4-HNE, carbonyls in proteins, oxidized DNA bases, tau protein, and cystatin C in CSF [40-49]. It is proposed that establishment of automatic systems including databases and accurate analyzes of various lipid mediators derived from enzymic and non-enzymic metabolism of neuronal membrane polyunsaturated fatty acids (eicosanoids, resolvins, neuroprotectins, isoprostanes, neuroprostanes, and isofurans) would facilitate identification of key biomarkers associated with acute neural trauma, neurodegenerative diseases, and neuropsychiatric disorders [5; 6; 48, 49]. A comprehensive lipidomics analysis using liquid chromatography-tandem mass spectrometry prostaglandins in rat hippocampal tissue indicates that systemic kainic acid administration produces large amounts of $PGF_2\alpha$ and PGD_2 and smaller amounts of other prostaglandins and hydroxyleicosatetraenoic acids [50]. This increase in eicosanoids can be prevented by intracerebroventricular administration of kainic acid receptor antagonists. Lipidomics, proteomics, and genomics can be used for the determination of lipid mediators in small tissue and biological fluid samples from normal human brain and brains from patients with neurodegenerative diseases [51]. Microarray analysis of tissue samples from brain regions from neurodegenerative diseases can provide information on candidate genes that modulate excitotoxicity, oxidative stress, and neuroinflammatory responses. This would facilitate the understanding of molecular mechanisms of the development of neurodegenerative diseases as well as information on molecular diagnostics and targets for drug therapy based on gene expression in body fluids such as CSF and blood [23, 52].

Multiplex quantitative proteomics methods, iTRAQ (isobaric Tagging for Relative and Absolute protein Quantification), in conjunction with multidimensional chromatography, follow- ed by tandem mass spectrometry (MS/MS), can be used to simultaneously measure relative changes in the proteome of cerebrospinal fluid (CSF) obtained from patients with AD, PD, and DLB compared to healthy controls [52]. iTRAQ studies have provided information on quantitative changes in AD, PD, and DLB as compared to controls; among more than 1,500 identified CSF proteins, 136, 72, and 101 of the proteins show quantitative changes unique to AD, PD, and DLB, respectively. Eight unique proteins have been identified by Western blot analysis, and the sensitivity at 95% specificity can be calculated for each marker alone and in combination. Several unique biomakers have been identified to distinguish among AD, PD and dementia with Lewy body (DLB) patients from each other as well as from controls with high sensitivity at 95% specificity. Based on these findings, it is suggested that a roster of proteins may be generated and developed into specific biomarkers that may eventually assist in clinical diagnosis and monitoring disease progression of AD, PD and DLB in larger and different human populations [52].

NEUROIMAGIMG AND DIAGNOSIS OF NEUROLOGICAL DISORDERS

Neuroimaging techniques, namely positron emission tomography (PET), fluorodeoxyglucose-positron-emission tomography (FDG-PET), single photon emission computed tomography (SPECT), magnetic resonance imaging (MRI), functional MRI (fMRI) and diffusion tensor imaging (DTI) are increasingly used to obtain useful information not only on metabolism and rates of generation of lipid mediators, but also on abnormalities in brain structure and function in mild cognitive impairment (MCI) and its conversion to dementia, as well as early neurodegenerative diseases (AD, PD, HD, and ALS) [53]. These procedures offer researchers and clinicians a new noninvasive window into the human brain and spinal cord.

PET utilizes incorporation of labeled radioligands (fatty acids and fluorodeoxyglucose) to determine of its distribution in various brain regions as a function of time. From measurements, fluxes, turnover rates, half-lives, and ATP consumption rates, *in vivo* generation of lipid mediators and fatty acid metabolism can be determined and imaged through PET. Based on the *in vivo* metabolism of various radioligands, PET can be used to image brain signaling and neuroplasticity in normal human brain and brain from patients with neurodegenerative diseases. Initial experiments on drug mediated stimulation of $[^{11}C]$arachidonic acid metabolism in animal models of AD and PD indicate that PET can be used to image various lipid mediators of arachidonic acid metabolism. Detailed investigations are required on the use of biomarkers, MRI, PET imaging and ultrasound measurement not only to judge the severity and progression of dementia during the course of various neurodegenerative diseases, but also to predict rates of cognitive decline and conversion of normal subjects to patient with clinical neurodegenerative disease [54, 55, 56]. Amyloid imaging using Pittsburgh Compound-B (PIB) has recently emerged as a non-invasive neuroimaging technique to visualize the accumulation of amyloid-β in living human brain. Based on this procedure it is reported that the accumulation of β-amyloid start during the preclinical stage of AD and reaches plateau phase before or during the mild cognitive impairment (MCI) stage supporting the view that amyloid imaging may be useful as a biomarker of AD, not only for the very early specific diagnosis, but also for monitoring the therapeutic effect with agents that reduce the accumulation of β-amyloid in brains from AD patient.[57].

DTI is based on the preferential movement of water protons within the brain along the axis of the axons. In contrast to MRI, this anisotropic diffusion can be used to generate information about the status of immature brain prior to myelination, during maturation, and in normal and disease states [58]. Recently, DTI has been used to reveal disruption of white matter microstructure in chronically injured brain and spinal cord. DTI provides reproducible quantitative measures, such as mean diffusivity and fractional anisotropy (FA) that reflect the underlying tissue properties of gray matter and white matter and may therefore become useful as developmental milestone. In addition, three-dimensional fiber tractography based on DTI can reveal the developing axonal connectivity of the human brain as well as aberrant connectivity in structural brain malformations. The information contained within the diffusion tensor data can be used to create 3-D mathematical renderings of white matter or tractography. Quantification of FA and mean diffusivity (MD) has been successfully used to demonstrate retrograde Wallerian degeneration (WD) of cranial corticospinal tract (CST) in cervical spinal cord injury (SCI). Significant reduction in the mean FA and increase in MD can be observed in the cranial CST in patients with SCI compared with controls, suggesting the occurrence of retrograde WD. Temporal changes in the DTI metrics suggest progressing degeneration in different regions of CST. These spatiotemporal changes in DTI metrics suggest continued WD in injured fibers along with simultaneous reorganization of spared white matter fibers, which may contribute to changes in the neurological status in chronic SCI patients [59, 60]. Thus, DTI provides an efficient tool for comprehensive, noninvasive, functional anatomy mapping of the human brain and spinal cord. Collective evidence suggests that in future neuroimaging methods can not only be used to obtain information on physiological and morphological changes, but following neuroprotective effects of drugs on brain and spinal cord tissues of normal subjects and patients with neurodegenerative diseases. For example, when neurological examination is normal, risk for dementia can be diagnosed by functional MRI. Similarly, when normal anatomic MRI scan fails to diagnose neurological deficits in traumatic brain injury (TBI), PET scanning can provide information on cortical dysfunction in TBI. Neuroimaging procedures can also provide insight at the molecular level when behavioral tests or anatomic brain imaging fails to provide useful information [58, 59, 60].

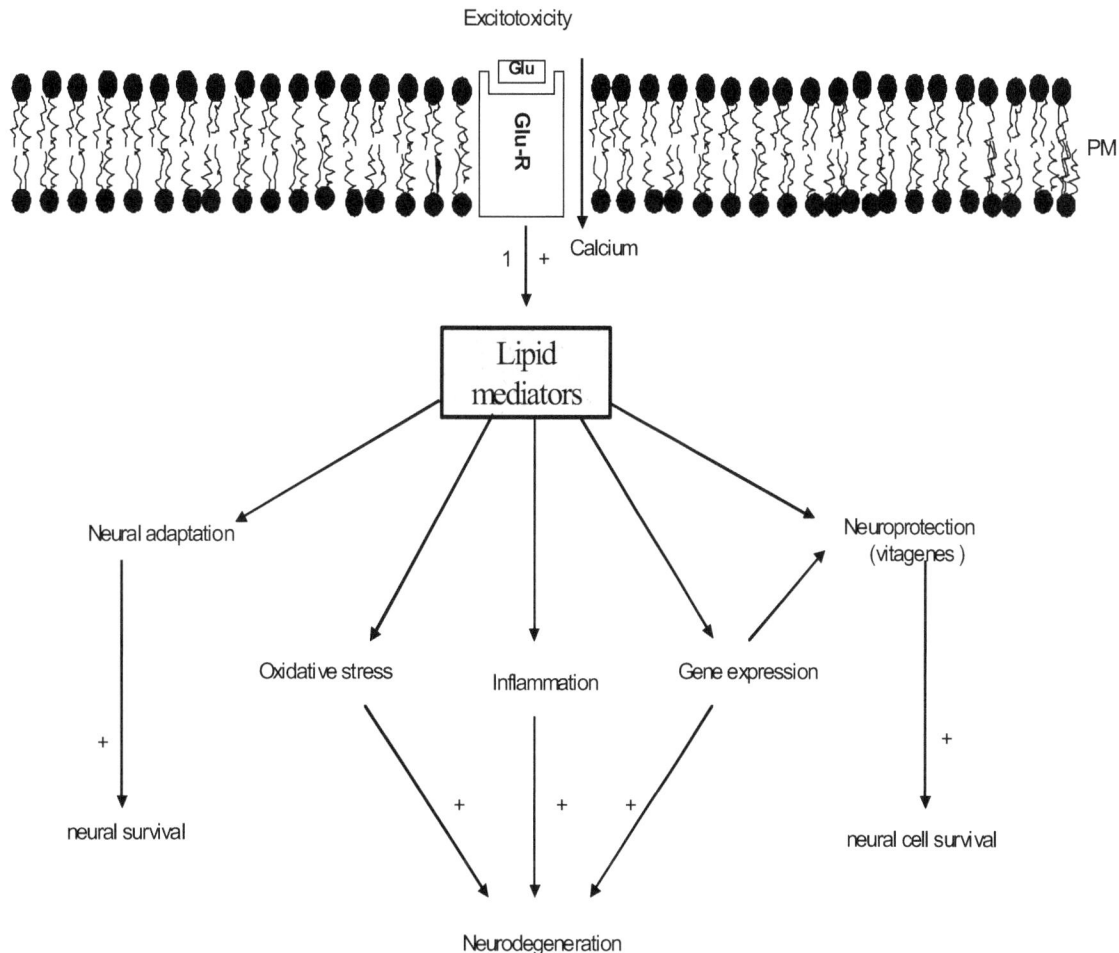

Figure 1: Diagram showing contribution of excitotoxicity, oxidative stress, inflammation, and gene expression in neurodegebneraion and neuronal survival. Plasma membrane (PM); glutamate receptor (Glu-R); glutamate (Glu); cytosolic phospholipase A_2; secretory phospholipase A_2; cyclooxygenase; lipoxygenase; epoxygenase (1); High levels of lipid mediators, such as eicosanoids; platelet activating factor (PAF); 4-hydroxynonenal (4-HNE); and isoprotanes (Isopros) cause oxidative stress (generation of ROS and activation of nuclear factor kappaB (NF-κB); inflammation (eicosanoids; tumor necrosis factor-α (TNF-α); interleukin-1β (IL-1β,); low levels of lipid mediators are associated with normal neural function. Positive sign (+) indicates increase in intentisty of neurochemical processes.

NEUROPROTECTION STRATEGIES AND NEURODEGENERATIVE DISEASES

Stress response signaling and neuroadaptation play an important role in neuronal survival and neuroprotection (Fig. 1). This approach can also be used for the treatment of neurodegenerative diseases. Sulfhydryl groups in proteins play a key role in redox sensing. It is well known that modulation of cellular redox state is crucial mediator of multiple metabolic, signaling, and accomplished by a complex network of longevity assurance processes that are controlled by vitagenes, a group of genes involved in preserving cellular homeostasis during stressful conditions [61, 62, 63]. Vitagenes encode for heat shock proteins (Hsp) Hsp32, Hsp70, the thioredoxin, and the sirtuin protein systems [61, 62, 63]. In particular, Hsp32 a heme oxygenase-1 (HO-1), has attracted considerable attention. It is recently shown that HO-1 induction, by generating the vasoactive molecule carbon monoxide and the potent antioxidant bilirubin, may represent a protective system potentially active against brain oxidative injury.

Accumulating evidence also evidence suggests that the HO-1 gene is redox-modulated and its expression appears closely related to conditions of oxidative and nitrosative stress.

Due to broad neuroprotective properties of the heat shock response, there is now strong interest in developing pharmacological agents capable of inducing such a response. These observations have led to new challenges and perspectives in neuropharmacology, as drugs mediating strong heat shock protein response can be used as neuroprotective agents in acute neural trauma, neurodegenerative diseases and neuropsychiatric diseases. Similarly, dietary antioxidants, such as polyphenols; ω-3 fatty acids, L-carnitine/acetyl-L-carnitine, carnosine, and curcumin (turmeric pigment) have recently been demonstrated to be neuroprotective through the activation of hormetic pathways, including vitagenes [61, 62, 63]. The hormetic dose-response challenges long-standing beliefs about the nature of the dose-response in a low dose zone, having the potential to affect significantly the design of pre-clinical studies and clinical trials as well as strategies for optimal patient dosing in the treatment of numerous diseases. Interest in the broad neuroprotective effects of endogenous (Hsp) and exogenous dietary antioxidants is significant. There is now growing interest in developing pharmacological agents that may induce neuroprotective responses in patients with acute neural trauma, neurodegenerative diseases, and neuropsychiatric disorders.

CONCLUSIONS

Collective evidence suggests that the intensity of cross talk (interplay) among the lipid mediators generated following excitotoxicity, oxidative stress, neuroinflammation, modulates the rate neurodegeneration in acute neural trauma, neurodegenerative diseases, and neuropsychiatric disorders [2]. Our emphasis on the cross talk among lipid mediators of excitotoxicity, oxidative stress, and neuroinflammation does not rule out the participation of other mechanisms involved in neurodegeneration. Lack of coordination among the above parameters may control the time taken by neural cells to die. For example, as stated above that in acute neural trauma, neuronal cell death occurs within hours to days, whereas in neurodegenerative diseases and neuropsychiatric disorders, neuronal damage take years to develop. For stroke and traumatic brain and spinal cord injuries, there is a therapeutic window of 4 to 6 h [2, 8] for partial restoration of many functions. In contrast, there is no therapeutic window for neurodegenerative diseases and neuropsychiatric disorders. Treatment of neurodegenerative diseases and neuropsychiateric disorders should start as soon as symptoms these diseases are detected.

REFERENCES

[1]   Farooqui, A.A.; Horrocks, L.A. *Int. Rev. Neurobiol.* **1994**, 36, 267.

[2]   Farooqui, A.A.; Horrocks, L.A. In *Glycerophospholipids in Brain: Phospholipases A₂ in Neurological Disorders* **2007**, Springer, New York.

[3]   Farooqui, A.A.; Ong, W.Y.; Horrocks, L.A.; Chen, P.; Farooqui T. *Brain Res. Rev.* **2007**, 56: 443.

[4]   Farooqui, A.A.; Ong, W.Y.; Horrocks, L.A. *Neurochemical Aspects of Excitotoxicity.* **2008,** Springer, New York.

[5]   Farooqui, A.A. *Hot Topics in Neural Membrane Lipidology.* **2009** Springer New York.

[6]   Farooqui, A.A.. *Beneficial effects of Fish Oil on Human Brain.* **2009** Springer, New York.

[7]   McIntosh, T.K.; Saatman, K.E.; Raghupathi, R., Graham, D.I.; Smith D.H.; Lee V.M.; Trojanowski, J.Q. *Neuropathol. Appl. Neurobiol.* **1998**, 24, 251.

[8]   Leker, R.R.; Shohami, E. *Brain Res. Brain Res. Rev.* **2002**, 39, 55.

[9]   Mattson, M.P. *Neuromolecular Med.* **2003**, 3, 65.

[10]   Bazan, N. G. *Cell. Molec. Neurobiol.* **2006**, 26, 901.

[11]   Bazan, N.G. *Curr. Opin. Clin. Nutr. Metab. Care.* **2007**, 10,136.

[12]   Bazan, N.G. *J Lipid Res.* **2009**, 50 Suppl, S400.

[13]   Serhan, C. N. *Curr. Opin. Clin. Nutr. Metab.* Care **2005**, 8,115.

[14]   Serhan, C. N. *Pharmacol. Ther.* **2005**, 105, 7.

[15]   Graeber, M.B.; Moran, L.B. *Brain Pathol.* **2002**,12, 385.

[16]   Bufill, E.; Blesa, R. *Rev Neurol.* **2006** 42, 25.

[17]   Phillis, J.; Horrocks, L.A.; Farooqui, A.A. *Brain Res. Rev.* **2006**, 52, 201.

[18]   Gallagher, S. *Psychopathology* **2004**, 37, 8.

[19]   Hornig, M.; Weissenböck, H.; Horscroft, N.; Lipkin W.I. *Proc. Natl. Acad. Sci.* USA.**1999**, 96, 12102.

[20]   Cohly, H.H.; Panja, A. Int. Rev. Neurobiol. **2005**, 71, 317.

[21]   Benes, F. M.*Brain Res Brain Res Rev 2000,* 31, 251.

[22]   Harrison, P. J. *Brain 1999,* 122, 593.

[23]   Facheris, M.; Beretta S.; Ferrarese, C. *J. Alzheimers Dis.* **2004**, 6, 177.

[24]   Wood, P.L. (1998). *Neuroinflammation: Mechanism and Management.* **1998**, Humana Press, Totowa, New Jersy.

[25]   Correale, J.; Villa, A. *J. Neurol.* **2004**, 251:1304.

[26]   Kidd, P.M. *Altern Med Rev.* **2005,** 10, 268.

[27]   Andersen, J.K. *Nat Med.* **2004**, 10, Suppl:S18.

[28]   Juranek I.; Bezek, S. *Gen. Physiol. Biophys.* 2005, 24, 263.

[29]   Shibata, N.; Kobayashi, M. *Brain Nerve* **2008**, 60, 157.

[30]   Dagda, R.K.; Zhu J.; Chu C.T. *Mitochondrion* **2009**, Epub ahead print.

[31]   Niizuma, K.; Endo H.; Chan PH. *J. Neurochem.* **2009**, 109, Suppl 1,133.

[32]   Niizuma, K.; Endo, H.; Nito, C.; Myer D.J.; Chan, P.H. *Stroke.* **2009**, 40, 618.

[33]   Wang, J.Y.; Wen L.L.; Huang Y.N.; Chen, Y.T.; Ku M.C. *Curr.Pharm. Des.* **2006**, 12, 3521.

[34]   Dwyer, B.E.; Takeda, A.; Zhu, X.; Perry, G.; Smith M.A. *Curr. Neurovasc. Res.* **2005**, 2, 261.

[35]   Migliore, L.; Fontana, I.; Colognato, R.; Coppede, F.; Siciliano, G.; Murri, L. *Neurobiol Aging.* **2005**, 26, 587.

[36]   Wilde, G.J.; Pringle, A.K.; Wright, P.; Iannotti, F. *J Neurochem.* **1997**, 69, 883.

[37]   Salmina, A.B. *J. Alzheimers Dis.* **2009**,16, 485.

[38]   Savitz, S.I..; Malhotra, S.; Gupta, G.; Rosenbaum D.M. *J. Cardiovasc Nurs.* **2003**, 18, 57.

[39]   Savitz, S.I..; Dinsmore, J.H.; Wechsler, L.R.; Rosenbaum, D.M.; Caplan L.R. *NeuroRx.* **2004**, 1, 406.

[40]   German J.B.; Gillies, L.A.; Smilowitz, J.T.; Zivkovic, A.M.; Watkins S.M. *Curr. Opin. Lipidol.* **2007** 18, 66.

[41]   Watson, A. D. *J Lipid Res.* **2006**, 47, 2101.

[42]   Bowers-Gentry R.C.; Deems, R.A.; Harkewicz R.; Dennis E.A. in *Functional Lipidomics* (Feng L.; Prestwich, G.D eds), **2006**, pp. 79-100, CRC Press-Taylor and Francis Group, Boca Raton.

[43]   Berliner, J.A.; Zimman, A. *Chem. Res. Toxicol.* **2007**, 20, 849.

[44]   Serhan, C.N. Prostaglandins Other Lipid Mediat. **2005**, 77, 4.

[45]   Adibhatla, R.M.; Hatcher, J.F.; Dempsey R.J. AAPS J. **2006**, 8, E314.

[46]   Milne. S.; Ivanova, P.; Forrester, J.; Alex Brown H. Methods. **2006**, 39, 92.

[47]   Morrow, J.D. Curr Pharm Des. **2006**, 12, 895.

[48]   Lu, Y.; Hong, S.; Gotlinger, K.; Serhan, C.N. ScientificWorld Journal. **2006**, 6, 589.

[49]   Perluigi, M.; Fai Poon, H.; Hensley, K.; Pierce, W.M.; Klein J.B.; Calabrese, V.; De Marco, C.; Butterfield, D.A. Free Radic. Biol. Med. **2005**, 38, 960.

[50]   Yoshikawa K, Kita Y, Kishimoto K, Shimizu T. J. Biol. Chem. **2006**, 281, 14663.

[51]   Butterfield DA, Perluigi M, Sultana R. Eur J Pharmacol. **2006**, 545, 39.

[52]   Abdi, F.; Quinn, J.F.; Jankovic, J.; McIntosh, M.; Leverenz, J.B.; Peskind, E.; Nixon, R.; Nutt, J.; Chung, K.; Zabetian, C.; Samii, A.; Lin, M.; Hattan, S.; Pan, C.; Wang, Y.; Jin, J.; Zhu, D.; Li, G.J.; Liu, Y.; Waichunas, D.; Montine, T.J.; Zhang, J. J. Alzheimers Dis. **2006**, 9, 293.

[53]   Weiner, M.W. (2009). Editorial: Imaging and Biomarkers Will be Used for Detection and Monitoring Progression of Early Alzheimer's Disease. J. Nutr. Health Aging 13:332.

[54]   Hampel, H.; Teipel, S. J.; Alexander, G. E.; Pogarell, O.; Rapoport, S. I.; Moller H. J. J. Neural Transm. **2002**,109, 837.

[55]   Rapoport, S. I. Prostaglandins Other Lipid Mediat. **2005**, 77, 185.

[56]   Masters, C. L.; Cappai R.; Barnham K. J.; Villemagne V. L. (2006). Molecular mechanisms for Alzheimer's disease: implications for neuroimaging and therapeutics. J. Neurochem. **2006**, 97, 1700.

[57]   Ishii, K. Rinsho Shinkeigaku. **2007**, 47, 915.

[58]   Rollin, N.K. Pediatr. Radiol. **2007**, 37, 769.

[59]   Guleria, S.; Gupta, R.K.; Saksena, S.; Chandra, A.; Srivastava, R.N.; Husain, M.; Rathore, R.; Narayana, P.A. J Neurosci Res. **2008**, 86, 2271.

[60]   Mukherjee, P., McKinstry, R.C. Diffusion tensor imaging and tractography of human brain development. Neuroimaging Clin N. Am. **2008,** 16, 19.

[61]   Calabrese, V.; Cornelius, C.; Dinkova-Kostova, A.T.; Calabrese, E.J. Biofactors. **2009**, 35, 146.

[62]   Calabrese, V.; Cornelius, C.; Mancuso, C.; Barone, E.; Calafato, S.; Bates, T.; Rizzarelli, E.; Kostova, A.T. Front Biosci. **2009**, 14, 376.

[63]   Farooqui, T.; Farooqui, A.A. Mech. Ageing Dev. **2009**, 130:203.

INDEX

hypoxia inducible factor 1α (HIF-1α), 51, 55, 56

I

IκB kinases, 16
immunoreactivity, 29, 30, 40, 41, 42, 43, 44, 45, 46, 76, 83
immunosuppressive, 2, 12
inducible nitric oxidase, 14
oxidative stress and neuroinflammation, 13, 132, 133, 134, 135, 138
inflammatory reactions, 1, 5, 15, 20, 74, 96
ischemia, 2, 5, 56, 92, 93, 94, 110, 127, 132
ischemic stroke, 9, 91, 92, 93, 98
isoprostanes, 17, 23, 24, 26, 27, 42, 135

K

kainic acid, 15, 39, 40, 135
keto-sphinganine, 42

L

L-Dopa, 63, 64, 65, 68, 109
leucine rich repeat kinase 2 (LRRK2), 134,
leukoaminochrome, 63
LeukoDOPAchrome, 69
leukotrienes, 1, 2, 3, 4, 6, 13, 14, 19, 20, 21
Lewy bodies disease, 26
lipid mediators, 1, 2, 3, 6, 8, 10, 11, 12, 13, 14, 15, 18, 19, 20, 39, 40, 41, 45, 46, 47, 48, 132, 133, 134, 135, 136, 137, 138
lipid peroxides, 40, 104
lipidomics, 15, 135
lipoxins, 1, 3, 4, 5, 8, 13, 14, 15, 20, 39, 121, 126, 127, 128, 135
lipoxygenases (LOX), 2, 8, 12, 20
15-lipoxygenase (15-LOX), 5, 6, 12
lyso-glycerophospholipids, 10
lysophosphatidylcholine (LysoPtd), 6, 16

M

magnetic resonance imaging (MRI), 51, 136
malondialdehyde, 23, 24, 27, 104
manganese superoxide dismutase (MnSOD), 27, 53
maresins, 121, 124, 126, 128
meingioma, 90
melatonin, 33
melanized neurons, 78
membrane attack complex, 72,73
methamphetamine (METH), 64
1-methyl-4-phenyl-1,2,3,6tetrahydropyridine (MPTP), 67
mild cognitive impairment, 23, 26, 51, 121, 128, 136
matrix metalloproteinase, 14, 16
microglial activation, 11, 50, 72, 74, 75, 76, 77, 78, 81, 83, 84, 122, 124
microhaemorrhage, 50
missense mutations, 32, 122, 124
mitochondrial complex I, 27
mitochondrial complex III, 50
mitochondrial dysfunction, 29, 39, 41, 47, 50, 51, 53, 54, 55, 57, 62, 65, 67, 68, 69, 77, 108, 127, 134

www.ingramcontent.com/pod-product-compliance
Lightning Source LLC
Chambersburg PA
CBHW041711210326
41598CB00007B/612